NANOMATERIALS FOR OPTOELECTRONIC APPLICATIONS

NANOMATERIALS FOR OPTOELECTRONIC APPLICATIONS

Edited by
Mohd. Shkir, PhD
Ajeet Kumar Kaushik, PhD
Salem AlFaify, PhD

AAP APPLE
ACADEMIC
PRESS

First edition published 2022

Apple Academic Press Inc.
1265 Goldenrod Circle, NE,
Palm Bay, FL 32905 USA
4164 Lakeshore Road, Burlington,
ON, L7L 1A4 Canada

CRC Press
6000 Broken Sound Parkway NW,
Suite 300, Boca Raton, FL 33487-2742 USA
2 Park Square, Milton Park,
Abingdon, Oxon, OX14 4RN UK

Library and Archives Canada Cataloguing in Publication

Title: Nanomaterials for optoelectronic applications / edited by Mohd. Shkir, PhD, Ajeet Kumar Kaushik, PhD, Salem AlFaify, PhD.

Names: Shkir, Mohd., 1982- editor. | Kaushik, Ajeet Kumar, editor. | AlFaify, Salem A., editor.

Description: First edition. | Includes bibliographical references and index.

Identifiers: Canadiana (print) 20200403478 | Canadiana (ebook) 2020040363X | ISBN 9781771889407 (hardcover) | ISBN 9781774638224 (softcover) | ISBN 9781003083948 (PDF)

Subjects: LCSH: Nanostructured materials. | LCSH: Optoelectronic devices. | LCSH: Nanotechnology.

Classification: LCC TA418.9.N35 N36 2021 | DDC 621.3815/2—dc23

Library of Congress Cataloging-in-Publication Data

Names: Shkir, Mohd., 1982- editor. | Kaushik, Ajeet Kumar, editor. | AlFaify, Salem A., editor.

Title: Nanomaterials for optoelectronic applications / edited by Mohd. Shkir, PhD, Ajeet Kumar Kaushik, PhD, Salem AlFaify, PhD.

Description: First edition. | Palm Bay, FL, USA : Apple Academic Press, 2021. | Includes bibliographical references and index. | Summary: "This book shines a spotlight on the significance and usefulness of nanomaterials for the development of optoelectronic devices and their real-life applications. It present an informative overview of the role of nanoscale materials in the development of advanced optoelectronic devices at nanoscale and discusses the applications of nanomaterials in different forms prepared by diverse techniques in the field of optoelectronic and biomedical devices. Major features, such as type of nanomaterials, fabrication methods, applications, tasks, benefits and restrictions, and saleable features, are well covered. Key features of Nanomaterials for Optoelectronic Applications: Provides an overview of and introduction to nanoscale materials Explains the features of 0D, 1D, 2D and 3D nanomaterials Exhibits the wide range of applications of nanomaterials in optoelectronics, photonics, biosensing, x-rays and x-ray detectors, medical imaging, visible light photodetectors, etc. Discusses the advances in miniaturized nanoscale devices for biomedical applications Describes the various preparation methods for advanced nanomaterials and their functionalization for fabrication of nanoelectronics devices Enlightens on the challenges and future prospects in nanoscale research This volume will be valuable for students, researchers, and scientists working in the field of nanotechnology, physics, chemistry, and materials science, providing a clear idea of the vast applications of nanomaterials in different forms prepared by diverse techniques in the field of optoelectronic and biomedical devices"-- Provided by publisher.

Identifiers: LCCN 2020053969 (print) | LCCN 2020053970 (ebook) | ISBN 9781771889407 (hardcover) | ISBN 9781774638224 (softcover) | ISBN 9781003083948 (ebook)

Subjects: LCSH: Optoelectronic devices--Materials. | Nanostructured materials.

Classification: LCC TK8304 .N36 2021 (print) | LCC TK8304 (ebook) | DDC 621.381/0450284--dc23

LC record available at https://lccn.loc.gov/2020053969

LC ebook record available at https://lccn.loc.gov/2020053970

ISBN: 978-1-77188-940-7 (hbk)
ISBN: 978-1-77463-822-4 (pbk)
ISBN: 978-1-00308-394-8 (ebk)

About the Editors

Mohd. Shkir, PhD

Mohd. Shkir, PhD, is an Assistant Professor at the Department of Physics, Faculty of Science, King Khalid University (KKU), Abha, Saudi Arabia, since 2013, where he played key role in the development of the Advanced Functional Materials and Optoelectronics Laboratory (AFMOL). He has published over 320 research papers in high-impact international and national journals, and his work has been cited over 3503 times. His h-index is 31, i10-index is 119, and RG score is 42.334. He holds an international patent on solar-cell fabrication system [ES2527976 (A1)—2015-02-02]. Dr. Shkir has received many awards, including the Award for Revolutionary Findings; DS Kothari Post-Doctoral Fellowship Award; IAAM Young Scientist Medal 2018, Singapore; MNRE-SEC-Research Associate; etc. His area of research includes nanosynthesis of different kinds of materials for biomedical, optoelectronic, photodetectors, radiation detection applications, etc. He had led many research groups in the past and currently leads a research group on "Investigation on Novel Class of Materials (INCM)" at KKU.

Dr. Shkir received his BSc and MSc degrees in physics from M.J.P. Rohilkhand University, Bareilly, India. He received his PhD degree in physics from Jamia Millia Islamia, New Delhi, India, in 2011. He performed his postdoc in the Crystal Growth Lab (CGL) at the Universidad Autónoma de Madrid, Spain, with Prof. E. Dieguez.

Dr. Shkir profile: https://scholar.google.com/citations?
user=Vs-NvR4AAAA J&hl=en

Ajeet K. Kaushik, PhD

Ajeet K. Kaushik, PhD, is an Assistant Professor of Chemistry at Florida Polytechnic University, Lakeland, Florida, USA. Dr. Kaushik is exploring nanotechnology, materials science, thin films, miniaturized sensors, electrochemistry, electro-chemical biosensors, nanomedicine, drug delivery to CNS, and biomedical engineering for health wellness. Prior joining Polytech, he has worked as an Assistant Professor to explore "magnetic nanomedicine along with drug deliver to the brain for the CND diseases management and miniaturized electrochemical sensing systems for personalized health wellness" at the Center of Personalized Nanomedicine at the Institute of NeuroImmune Pharmacology, Department of Immunology and Nano-Medicine, Herbert Wertheim College of Medicine, Florida International University (FIU), Miami, USA. He is the recipient of various reputed awards for his service in the area of nanobiotechnology for health care. His excellent research credentials are reflected by his four edited books, 100 international research peer-reviewed publications, and three patents in the area of nanomedicine and smart biosensors for personalized health care. In the course of his research, Dr Kaushik has been engaged in design and development of various electro-active nanostructures for electrochemical biosensor and nanomedicine for health care. His research interests include nanobiotechnology, analytical systems, design and develop nanostructures, nanocarriers for drug delivery, nano-therapeutics for CNS diseases, on-demand site-specific release of therapeutic agents, personalized nanomedicines, biosensors, point-of-care sensing devices, and related areas of health care monitoring.

He received his PhD (chemistry and nano-enabled biosensors) in collaboration with the National Physical Laboratory and Jamia Milia Islamia, New Delhi, India.

His Google Scholar profile:
https://scholar.google.com/citations?user=RYH8Z_4AAAAJ&hl=en
Website: https://akaushik3.wixsite.com/nanocare

Salem AlFaify

Salem AlFaify, PhD, is a Professor of Physics, a leader of the "quantum functional materials for advanced applications" (QFMAA) research group, and a leading researcher at the Advanced Functional Materials and Optoelectronics Laboratory (AFMOL) at the Department of Physics, Faculty of Sciences, King Khalid University (KKU), Saudi Arabia. He was formerly President of the Saudi Physical Society (SPS) from 2013 to 2016. He has authored and co-authored more than 220 articles in peer-reviewed and well-known ISI journals. He works with collaborators and researchers with mutual interest from many institutes and universities around the world. His research interest is primarily in the area of the condensed matter physics at the nanoscale, in particular, the correlation nature of the nano-quantum structures and their properties and applications. In addition to the growth of varying forms of nanostructured materials and their basic characterization by XRD, SEM, TEM, etc., he is interested in utilizing the powerful techniques related to accelerators physics such as the ion beam analysis (IBA) and synchrotron radiation to fundamentally investigate the essence of the nanomaterials and understand and engineer their novel properties for modern and futuristic applications.

Dr. AlFaify obtained his PhD in condensed matter physics/nanomaterials in 2011 from Western Michigan University, Kalamazoo, MI, USA. He was awarded a thesis appointment scholarship from the Department of Energy (DOE), USA, to conduct his PhD research project at the Center for Nanoscale Material (CNM) at Argonne National Laboratory (ANL), one of the prominent national laboratories of the DOE, USA, operated and managed by the University of Chicago, Chicago, IL, USA.

Contents

Contributors

Md. Mottahir Alam
Department of Electrical and Computer Engineering, Faculty of Engineering and Technology,
King Abdulaziz University, Jeddah 21589, Saudi Arabia

Salem A. Alfaify
Advanced Functional Materials and Optoelectronics Laboratory (AFMOL), Department of Physics,
King Khalid University, Abha 61413, Kingdom of Saudi Arabia

H. Elhosiny Ali
Advanced Functional Materials & Optoelectronics Laboratory (AFMOL), Department of Physics,
Faculty of Science, King Khalid University, Abha 62529, Saudi Arabia
Physics Department, Faculty of Science, Zagazig University, 44519 Zagazig, Egypt

Abdullah Almohammedi
Department of Physics, Faculty of Science, Islamic University of Madinah, Prince Naif bin Abdulaziz,
Al Jamiah, Madinah 42351, Kingdom of Saudi Arabia

Sajid Ali Ansari
Department of Physics, College of Science, King Faisal University, P.O. Box 400, Hofuf,
Al-Ahsa 31982, Saudi Arabia

Kamlesh V. Chandekar
Department of Physics, Rayat Shikshan Sanstha's, Karmaveer Bhaurao Patil College, Vashi,
Navi Mumbai 400703, Maharashtra, India

Meenakshi Choudhary
Swiss Institute for Dryland Environmental and Energy Research, Blaustein Institutes for Desert Research,
Ben-Gurion University of the Negev, Israel

Penny P. Govender
Department of Chemical Science-DFC (formerly Department of Applied Chemistry),
University of Johannesburg, 17011, Doornfontein, 2028 Johannesburg, South Africa

Chaudhery Mustansar Hussain
Department of Chemistry and Environmental Science, New Jersey Institute of Technology, Newark,
NJ 07102, USA

Mohd Imran
Department of Chemical Engineering, Faculty of Engineering, Jazan University, Jazan 45142,
Saudi Arabia

Tien-Chien Jen
Department of Mechanical Engineering Science, University of Johannesburg, South Africa

Ziaul Raza Khan
Department of Physics, College of Science, University of Hail, Hail 2440, Kingdom of Saudi Arabia

Mohd. Taukeer Khan
Department of Physics, Faculty of Science, Islamic University of Madinah, Prince Naif bin Abdulaziz,
Al Jamiah, Madinah 42351, Kingdom of Saudi Arabia

Vinod Kumar
Department of Materials Engineering, Ben-Gurion University of the Negev, Beer-Sheva, 84105, Israel

Bindu Mangla
Department of Chemistry, J.C. Bose University of Science and Technology, YMCA, Faridabad, India

M. Aslam Manthrammel
Advanced Functional Materials & Optoelectronics Laboratory (AFMOL), Department of Physics, Faculty of Science, King Khalid University, Abha 62529, Saudi Arabia

Nazish Parveen
Department of Chemistry, College of Science, King Faisal University, P.O. Box 380, Hofuf, Al-Ahsa 31982, Saudi Arabia

Mohammad Shariq
Department of Physics, Faculty of Science, Jazan University, Jazan 45142, Saudi Arabia

Mohd. Shkir
Advanced Functional Materials & Optoelectronics Laboratory (AFMOL), Department of Physics, College of Science, King Khalid University, Guraiger, Abha 61413, Kingdom of Saudi Arabia

Sudheesh K. Shukla
Department of Chemical Science-DFC (formerly Department of Applied Chemistry), University of Johannesburg, 17011, Doornfontein, 2028 Johannesburg, South Africa

Abhay Kumar Singh
Department of Mechanical Engineering Science, University of Johannesburg, South Africa

Rui Wang
School of Environmental Science and Engineering, Shandong University, Jinan, Qingdao 266237, China

Abbreviations

AAM	anodic alumina membrane
AAO	anodized aluminum oxide
ALD	atomic layer deposition
BHJ	bulk heterojunction
BNNT	boron nitride nanotube
BP	black phosphorus
CB	chlorobenzene
CBD	chemical bath deposition
CBs	conduction bands
CdS	cadmium sulfide
CEAS	cavity-enhanced absorption spectroscopy
CIGS	copper indium gallium diselenide
CIS	copper indium sulfide
CNT	carbon nanotube
CPs	conjugated polymers
CQDs	carbon quantum dots
CV	cyclic voltammetry
CVC	chemical vapor condensation
CVD	chemical vapor deposition
D–A	donor–acceptor
DIO	diiodooctane
DOS	density of states
e–h	electron–hole
EIS	electrochemical impedance spectroscopy
EL	electroluminescence
EM	electromagnetic
EMA	effective mass approximation
ENs	engineered nanomaterials
EQE	external quantum efficiency
FETs	field-effect transistors
FF	fill factor
FWHM	full-width at half maximum
GO	graphene oxide
Hap	hydroxyapatite

HBM	hyperbolic band model
ITO	indium tin oxide
ITO	indium transparent electrode
LAS	laser absorption spectroscopy
LDF	laser Doppler flowmetry
LED	light emitting diode
MBE	molecular-beam epitaxy
MC1DNM	multicomponent 1D nanostructured materials
MOFs	metal-organic frameworks
MRI	magnetic resonance imaging
MS	magnetron sputtering
MW	multiwalled
NCs	nanocarbons
NCs	nanocrystals
NFA	nonfullerene acceptor
NIOS	noninvasive optoelectronic sensors
NIR	near-infrared
NPs	nanoparticles
NRs	nanorods
NSMs	nanostructured materials
NTs	nanotubes
NW	nanowire
OA	oleic acid
OCT	optical coherence tomography
OLEDs	organic light-emitting diodes
OPV	organic photovoltaic
OS	optoelectronic sensors
PANI/PEO	polyaniline/polyethylene oxide
PCE	power conversion efficiency
PDA	prostate-specific antigen
PDs	photodiodes
PEG	polyethylene glycol
PL	photoluminescence
PLD	pulsed laser deposition
PMC	polymer matrix nanocomposite
PPY	poly(pyrrole)s
PVD	physical vapor deposition
QCLs	quantum cascade lasers
QD	quantum dot

QE	quantum efficiency
SCMs	supplementary cementitious materials
SILAR	successive ionic layer adsorption and reaction
SMM	single-molecule microscopy
SNCs	silica-based nanocapsules
STM	scanning tunneling microscope
TDA	temperature-dependent aggregation
UPD	underpotential deposition
USCVD	ultrasonic spray chemical vapor deposition
UV	ultraviolet
VBs	valence bands
VD	vapor deposition
VLS	vapor–liquid–solid
VOCs	volatile organic compounds
WH	Williamson–Hall
XRD	X-ray diffraction

Preface

The currently written textbook is intended for students, researchers, and scientists working in the field of nanotechnology in such areas as physics, chemistry, nanotechnology, and materials science. The book will be helpful to master's students as well as undergraduate and doctorate students who are looking for careers in nanoscience and thin films technology.

The key purpose of this book is to provide a clear idea about the applications of nanomaterials in different forms prepared by diverse techniques in the field of optoelectronic and biomedical devices.

Special care has been taken to present the topics in a very simple manner. This book contains nine chapters that broadly deal with the topics concerned with nanomaterials and thin films, their forms, experimental techniques, and their physical properties at the nanoscale.

The key features of the book are:

1. Shows the scenarios of nanotechnology in optoelectronic device applications and developments.
2. Articulates the various preparation methods for advanced nanomaterials and their functionalization for the fabrication of nanoelectronics devices.
3. Demonstrates the consequence of economizing and combining 0D, 1D, and 2D nanomaterials for the future of electronics.
4. Establishes the wide range of applications of nanomaterials in optoelectronics, photonics, biosensing, X-ray, γ-ray detectors, medical imaging, visible light photodetectors, etc.
5. Presents innovative and new notions for evolving novel nanomaterials for nanodevices for the development of modern technological systems.
6. Discusses the imminent projections and tasks in developing nanoscale devices for safety of society health and advanced technology.

Chapter 1 reviews nanomaterials and their applications. Nanoscience and their optoelectronics applicability are one of the emerging scientific areas in materials science and technology. Considering this aspect, this chapter provides an introduction of nanoscience, including different forms of nanomaterials as well as nanotechnology. It covers the historical background of nanomaterials and discusses various kinds of the naturally occurring nanoforms. A detailed description of nanomaterials and their classifications is given, including zero-dimension, one-dimension, three-dimension, interpretation of nanostructures, nanoparticles, nanowires, nanotubes, nanolayers, and nanoporous properties. Descriptions of synthesis strategies of nanomaterials with chemical, physical, and biological approaches of various methods have been discussed. Additionally, the importance of nanomaterials and the significance of nanoscience have also been covered. The potential optoelectronic photo-voltaic cells application of different kinds of nanomaterials has been described along with their current status and working mechanisms, such as silicon and thin films solar cells, multijunction solar cells, organic solar cells, dye-sensitized solar cells (DSSCs), quantum dots-sensitized solar cells (QDSSC). It also includes carbon nanotubes, graphene, and transition metal dichalcogenides based optoelectronics applicability. At the conclusion, the shortcomings of the nanomaterials have also been incorporated.

Chapter 2 discusses 0-dimentional nanostructures and applications. Semiconductor nanomaterials are one of the most important components of advanced optoelectronics devices. Size-dependent properties of semi-conductor nanomaterials have provided a broad range of modification in device sizes, power consumption, and efficiency enhancement for the advancement of optoelectronic devices. Zero-dimensional semiconductor nanomaterials (0D NMs) can be promising candidates, due to their unique physical, electronic, and surface properties, for meeting these objectives in advancement of optoelectronic devices. The biggest challenge of the 21st century is the huge amount of energy demand and supply. Energy-harvesting devices, such as solar cells, low power consumption, and efficient lightning devices, light emitting laser diode, laser-controlled devices, energy-stored devices based on 0D nanomaterials, will play a major role to resolve this toughest demand of energy in the future. Therefore, 0D nanomaterials have attracted intense attention of materials researchers to develop various types of 0D nanomaterials with suitable, optical, electronic, and electrical properties for manufacturing of advanced optoelectronic devices applications. This chapter describes the classification of nanomaterials and growth techniques, which is discussed in brief for nanomaterials. The chapter then focuses on the development of 0D nanomaterials by using various techniques and demonstrates the potential applications of 0D nanomaterials for optoelectronic applications in recent trends.

Chapter 3 particularizes one-dimensional (1D) nanomaterials and their applications. Nanotechnology enables the control of size, shape, and crystallographic orientation of nanomaterials at nanoscale range. Among various nanoshapes, 1D nanostructures have their specific advantages as compared to isotropic nanoparticles. One-dimensional nanomaterials are an ideal system for exploring a large number of novel phenomena at the nanoscale level and for investigating the dimensional as well as size-dependent properties for the different applications, including optoelectronics. In recent years, integration of optical switches or interfaces, based on 1D nanomaterials with tailored geometrics, has made significant advancement. One-dimensional-based optoelectronic devices can be configured either as a resistor whose conduction could be altered by a charge transfer process or as a field effect transistors (FETs). Functionalization of the structural surfaces offers numerous promising opportunities for intensifying optoelectronic competences. This chapter provides a comprehensive appraisal of the state-of-the-art research activity on the synthesis and functionalization of 1D nanomaterials along with their respective optoelectronic applications.

Chapter 4 is dedicated to discussing two-dimensional (2D) nanomaterials and their applications. The authors present a general classification, synthesis, and fabrication of nanoparticles. Subsequently, we focus more attention on the classification of nanomaterials in 2D for optoelectronic applications. Two-dimensional materials like nanocomposite materials have become the most powerful semiconductor materials in the field of optoelectronic devices due to their extraordinary properties. Due to the layer dependent and properly shaped bandgaps, photodetectors, depending on different 2D components, are intended and produced in a rational manner. Using the distinctive characteristics of 2D components, many unexpected physical events of collisions relying on 2D components can be achieved when separate 2D components are placed together. This allows heterojunctions more famous than 2D components themselves, and the development of 2D components for humans is easier than ever before. In this chapter, we focused on the polymeric nanocomposite as it is one of the important famous types of nanocomposites.

Chapter 5 deals with three-dimensional (3D) nanomaterials and their uses. In this chapter, the authors provide the general discussion on 3D. Over the last few decades, various materials, especially zero-, one-, and two-dimensional, have attracted much attention due to their various exceptional properties as compared to the bulk materials. However, these materials had various issues including structural and assembly inhomogeneity. Though, constructing 3D using simple techniques have become foundations of nanotechnology and nanoscience, which is due to the unique 3D structure. In addition, the development of three-dimensional nanomaterials has one of the unique assemblies that have been used for various applications from energy to environmental applications in the past few years. In this chapter, we summarize the synthesis of various three-dimensional material and its optoelectronic applications.

Chapter 6 deals with the advances of nanostructured thin films and their optoelectronic applications. This chapter emphasizes the basics and synthesis of nanostructured thin films using different techniques. The properties that were specific to the material turn out to be size dependent in nanoscale. Thus, the structural, electrical, optical, and morphological characteristics of the sample show different behaviors, which can be highly tuned for the device-oriented properties depending upon the growth mechanism involved. In this chapter, we will deal more about such changes regarding the samples grown in thin film form. Fundamentals behind various vacuum and wet-chemical techniques for the nano thin-film synthesis are discussed. The chapter also covers the substrate selection as well as the importance of tuning the synthesis procedures for device-oriented application. The essential optical characterization details have been discussed in briefed at the end of the chapter to determine the band structure of semiconductor thin film samples.

Chapter 7 discusses fabrication methods of thin film polymer solar cells. In the last decade, organic solar cells have emerged as one of the most prominent candidates for the conversion of sun light into electricity and have already achieved more than 17% efficiency. This chapter provides an insight into the organic photovoltaics technology including device architecture, materials, device physics, and fabrication methods. We believe that this chapter will be worthwhile for the beginning researcher in organic photovoltaic technologies.

Chapter 8 deals with the optoelectronics for biomedical applications and covers the basics of optoelectronics and the general optoelectronic properties applied in the developments of equipment used in the biomedical fields. This chapter starts with the introduction of photonics and electromagnetic wave. Semiconductor energy-band diagrams and their interaction with the light have been mentioned afterward. Some of the important optoelectronic materials are also introduced and their properties and applications are explained, especially as they pertain to the biomedical fields. Recently explored optoelectronic materials such as conducting polymers, phosphorene, carbon nanotubes, and other metal oxides nanoparticles are also mentioned in the optoelectronic materials section. Optoelectronic devices starting from simple diode to complex solar cells have been added and presented as well. Special attention has been given to the application of optoelectronics for biomedical equipment. The well-known optoelectronic sensors have been described and classified according to their nature of invasiveness and noninvasiveness. The design and the components of these sensors as well as working principles are also part of the discussion carrying out in this chapter. Other familiar techniques such as endoscopic imaging techniques and other important medical equipment have also been included. Some of the important diagrams or schemes has been included and explained according to their function. The future of optoelectronics and its application into biomedical field have been discussed. The future development in optogenetics and retinal prosthesis has been included in the biomedical application section as well as in future prospective section. A conclusion of all the components is covered in this section, which is the closing of this chapter.

Chapter 9 deals with the future prospects and challenges in the field of nanomaterials and applications.

This book gives an overview of the usefulness of nanostructured materials of different forms for the progress of optoelectronic devices. Major features, like type of nanomaterials, fabrication methods, applications, tasks, benefits and restrictions, and saleable features, are well-covered.

The authors, from different fields such as nanotechnology, biochemistry, chemistry, material science, and bioengineering, solar cells, thin films, have taken care for writing their respective chapters.

The editors sincerely acknowledge their teachers/guides, friends, and faculty members from diverse departments for their valuable suggestions and also encouragement while preparing the whole manuscript. The editors are also gratefully thankful to publisher of book and journals cited in the bibliography in the respective chapters.

CHAPTER 1

Introduction to Nanomaterials and Their Applications in Optoelectronics

ABHAY KUMAR SINGH and TIEN-CHIEN JEN

Department of Mechanical Engineering Science,
University of Johannesburg, South Africa

ABSTRACT

Nanoscience and their optoelectronics applicability is one of the emerging scientific areas in the materials science and technology. Considering this aspect, this chapter deals with an introduction of nanoscience which include different forms of nanomaterials as well as nanotechnology. Various kinds of the naturally occurring nanoforms are also described including nanomaterials historical background. A detail description on nanomaterials and their classifications including zero-, one-, three-dimension, interpretation of nanostructures, nanoparticles, nanowires, nanotubes, nanolayers, and nanoporous have also been included. To describe nanomaterials synthesis, strategies such as chemical, physical, and biological approaches have been discussed. Additionally, the importance of nanomaterials and the significance of nanoscience have also been covered. The potential optoelectronic photovoltaic cells application of different kinds of nanomaterials has been described with their current status and working mechanisms, such as silicon and thin-films solar cells, multijunction solar cells, organic solar cells, dye-sensitized solar cells, and quantum dots-sensitized solar cells including carbon nanotubes, grapheme, and transition metal dichalcogenides based optoelectronics applicability. At the end, shortcomings of nanomaterials have also been addressed.

1.1 INTRODUCTION

The atomic level investigation of matter provides an extensive scope to every areas of the science such as physics, chemistry, engineering, biotechnology, and medicine. In the modern world, with the promising tools, various materials are prepared by alternating either their form or shapes. The alternation or modification in forms or shapes may provide various changes in their physical and chemical properties [1]. Considering this worldwide concern, several efforts have also been made about the energy-related climate change where spiralling rate of fossil fuels can be controlled. Moreover, interests in the renewal energy are rising drastically due to their worldwide demand. These materials can also efficiently protect the environment from the pollution. All these problems are encountered increasingly due to industrialization and urbanization of the human society in a few decades.

In general, all the existing materials can be classified into two key groups, such as natural and artificial materials. Organic matter, mineral matter, and living matter fall in natural category, while artificial materials can be manufactured through the synthesis process under the well-defined conditions. Both kinds of materials have chemical compositions and structures with the specified physical properties or functions. More specifically, the artificial materials are developed in a particular environment for the specific properties related to their precise field of applications. Recent progress in this area occurred through the size effect by transformation of bulk to nanoscale material. There are number of approaches available to fabricate multidimensional materials with the diverse properties to enhance interface performance [2].

Thus the advanced nanomaterials can be defined in various ways, based on their availability and usability. With the usual straightforward explanation of the materials, these are associated from the progressive technologies and perspective to derive direct/indirect benefits for the highly specialized outcomes in multidisciplinary areas. As per the US National Institute for Standards and Technology descriptions: "The materials with unique functionalities have been identified and developed as large quantities enough for innovators and for manufacturers to investigate and authenticate in order to increase new products for consumers" [3]. An alternative definition of the advanced materials comes from the Technology Strategy Board; materials those are designed for some targeted applications with better properties. They may not only be new materials but can also be alloys or composites of graphene or high-temperature superconductors with new or superior structural (hardness, strength, and flexibility) and functional properties (electric, electronic, magnetic, optical) [4, 5]. On the basis of the recent investigations of such materials, the possible key utility is schematically represented in Figure 1.1.

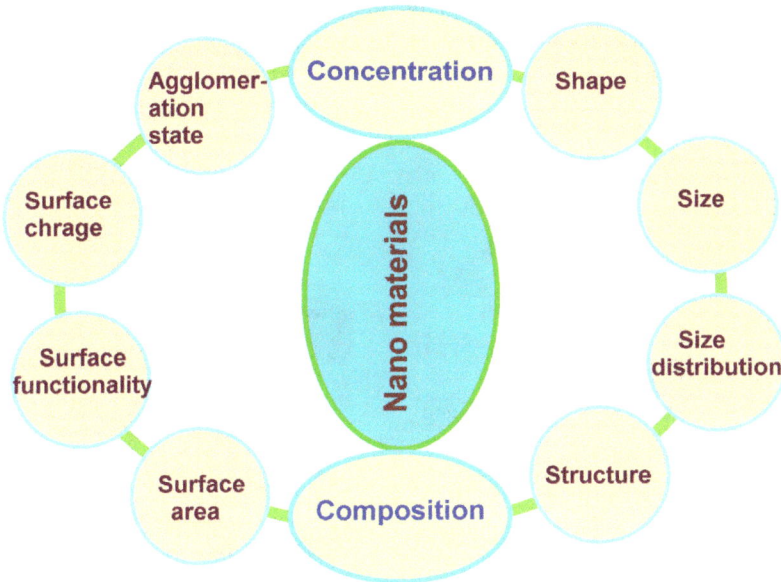

FIGURE 1.1 Normal characteristics of nanomaterials.

The importance of advanced nanomaterials is mainly governs from their basic properties. Mainly by the controlling factors of pioneering processes, these materials can play a significant role in their specific applications. Therefore, it is essential that the required material properties should be decided before designing and developing a new innovative composition. As an example, most of alloys with high strength are usually used for the stronger, lighter, and safer vehicles. Hence using material possessing to improve properties may open up a new way in the modern technologies as a better performing tool in energy-storage devices (batteries/supercapacitors), electronic inks (multi-printing), high-voltage transmission lines, and healthcare-associated applications. The advance nanomaterials key feature is briefly mentioned in the schematic represented in Figure 1.2. As per their characteristics, they are mainly separated into four categories: chemical, physical, mechanical, and dimensional. Further each category has its own subdivisions according to their key properties. The advanced nanomaterials structure is classified into two key classes; the atomic level and the microscopic level. In the atomic level, atoms are arranged in different ways, which exhibit diverse activities as compared to the bulk molecular level. As an example, property of graphite differs from the graphene. While at the microscopic level, the granular arrangement of materials as small grains can be identified through microscopy. This provides evidence about alternation in optical property of the developed nanomaterial [6].

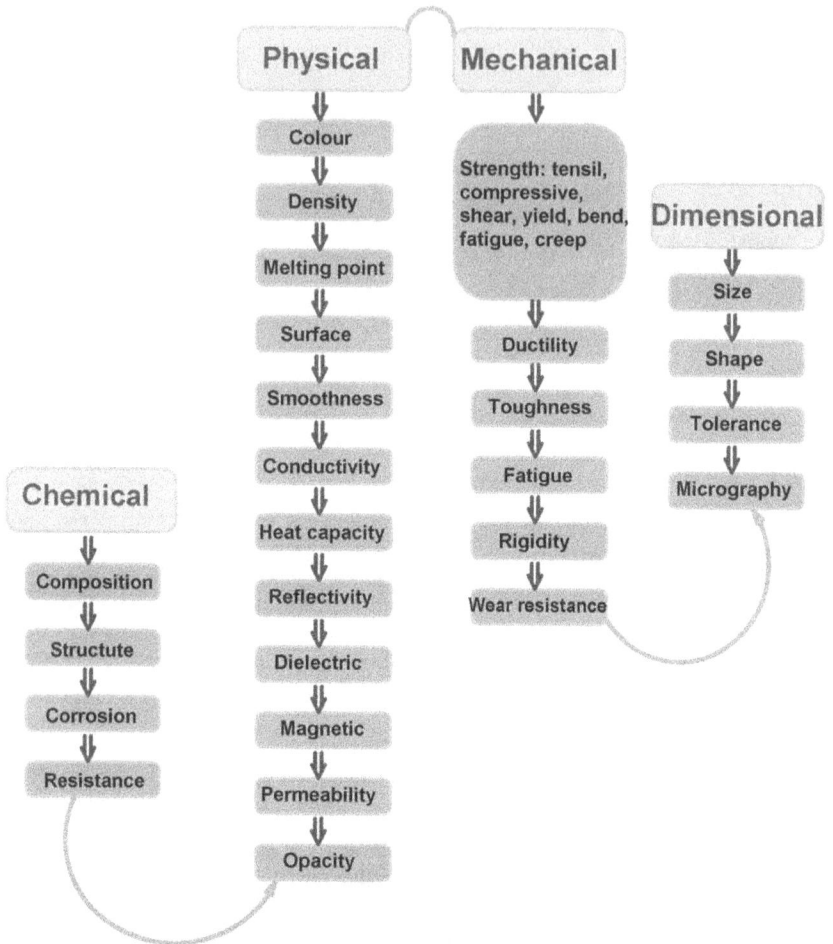

FIGURE 1.2   Schematic for the nanomaterials selection.

1.1.1   *AN INTRODUCTION OF NANOSCIENCE*

Nanoscience is a relatively new discipline in materials science which is concern with unique properties, due to assembly of atoms or molecules on the nanoscale. Nanoscience essentially deals with the objects/particles and its phenomena at an extremely small scale, typically in the range 1–100 nm. The word "nano" refers to a scale size in the metric system. It is equal to one-billionth of the base unit, around 100,000 times smaller than human hair diameter. The dimension of nanometer is 10^{-9} m (1 nm = 10^{-9} m) in the atoms and molecules world (the size of H atom is 0.24 nm and, therefore, line up of 10 hydrogen atoms is 1 nm). Hence nanoparticles contained 100 to 10,000 atoms. Their particles size is roughly in the range of 1–100 nm building blocks of the materials.

1.1.2 AN INTRODUCTION OF NANOMATERIALS

Nanomaterials are formed owing to building blocks of nanoparticles; therefore, they can be expressed as a set of substances where at least one-dimensional should be less than 100 nm. However, nanomaterials are also classified for the quite larger particles size in the range of 0.3–300 nm, such as in environment, health, and consumer protection due to their atomic/molecular levels organizations. Classification of larger particles size range of nanomaterials provides a path for the intensive research and their deeper understanding. In addition, whether a specific nanomaterial can be useful for the human health or not: under what particle size range. Typically fullerenes, carbon nanotubes, and graphene nanocarbons are the excellent examples of nanomaterials. The particles size of different kinds of fabricated nanomaterials, a few natural and biological species, is illustrated in Figure 1.3.

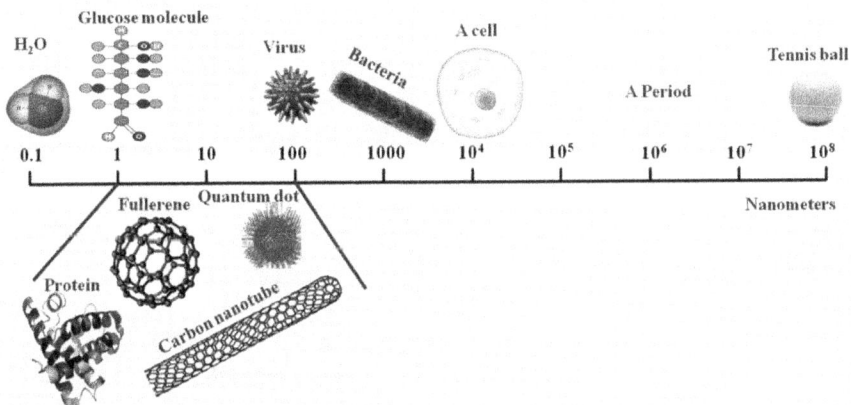

FIGURE 1.3 Size of the various objects nanomaterials and biomolecules (graphics were taken from Google images).

1.1.2.1   DEFINITION OF NANO-OBJECT

Materials those confined in one-, two-, or three-dimensions at the nanoscale including nanoparticles (all three dimensions at the nanoscale), nanofiber (two-dimensions at the nanoscale), nanoplates (one-dimension at the nanoscale), and nanofiber which is divided into nanotubes (hollow nanofiber) nanorods (solid nanofiber), and nanowire (electrically conducting or semiconducting nanofiber) are classified as nano-object. Though, term nano-object is not very popular in modern world.

1.1.2.2   DEFINITION OF PARTICLE

Particles are defined as a minute piece of matter that established the physical boundaries. This minute piece of the mater can move as a unit. Usual particle definitions impact on the nano-objects is follows as:

- **Definition of nanoparticle:** Nanoparticles are the nano-object having all three external dimensions at the nanoscale. Nanoparticles can be in the amorphous or the crystalline form, while, their surfaces can act as carriers for the gases or liquid droplets.
- **Definition of nanoparticulate:** This is defined as a collection of nanoparticle; they reflect a collective behavior.
- **Definition of agglomeration:** Agglomeration reflects a group of particles which held together under the weak van der Waals, such as forces including some electrostatic forces and surface tension. The agglomeration of the particles usually retains a high surface-to-volume ratio.
- **Definition of aggregation:** Aggregation of particles is defined as a group of particles held together by strong forces those associated from the covalent or metallic bonds. The aggregation of the particles also retains a high surface-to-volume ratio.

1.1.2.1   INTERPRETATION OF NANOTECHNOLOGY

Nanotechnology is interpreted as the formation of structures and their functionality designed at the atomic or molecular scale having at least one-dimension in nanometers. Their size-dependent feature allows them form novel structures with the significantly improved physical, chemical, and biological properties. Hence, in general, nanotechnology can be defined as the research and development which involve measurement and manipulation of matter at the atomic, molecular, and supramolecular levels, whereas, at least one-dimension should be in the range of 1–100 nm.

In the case of materials characteristics structural features, the sizes between isolated atoms and bulk materials intermediate range approximately 1–100 nm. The nano-object often shows physical distributions substantially different from both the atoms and bulk materials. Thus, the term "nanotechnology" in public domain is largely used as a reference for both nanoscience and nanotechnology. Therefore, nanoscience can be described as the science dealing with physics, chemistry, materials science, and biology with the manipulation and characterization of matter between the molecular and micron size length scales. However, nanotechnology deals with emerging engineering discipline by employing methods from nanoscience to create the desired products.

1.2   NATURAL OCCURRENCE OF NANOMATERIALS

Nanostructures substances are abundant in nature. Nanoparticles are distributed widely in the universe. It is considered that nanoparticles are the building blocks for the planet formation. In the biological world, several natural structures including proteins and the DNA diameter around 2.5 nm, viruses (10–60 nm) and bacteria (30 nm–10 μm), all these are obeying definition of nanomaterial, with other mineral or environmental origin substances, such as fin fraction of desert sand, smog, fumes originating from volcanic activity, forest fires, and certain atmospheric dusts. Generally biological systems built up inorganic–organic nanocomposite structures to improve the mechanical, optical, magnetic, and chemical sensing properties. These systems can also form nacre (mother-of-pearl) for the mollusk shell in biologically lamellar ceramic. The formed biologically lamellar ceramic exhibits structural robustness despite of their brittle nature of the constituents. Such nanomaterials can be evolved and optimized through the evolution over millions of years into sophisticated and complex structures. Usually, in the natural nanosystems formation process begins from the bottom-up approach at the molecular level those involving self-organization concepts. This is highly successful concept for building the larger structural and functional components. The functional nanosystems are wildly characterized from the complex sensing, self-repair, information transmission, storage, and other functions based on molecular building blocks. The most common examples are complex structures of the teeth, like shark teeth is the composite of biomineralized fluorapatit and other organic compounds. Such structures have a unique combination of hardness, fracture toughness, and sharpness. Evolution has not been done on that scales also work reported on much smaller scales too, like production of finely honed nanostructures, parts less than a millionth of a meter across which is smaller than 1/20th of the width of a human hair: these nanostructures are helpful in animals climb, slither, camouflage, flirt, and thrive. This kind of sensory patch for the amphibian ears microscopic image visualization has been reported by the investigators [7]. According to this investigation, amphibian ears consist of a single bundle of stereo cilia under projection of the epithelium papilla. These papillae can act as nanomechanical cantilevers to measure the deflection in a smaller range (3 nm). In the more convenient way, it can be correlated to many of the shimmering colors in butterfly's wings produced not with pigments but also with nanostructures. The butterfly's wings scales are patterned with nanoscale channels, ridges, containing protein cavities of chitin, as shown in Figure 1.4(a). Therefore, we can express it in this way; due to absorption of some light from different wavelengths reflected from the particularly shaped nanostructure. These are physically bend and scattered light in the directions appeared in our eyes. Such scattering can also create iridescent in human eyes due to color changes with the angle. When this radiation falls in the infrared wavelength range and hits to chitin nanostructures, then their shape changes because of the expansion of size of molecular species. Figure 1.4(b) shows the peacock feather impressive colors where barbs project the main feather stem and barbules (~0.5 mm long) to each side of the barb generate the typical "shimmer" of iridescence.

The arthropods also uses compound eye with the nanoscale features to enhance their visual sensitivity. Like an insect's compound eye, which consists of around 50–10,000 individual facets, their eyes are studded with an array of nanoscale protuberances, this is called "corneal nipples." Corneal nipples are tiny structures in the range 50–300 nm cut the glare that reflects off the insect eye. The nanostructures of the nipple pattern on their moth eyes have also provided a new idea for antireflective coatings for the solar cells. The single-molecule precision of male silk moth however pheromones from female moth emitted up to 2 miles away. In this order, spider silks are also identified as one of the toughest materials known to man, stronger than steel and their webs can withstand gusts of wind. This kind of strength arises due to nanometers of thin crystal proteins, which are stacked with hydrogen bonds. This allows the silk to stretch and flex under pressure.

The described examples are only a few of the countless how nature employs nanotechnology in different ways. Of course, in this area we are with the most important technology, which being the human body itself. That contains billions of nanoscale machines. Therefore, it is worth to interpret the fascinating technological advances in nanoscale materials. The most important issue investigators are still struggling to build nanotechnology-based devices that can be close to nature. Herein we are introducing a few key organic naturally occurring materials.

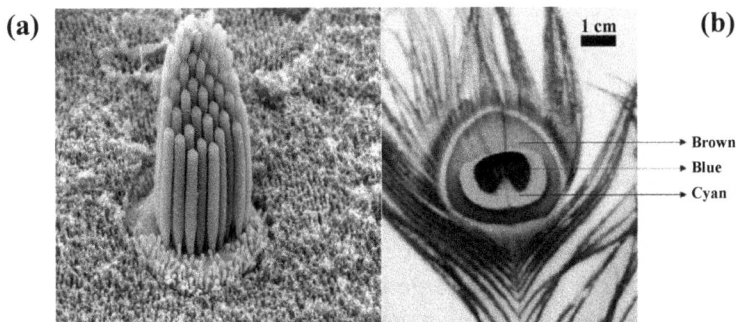

FIGURE 1.4 Natural nanotechnology: (a) sensory patch in amphibian ears (http://scinerds.tumblr.com/post/35542105310/stereocilia-stairsteps); (b) peacock feather showing barbules, this is representing a photonic lattice (Reprinted with permission from Ref. [186]. © 2010 Springer).

1.2.1 NATURAL EROSION AND VOLCANIC ACTIVITY

Due to natural erosion and volcanic explosions, the mineral worlds are produced in the form of nanoparaticles. The ashes that are released during volcanic eruptions can reach temperatures more than 1400 °C. When these ashes are realized in atmosphere and come into surrounding water contents, then due to the chemical interaction with the environment, this results in deposition of a wide range of nanoparticles, which may have negative health effect. These particles have diameter typically in the size range 100–200 nm which will be readily suspended in the air. Their suspension in the atmosphere could cause the inhalation of nanoparticles that originate from volcanic ashes which leads to serious respiratory disorders following the deposition of these particles in the upper, tracheobronchial and alveolar regions of the respiratory tract.

1.2.2 COLLOIDS

In the common day-to-day life, we are experiencing the naturally occurring liquid colloids, such as milk, human and other animals' blood, and aerosols are examples of the natural colloids. In all the naturally existed substances, nanoparticles are dispersed in the liquid or gas medium without forming a solution, but they form a colloid. All such naturally formed colloids/or nanoparticle materials have significant characteristic of scattering light which creates their color (e.g., milk and blood). Thus the naturally existed substances color arises due to the scattering of light by the nanoparticles of constituents.

1.2.3 CLAYS

Clays are the minerals that contain layered nanostructured of silicate materials. Predominantly the crystal structures of clays have been characterized as two-dimensional. As an example, mica is one of the most explored materials [8]. The substance, mica, has a large number of silicate sheets which held together with the relatively strong bonds. However, the montmorillonite and smectic types of clay have relatively weak bonds between the conjugative layers. Their each layer consists of two sheets of silica which held together through the cations, such as Li^+, Na^+, K^+, and Ca^{2+}. Such cations presence is necessary to compensate for the overall negative charge of the single layers. The formed layers diameter are around 20–200 nm laterally into aggregated form called tactoids and their thickness can be about 1 nm or thicker. The clays' fine nanostructure gives them adequate properties. Like nanostructured clay swells to several times of the original volume when water is added to it. This is due to opening of the layered structure under the influence of the water molecules that replace the cations. The swelling of the clay is a significant property particularly for the soil stability; it can be used for the road construction.

1.2.4 NATURAL MINERALIZED MATERIALS

The natural materials, such as shells, corals, and bones formed calcium carbonate crystals with other natural materials under the self-assembling process. As an example, polymer forms fascinating three-dimensional structures. Also shell is grown layer-by-layer coating of protein through the chitin support in a polysaccharide polymer. Further, proteins can act as a nano-assembly system for the controlled calcium carbonate crystal growth. Each grown crystal is almost retained a honeycomb-like matrix with the protein and chitin. The relatively flexible enclosed structures are one fundamental requirement for the mechanical properties of the shell, for example, mitigating cracking. Their typical crystal size is around 100 nm. This is a direct consequence of the mollusk shell which has extraordinary physical properties, such as strength and resistance of compression.

1.2.5 NANOSCIENCE UTILITY FOR THE BIOLOGICAL WORLD

Despite of various kinds of the utility of nanomaterials in the biological world, here we will discuss the two most significant examples of nanoscience in biological world.

1.2.5.1 LOTUS EFFECT

Athough it had been well recognized for a long time that water repellence of lotus, the scientific basis was understood in 1997 by the two botanists Wilhelm Berthelot and Christophe Neinhuis of the University of Bonn Germany. They examined leaf surfaces of lotus using a scanning electron microscope and resolve surface nanostructures in the range of 1–20 nm [9]. Figure 1.5(a) exhibits nonwettable lotus plant leaf. These leafs are having self-cleaning property due to the super hydophobicity property. This process occurs due to convex papillae on the surface of leaves which is coated from crystal wax with dimension approximately around 10–100 nm, as exhibited in Figure 1.5(b). The water drop picks up the dirt particle when it rolls off the leaf surface, as a consequence leafs are showing a self-cleaning process (see Figure 1.5(c)). Similarly, several other plants leaves are also showing self-cleaning process.

With such leafs papilla, it could largely reduce the contact area of water droplets on the leaf surface. Each epidermal cell forms a micrometer-scale papilla, which contains a dense layer of with waxes superimposed on it. The every papillae braches contains nanostructures on their surface. As an example, the lotus leaves are almost spherical, water droplets simply rolls off on it, and not come to rest on the surface even tilted slightly. This is usually referred to as the Lotus effect. Hence self-cleaning effect of the lotus flower leafs surfaces are attributed due to the combination of micro- and nanostructures formation. In this combined structures-induced hydrophobic properties provide the water and dirt-repellent actions. In the past few decades, a large number of companies realize resembling the surface morphology and chemistry of the lotus leafs, such as paint, glass surface, ceramic, tiles with the dirt-repellent properties.

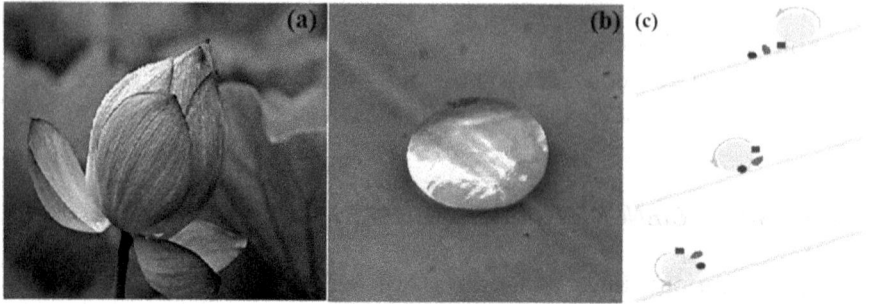

FIGURE 1.5   (a). Lotus (*Nelumbo nucifera*) plant (https://theponddigger.com/water-lotus/); (b) spherical water droplet on a nonwettable lotus plant leaf, (c) schematic of a drop which picks up the dirt particles as it rolls off on the surface for self-cleaning.

1.2.5.2   GECKOS TECHNOLOGY

In the animal world, geckos are known for the sticky toes that give them the ability to climb up walls when they hang upside down on ceiling and walk on a leaf. They have this ability due to special attached elements at nanoscale. As shown in Figure 1.6(a) and (b), a gecko's toes contain a billion of tiny adhesive hairs, their dimensions for both width and length are around 200 nm. The existence of such hairs provides a direct physical contact with the surface. The spatula-shaped hairs ends also make their adhesion stronger. The biological industry researchers imitate these structures with these properties to develop artificial dry adhesive systems. Considering their potential applications for the reusable adhesive fixture to strengthen the duct tape, this can be removed as easily as a sticky note.

Other promising examples of the naturally occurring materials, such as cotton, spider's silk, and opals can also consider mentioning due to their nano features and adequate properties. Specifically cotton has nanoscale arrangement of cellulose fiber with high strength, durability, and absorbance. Whereas, spider silk can have much higher strength (around five times) than that of steel, which made them the natural supramolecular organization of fibrolite nanoscale. The precious stone opal consists of silicon dioxide in 150–300 nm diameter spheres with hexagonal or cubic close-packed lattice. Such ordered silica spheres can produce the internal colors owing to interference and diffraction of light passing through the microstructure of the opal. This kind of realization of the nature can provide us a model to improve the engineering design. Such engineering research created field is called "biomimicking" or bio-inspired material science. Thus, it has been possible to process several types of bio-inspired nanostructures at the biological nanomaterials. Table 1.1 summarizes some potential technological applications of such materials.

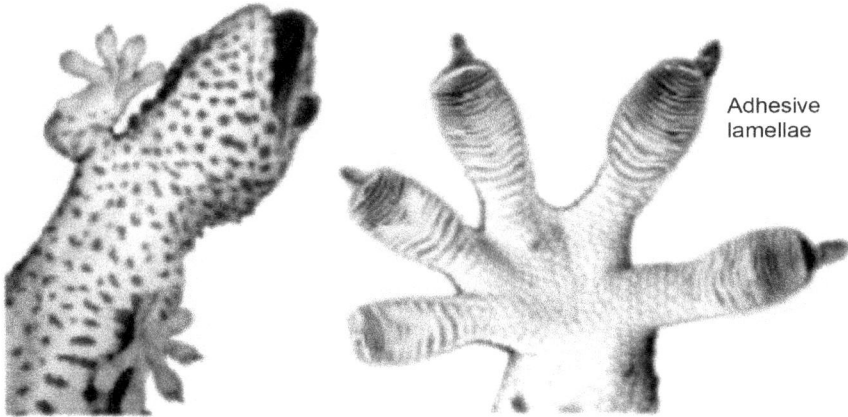

FIGURE 1.6 Gecko's adhesive system structure: (a) ventral view of a tokay gecko (*Gekko gecko*); (b) sole of the foot which exhibiting adhesive lamellae (Images has been taken from Google images).

TABLE 1.1 Adequate Properties of the Bio-Inspired Materials

Natural System/Materials	Bio-Inspired Properties
Substructure of nacre	Low-density, high-strength composites
Wood, ligaments, and bone	High-strength structural material
Eels and nervous system	Electrical conduction
Spider silk	High-tensile strength fiber
Butterfly and bird wings	Photonic crystals
Deep-sea fish and glow worms	Photoemission
Moth eye	Antireflective
Human brain	Artificial intelligence and computing
Gecko's feet	Adhesion
Lotus leaf, human skin, fish scales	Hydrophobic surfaces, self-cleaning
Shark skin	Drag reducing
Gravitropism	Gravity
Electrotropism	Electric field
Hydrotropism	Gradient of water in soil

1.2.6 HISTORICAL BACKGROUND

Thousands BC ago, people used the natural fabrics, such as flex, cotton, wool, and silk, and processed them as products. Speciality of these fabrics is that they developed the typical nanoporous materials with a network of pores in the particle size range 1–20 nm. The nanoporous natural fabric possesses high utilitarian properties of absorbing sweat, quickly swelling, and getting dried soon. In the ancient times, people also mastered in making bread, wine, beer, cheese, and other foodstuffs, but fermentation processes at the nano level were critical and challenging for them. In the Romans pre-Christian era, human had introduced the metals with nanometric dimensions in glass making. A cup which describe the death of King Lycurgus around 800 BC years ego also contains the nanoparticles of silver and gold [10]. The unique feature of this cup is that when a light source is placed inside the cup then its color changes from green to red. It was due to the presence of nano-sized particles of silver (66.2%), gold (31.2%), and copper (2.6%) embedded in the glass stoichiometry. The light absorption and scattering of these nanoparticles created different colors. Similarly, stained-glass windows of the great medieval cathedrals also contain metallic nanoparticles.

Colors of certain Mayan paintings are also due to the existence of metallic nanoparticles in their constituent [11]. In the eighth century, Mayan artisans had introduced the unique pigment which we know as Maya Blue. They had endured their lively blue tones during more than 12 centuries of the harsh jungle environment. The Maya Blue was neither an ordinary organic dye nor a simple mineral. In fact, it is a hybrid organic–inorganic nanocomposite with the organic dye molecules protected under palygorskite, which is a complex natural clay. The Asian countries (like India and China) art history is also having several examples to use nanotechnology in ancient time. A fascinating photography technique was developed in the 18th and 19th centuries by the interpretation of silver nanoparticles. In the 21st century, nanostructured catalysts have also been introduced. The precipitated and fumed silica nanoparticles were manufactured in the early 1940s and they were sold in the United States and Germany as substitutes for ultrafine carbon black for rubber reinforcements. In this order, the nano-sized amorphous silica particles were found large-scale applications in many daily-life consumer products, such as nondairy coffee creamer for the automobile tires, optical fibers, and catalyst. Moreover, nanoparticles size allows us to include colloids with soils which are used over a hundred years.

Around 1857, Faraday had introduced his experiments to use the colloidal gold. In this introductory lecture in Royal Society, Faraday had presented a purple color slide and stated that it contained "gold reduced in exceedingly fine particles, which becoming diffused, produced a ruby red fluid," "in his various preparations of gold were demonstrated, whether ruby, green, violet or blue, etc., consists substance in a metallic divided state." In order to explore such nanomaterials Faraday had postulated correctly the physical state of colloids. In addition, he had also described how a gold colloid would change color (turning blue) due to addition of a salt. Afterward, the colloidal science has been investigated widely. Further, to define the better description of these materials in the early 20th century, Gustav Mie had presented the Mie theory based on the mathematical explanation of the light scattering. As per this description, a relationship between metal colloid size and optical properties of solutions was established. To contribute to this, the Nobel Prize winner for Quantum Electrodynamics, Richard Feynman (Figure 1.7) had commented as, "Nature has been working at the level of atoms and molecules for millions of years, so why do we not?," After his call in a lecture in 1959, the nanotechnology has made tremendous progress not only in technical disciplines but also in medicine and pharmaceutics [12]. Considering all these developments in nanoscience and nanotechnology, world began speculating on the possibilities and potentials of nanometric materials and on the fact that the manipulation of individual atoms could allow us to create very small structures. Those properties would be entirely different from larger particles size structures for the same composition. In addition, they are in even more radical proposition. In principle, it was concluded that the possibility to create "nanoscale" machines is through a cascade of billions of factories. Therefore, several factories can be produced smaller scaled versions of machine hands and tools. Together with the above-discussed term and conditions, various additional factors may also impact at the nanoscale level. More specifically as Richard Feynman suggested that, when scale got smaller and smaller the gravity would become more negligible; hence, the van der Waals attraction and surface tension would become very important. He also delivered the first academic talk, where a main tenet of nanotechnology was proposed, the direct manipulation of individual atoms (molecular manufacturing). Thus, Richard Feynman is considered as the "Father of Nanotechnology," besides he had never explicitly mentioned the term "Nanotechnology."

For the utility of nanomaterials in various common purposes, their evolution of integrated chips can also be considered as a part of history of nanotechnology. The first promising electronic transistor was invented in 1947 based on a bulk macro-object. Since high-processing electronics devices keep demanding with the miniaturization, therefore, the dimensions of the transistor have been reduced considerably in the last five decades. The major achievement of the miniaturization of transistor up to the size 90 nm has been introduced in year 2002 [13]. In modern days, a single transistor in an Intel and other commercially produce high-performance core processor is around 45 nm. Further, in order to miniaturization with the Moore's law transistor would be as small as 9 nm is expected by the leading companies in the upcoming years. Though, this dimension is below the fabrication capabilities of last-generation tools used in the microelectronic industry, therefore, an extensive innovation is required. To fulfil this future requirement, numerous novel approaches such as quantum computing and molecular engineering are under investigation to achieve the workable transistor of this size.

FIGURE 1.7   Richard Feynman (image taken from the Google images).

Material science/engineering is also under an intensive examination to fulfil the future challenges with designing the suitable nanomaterials compositions for the specific use. Though several materials were produced at the nanoscale levels but yet not properly characterized due to the lack of analytical tools at this level. As an example, the process of anodizing was first reported in the early 1930s. This innovation was one of the most significant processes for the industrial use to protect aluminum from corrosion. From this process deposited thin films protected by the oxide layer on the aluminum surface. The interesting thing with this is that inventors were not aware that the protective layer is actually a nanomaterial, where the anodic layer is composed of hexagonally close-packed channels with diameter ranging from 10 to 250 nm or greater. The first time "nanotechnology" was used by the Norio Taniguchi in 1974 in an *International Conference on Precision Engineering*. As per his definition "production technology to get extra high accuracy and ultra-fine dimensions, that is, the preciseness and fineness on the order of 1 nm (nanometer), 10^{-9} m, in length." Particularly, the invention and development of nanotechnology has been enabled from the two analytical tools, which revolutionized the imaging (and manipulation) of surfaces at the nanoscale. These are the equipment, scanning tunneling microscope (STM) and atomic force microscope. These two equipment are capable of visualizing the surfaces at an atomic level resolution. These two instruments were invented by the IBM Zurich researchers, Binning and his co-workers. After the invention of these versatile tools, practically doors were opened for the nanoworld scientists. With the advancement of the STM, scientists also demonstrated that this tool not only can image surfaces with atomic resolution but can also able to move individual atoms. This is the first time that STM was realizing the Feynman's vision of atom-by-atom fabrication. The basic idea about this definition was explored with a deeper understanding by the Eric Drexler in the 1980s. He had also promoted the technological significance of nanoscale phenomena and devices through speeches and the books. The described investigations could also be useful for the various engines of creation. Thus coming era of nanotechnology and nanosystems can be useful to fabricate molecular machinery and to simulate molecular computation [14, 15]. The development of nanoscience and nanotechnology had also provided the momentum for the birth of cluster science in the late 1980s. This branch of physics deals with the synthesis and properties enhancement of metallic and semiconductor nanocrystals, which led to fast increasing number of metal and metal-oxide nanoparticles and quantum dots. In the last decade, United States National Nanotechnology Initiative was founded to coordinate Federal Nanotechnology Research and Development. In brief, the milestones related to the revolution of nanoscience and nanotechnology from ancient to modern era are summarized in Table 1.2.

TABLE 1.2   Key Innovations in the Area of Nanoscience and Nanotechnology

Year	Key Inventions
400 BC	Democritus of Abdera gave reasoning about atoms and matter
500 AD	Glazes artisan in Mesopotamia, Mayan paintings
1857	M. Faraday introduced the colloidal dispersion of gold that is stable for almost a century prior to destroyed during World War II
1931	Invention of electron microscope by Max Knoll and E. Ruska
1959	R. Feynman gave the idea "There's Plenty of Room at the Bottom" which described the molecular machines' at the atomic level building with the precision
1974	First time term nanotechnology was used by N. Taniguchi for the fabrication methods below 1 µm
1977	E. Drexler had first time introduced the concept of molecular nanotechnology at MIT, USA
1981	Invention of invented scanning tunneling microscope by G. Binnig and H. Rohrer (IBM), they had received Nobel Prize of Physics in 1986.
1985	Year of fullerene discovery by R. F. Curl Jr., H. W. Kroto, and R. E. Smelly, they had won Nobel Prize of Chemistry in 1996.
1986	G. Binning, C. F. Quate, and Ch Gerber (IBM) was invented of atomic force microscopy and Eric Drexler introduced the concept; engines of creation: the coming era of nanotechnology, in first book on nanotechnology
1991	S. Iijima was discovered the carbon nanotube (CNT)
1998	C. Dekkar and co-workers had invented the carbon nanotube transistor.
2000	S. Hell was discovered the stimulated emission depletion (STED)
2001	Discovery of the fastest silicon transistor by Intel Corporation, which switches on and off 1.5 trillion times per second
2004	A. Giem and K. Novoselov had invented the graphene and they won Nobel Prize in 2010
2004	Intel had introduced the Pentium 4 "PRESCOFT" Processor based on 90 nm technology
2006	E. Betzig and W. Moerner had discovered the single-molecule microscopy (SMM), this led to the discovery of nanomicroscopy, surpassing the limits of optical microscopy. The STED and SMM techniques super resolved fluorescence microscopy has emerged to study synapses in Alzheimer's and Huntington's disease. To gain a better understanding of protein development in embryos, for this E. Betzig, S, W. Hell, and W. E. Moerner had received Nobel Prize of Chemistry in 2014.

1.3   NANOMATERIALS AND THEIR CLASSIFICATION

In the materials world what is made of, the newly found and/or deveoped materials are of hugely interested and are constantly important. These nanomaterials are also intrinsically complicated. Such materials in general are comprised of a large number of atoms and molecules and their properties are determined by complex heterogeneous structures. These materials are historically heterogeneous and play crucial roles in determining properties that have been determined largely empirically and manipulated through the choice of the compositions of the starting materials and the conditions of processing. The nanomaterials imply those who meet at least the following criteria:

- Materials that consist of particles with one or more external dimensions in the size range of 1–100 nm for more than 1% of their composition.
- The internal/surface structures in one or more dimensions in the size range of 1–100 nm.
- More specific surface-to-volume ratio >60 m^2/cm^3, excluding materials consisting of particles with a size <1 nm.

Nanomaterials can be nanoscale in zero dimension (quantum dots), one dimension (e.g., surface films), two dimensions (e.g., strands or fibers), or three dimensions (e.g., precipitates, colloids).

1.3.1 ZERO-DIMENSIONAL MATERIALS

These materials, electrons, are confined in all the three directions at the nanoscale level. The overall dimensional confinement of these materials gives the properties at the level of zero dimensions (0D) which are more or less identical. In the past decades, the field of 0D materials has been developed and investigated widely toward their applications. The pure nanoparticles have a uniform sized particles, which are in arrays and it is called quantum dots. These uniform-sized nanoparticle are heterogeneous particles as arrays, those belong to basic core shell of the quantum dots, for examples, onions, hollow spheres and nanolenses are the 0D materials. The 0D materials quantum dots, electrons are not able to escape from their regions because of their confinement in all directions. Therefore, quantum dots can only exist inside an "infinitely deep potential well." Thus, there is negligible possibility of the delocalization of electrons. The 0D materials have predominately similar length and width. The 0D materials can have amorphous or crystalline nature.

Usually, semiconductor nanoparticles are described as quantum dots. They are reflecting the promotion of an electron from the valence to conduction bands, under the bulk lattice pattern, where an electron–hole pair (exciton) should be produced, according to quantum confinement theory. Therefore, the Bohr exciton radius (r_β) typically formed physical separation between the electron and the hole. This is not only applicable for the semiconductors but also for nanomaterials due to their compositions. The 0D materials diameter (L) of the nanocrystal and that can also be in a similar order of extent with the r_β for the more confined state. The 0D materials exciton quantum confinement effect can produce discrete energy levels at a small dimension. Therefore, quantum dots have a significant change in band gap and dimensions of a nanocrystal, which can vary by bringing in/out of the single atom. The discrete energy (E_n) of different 0D nanomaterials can be described from the following equation:

$$E_n = \left[\frac{\pi^2 h^2}{2mL^2}\right]\left(n_x^2 + n_y^2 + n_z^2\right) \tag{1.1}$$

where h is Planck's constant, m is the mass of an electron, amu, L is the orbital perimeter (nm), and n is the dimensionless dimensional coordinates.

According to the above-described equation, the 0D nanomaterials band gap values can be easily adjusted for all kinds of usual semiconducting crystals; simply altering the diameter of the quantum dot by controlling the size in comparison to r_β with dimensions [16].

Electronic properties of the quantum dots are interestingly decreased when particle size is below 10 nm and the electronic transition energy of the same material can increase up to 2 eV relative to the bulk material [17]. The first time fluorescent carbon quantum dots (CQDs) were discovered during the purification of the single-walled CNTs from the arc-discharge method [18]. The CQDs became a very important material for the 0D nanomaterial, due to their unique optical, electronic, and physicochemical properties. The 0D nanomaterials are considered as an active advanced material for the variety of applications, such as biological sensing, imaging, drug delivery, optoelectronics, photocatalysis, and photovoltaic processes. These materials exhibit easy tunability of photoluminescence, exceptional photo-induced electron transfer, and highly order light harvesting, which makes them significant for the solar energy conversion [19].

1.3.2 ONE-DIMENSIONAL MATERIALS

In the past decade, 1D structured materials (thin film or manufactured surfaces) have been widely studied. They are having two dimensions at nanoscale and one dimension outside the nanoscale. Such electrons confinement is in only two dimensions that restrict their movement. Similar to 0D materials, they can also be crystalline (single or poly), metallic, ceramic, and polymers. The nanowires, nanotubes, nanoribbons, nanoscrolls, nanobelts, nanofibers, nanorods, and nanofilaments are the best examples of 1D structured materials. Usually these materials exist in pure form; however, in some cases, they can occur in impure form (from doping in semiconductors). Such materials are available as individual material and they can be implanted with other materials too. The 1D structured materials are having significant factor in terms of greater length than width [16]. As per quantum confinement theory, the discrete energy (E_n) of 1D nanostructured materials can be obtained by the following relationship:

$$E_n = \left[\frac{\pi^2 h^2}{2mL^2} \right] \left(n_x^2 + n_y^2 \right) \tag{1.2}$$

It is well known that 1D structured materials are the versatile candidate for the novel systems due to their promising features, such as easy functionalization, size, and dimensionality. They are also useful for the energy-associated applications. They can offer a variety of benefits like easy electrical transport with direct current pathways, shorter ion diffusion length, and volume expansion as compared to nanoparticles [20]. Moreover, the porous 1D nanostructured materials can have more advantages as compared to usual 1D nanostructured materials under the organized porosity at the level of the nanoscale. Innovation of this kind of porous materials opens pathways for applications in a variety of sectors, for example, conversion and storage of energy, gas sensing and storage, adsorption, catalysis, and biosensing. The 1D architecture and porous properties of nanomaterial's combination offers flicker that may contribute in their performance toward better energy storage. The porous 1D nanostructure materials have hollow 1D geometry (also named a tubular nanostructure), hierarchical porous 1D architecture, nanoparticles embedded 1D porous configuration, and 1D porous nanoarray. These materials are available nowadays for their effective utilization. The porous nanomaterials can have high surface area with the shortest ion diffusion length, which can act as a host for fillers than nonporous 1D structures [21].

1.3.3 TWO-DIMENSIONAL (2D) MATERIALS

In 2D nanostructured materials, electron confinement is only in one dimension. The particle size lies at the nanoscale in one dimension; this indicates the surplus movement of electrons surrounded by the other two dimensions. They always appear in the plate-like shapes, such as nanosheets, nanolayers, nanofilms, and nanocoatings. Similarly to 1D materials, 2D materials are also in crystalline or amorphous forms as compared to metallic, ceramic or polymeric materials. The variety of chemical compositions has been extensively used to make single or multilayer sheet-like structures. Two-dimensional nanostructured materials can also be deposited on various kinds of substrates like metals and ceramics with the incorporation of other surrounding materials like carbon. The 2D nanostructured materials length is also more than the width like 1D materials and their electrons are also confinement in delocalization state [16]. Two-dimensional nanostructured materials quantum confinement discrete energy (E_n) can be illustrated by the equation

$$E_n = \left[\frac{\pi^2 h^2}{2mL^2} \right] \left(n_x^2 \right) \tag{1.3}$$

In the beginning, the research of 2D nanostructured materials did not receive much attention prior the discovery of graphene. The separation of graphene from graphite in 2004 boosted the development of the 2D nano-structured materials. Innovation of graphene and graphene-based materials have provided more evidence for fundamental research toward their unique properties and testing for the practical utility to the electronic and energy storage devices [22]. The 2D nanostructured material, grapheme, is distinct to graphite. The ultrathin material of graphene was the focus toward many research fields due to their tremendous favorable properties. These thin sheets of 2D materials offer extensive assortment of essential building blocks for next-generation electronic devices. Similarly, boron nitride (h-BN), black phosphors, and transition-metal dichalcogenides (TMDCs) are the recently identified as emerging fields for the 2D nanostructured materials beyond the graphene. Due to the existence of vertical confinement of electrons and holes, they may provide an exotic physics phenomena in the limit of monolayer [23]. Additionally, their naturally existing thin-layered structure provides easy circumstances to achieve better geometric dimension and the formation of useful monolayer for the modern electronic materials. They also have advantages on low power consumptions, lightweight, and flexibility [24]. For example, graphene-based materials can perform as an outstanding conductor [25]. Due to their wide band gap property, like h-BN can be used for the gate dielectric or deep ultraviolet emitters [26]. The TMDC-based semiconductors have also shown their potentials for many applications. This is due to their numerous superior quantum efficiency (QF) optical/optoelectronic applications with the elevated one-off ratio [27]. In the recent years, investigators have developed a few potential groups of the 2D materials from elements across the periodic table; those have shown better conductivity, flexibility, improved strength with the easiest chemical tunability in terms of energy-associated applications, electromagnetic shielding, and environmental remediation [28]. In the broader way, scientists have classified the 2D nanostructured materials in different groups based on the reported facts [29–34]; it is given in Table 1.3.

TABLE 1.3   Classification of the 2D Nanostructured Materials

2D Nanostructured Materials Groups	Key Classification
MXenes (Ti_3C_2, Ta_4C_3)	Around 30 MXenes were developed based on Sc, Ti, Zr, Hf, V, Nb, Ta, Cr, and Mo
Xenes (borophene)	The arrangement of boron atoms in sheets under the honeycomb pattern is called borophene, which is a metal-like conductor. Boron can easily forms the polymorphs because of its electron-deficient nature
TMDCs (MoS_2, WS_2, ReS_2)	When transition metal atom is sandwiched between two chalcogenide atoms, then between them a TMDC layer is formed. A weak van der Waals force acts to holds the TMDC layer. Due to covalent linkages, several stacks of polytypes and polymorphs occur in a single TMDC layer. The 1T, 2H, and 3R are the usual structural polytypes of TMDC that denote to one tetragonal (1T), two hexagonal (2H), and three rhombohedral (3R) symmetries, respectively.
Nitrides (GaN, BN, Ca_2N)	The boron and nitrogen atoms alternative arrangement in a honeycomb pattern containing a layered structure that retains its large band gap and dielectric properties. Usually they are insulators
Organic materials (covalent organic frameworks, 2D polymers)	Arrangement of some crystalline organic compounds as stacked molecular sheets

1.3.4   THREE-DIMENSIONAL MATERIALS

Usually bulk nanomaterials are known as 3D nanostructured materials. The 3D nanostructured materials have free nature of the particles. This reflects

3D nanostructured materials particles do not have quantized behavior during the motion. In this kind of nanomaterials, porosity plays a vital role. Because such nanometrials are formed due to building blocks of the particle, which are simply brought together to form hierarchical 3D nanoarchitectures. This can be described by using the term of nanocrystalline structure and 3D nanostructured materials are composed of numerous nanosized crystals in multiple arrangements containing different orientations. The 3D nanomaterials can be in the form of nanoparticles dispersions, bundles of nanowires, nanotubes, and nanosized multilayers.

In this decade, an intensive research that is in progress is made toward porous 3D nanomaterials and 3D porous interconnected graphene-based materials for the future energy applications. Owing to their no agglomeration property with high specific surface area, which can have powerful mechanical strengths and quick mass and electron-transfer kinetics, as a consequence, the 3D porous arrangement and outstanding fundamental properties of graphene [35, 36] can be achieved. Like graphene, the metal-organic frameworks (MOFs) porous coordination of polymers are also consisting of metal ions or clusters connected to each other by the organic linkers containing a 3D porous interconnected structure under the uniform cavities with long-range order. In most of the compounds, such as carboxylates, phosphonates, sulfonates and heterocyclic, the rings act as the organic linker and bridges of the inorganic secondary building units. It is possible to tune the size, shape, and the functional properties of the newly forming MOFs by altering the organic and inorganic materials. In the presence of porous materials, such as zeolites and carbon, the MOF has various distinct properties. But the encapsulation of molecules into the MOF is different from these materials due to their well-defined host–guest behavior. Hence, MOFs may act as a very good host to welcome the guest molecules in a well-organized arrangement all over a lattice in such materials [37, 38].

Thus to design and develop the heterogeneous nanostructures, it is important to choose a pure source material to form an appropriate nanostructure for the specific purpose. Based on the structural complexity of different heterogeneous nanomaterials, they can be classified into four major categories; zero-, one-, two-, and three-dimensional materials [39]. Their multi nanocomponents combinations were investigated with their heterogeneous nanostructured materials. Each of them tailored to address different uses, such as high-energy density, high conductivity, and excellent mechanical stability. Hence, ensuing composite materials that can exhibit synergic properties are very important in providing a lot of benefits to act as a promising material [39].

1.3.5 INTERPRETATION OF NANOSTRUCTURES

Nanostructures material can be one-, two-, or three-dimensions assembled at the nanometer scale having patterns, such as nanosphere, nanotubes, nanorod, nanowire, and nanobelt [40]. Nanostructured materials can have variety of structures as shown in Figure 1.8. The nanocrystalline is defined as the crystalline that possesses crystalline order at the nanoscale size. If the 3D nanostructured material particles are extrinsic in one-dimension then it is called quantum well and if in 2D then it is called quantum wire. When all the three dimensions of the material are at nanoscale then it is called 1D quantum dot. The nanostructured materials can also have several structures with the improved optical, magnetic, electrical, and other properties emerge as described above. The most important aspects of the nanostructured materials are the creation of functionality, devices, and systems through control of matter on the nanometer length scale (1–100 nm) and the exploitation of novel phenomena. The term "nanostructured materials" demonstrates two basic things: (1) at least some of the properties determining heterogeneity in materials occur in the size range of nanostructures (1–100 nm) and (2) these nanostructures might be synthesized and distributed (or organized), at least in part, by design. The nanostructure material mainly focused on the following points;

1. What are novel aspects which make nanostructures interesting?
2. Adopting what type of the process can they be synthesized?
3. Under what circumstances they can be introduced into materials?
4. What kind of relationships exists between compositions and structures to control the matrices and interfaces properties of the developed materials?

Hence the "Materials by Design" is coming with a goal to define material science and their applications. It remains challenging and difficult task due to a vast number of researchers are still focusing on explaining the basic questions. As per nanostructured materials concept described above, all the conventional materials, such as metals, semiconductors, glass, ceramics, or polymers can be formed structures at the nanoscale dimension levels. In principle, the nanomaterials spectrum can be inorganic to organic and crystalline to amorphous particles. These spectrums can be in the form of single particles, aggregates, powders, or dispersed in a matrix, over colloids, suspensions and emulsion, nanolayers, and nanofilms, supramolecular structures likedendrimers, micelles, or liposomes.

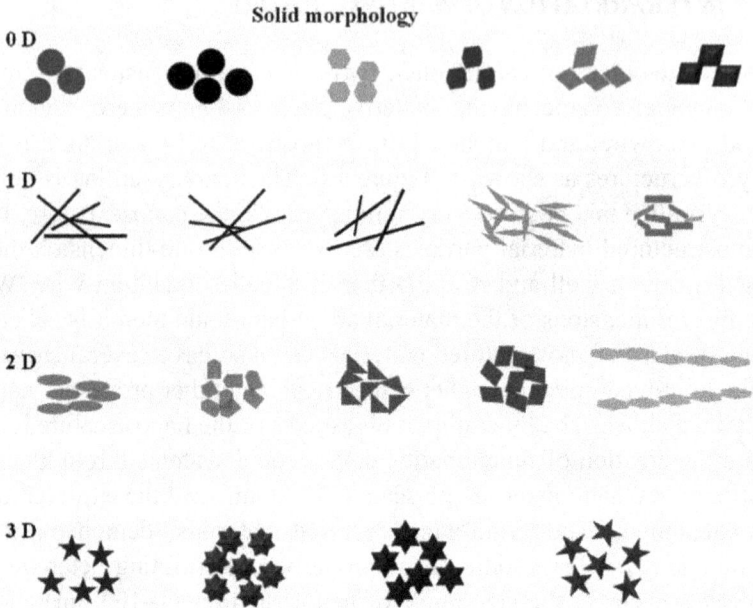

FIGURE 1.8 Typical morphologies of solid inorganic nanoparticles with 0D, 1D, and 2D shapes and 3D complex structures.

1.3.6 INTERPRETATION OF NANOPARTICLES

Nanoparticles are the intermediate state of solid particles between the macroscopic and atoms/molecules. Nanoparticles can have a large number of atoms or molecules which constitute a variety of sizes and morphologies, such as amorphous, crystalline, and spherical, needles. Therefore, nanoparticles can relate to different scientific fields, for example, chemistry, physics, optics, electronics, magnetism, and mechanism of materials. Some of nanoparticles have already been well studied in their practical from. The past studies reveal the small size impacted on large surface effect and their quantum tunnel effects on the nanoparticles special physical properties which can be used for the variety of applications. In this order, nanoparticles are available commercially in the form of dry powders or liquid dispersions. Several nanoparticles combinations of aqueous or organic liquid form a suspension or paste. For that alternation, it is required to use chemical additives (surfactants, dispersants) to obtain a uniform and stable dispersion of particles. This requires further processing of particles dispersion and nanostructured powders which can be used to fabricate surface coatings. It is worth noting that nanomaterials-based devices may or may not retain the nanostructure of the particular raw materials. The industrial scale production of nanoparticulate materials such as carbon black, polymer dispersions (or micronized drugs) have been well established for a long time. Metal-oxide nanoparticle is another important class of nanomaterials that includes silica (SiO_2), titania (TiO_2), alumina (Al_2O_3), and iron oxide (Fe_3O_4, Fe_2O_3), along with other compound semiconductors (e.g., cadmium telluride, CdTe or gallium arsenide, GaAs), metals (e.g., Ag and Au), and alloys can also include in this category which have been commercialized.

1.3.7 INTERPRETATION OF NANOWIRES AND NANOTUBES

Nanostructures such as nanowires, nanotubes, and nanorods can be obtained from different materials, for example, metals, semiconductors, carbon. The carbon nanotube is one of the most potential nanostructures in the linear arrangement that offers a variety of modification in the form of single or multiwalled filled or surface modified. In the modern carbon era, nanotubes are usually produced from the chemical vapor deposition (CVD) methods, which could be up to several tons per year.

1.3.8 INTERPRETATION OF NANOLAYERS OR NANOCOATINGS

In the area of nanotechnology, nanolayers are another important topics. The nanolayers can be directly related to nanoscale engineering of surfaces. A vast range of functionalities and new physical effects, such as magnetoelectronic or optical, can be achieved by using nanolayers. At the nanoscale, design of surfaces and layers is often desirable to optimize the interfaces for different material classes to obtain the required special properties.

1.3.9 INTERPRETATION OF NANOPOROUS MATERIALS

In the range of nanometer, nonporous materials can be defined as per their pore sizes for the special interest in term of broad range of commercial applications. The nanoscale materials can also have outstanding properties, such as thermal insulation, controllable material separation, and applicability in the form of templates or fillers. Aerogel is one of the important classes since nanoparticle materials can be produced by sol–gel chemistry. These material potential applications are not limited to the described areas; they can have a broad range of applications including catalysis, thermal insulation, electrode materials, environmental filters and membranes as well as controlled release of drug carriers.

1.4   STRATEGIES FOR THE NANOMATERIALS SYNTHESIS

Synthesis of nanoparticles is an important issue in materials science and engineering. There are many synthetic approaches that are adopted for the synthesis of nanostructured materials. There are two important classes of controlled syntheses, namely, size-controlled synthesis and shape-controlled synthesis process of nanomaterials. Two main approaches are utilized to fabricate nanomaterials: top-down approach and bottom-up approach. In top-down approach, a bulk matter with the intention to reduce to smaller sized nano-dimensional particles, while, in bottom-up approach a material is derived from its bottom stage, atom by atom, molecule by molecule, or cluster by cluster, as illustrated in Figure 1.9. Usually, there are four major processes employed for the synthesis of nanostructured materials: gaseous phase, vapor deposition, wet chemistry, and grinding. The schematic of various synthetic processes is illustrated in Figure 1.10.

1.4.1   *CHEMICAL APPROACHES*

Chemical synthesis process of nanomaterials is one of the most important approaches at the nanoscale to create the scientific and technical essential properties. The adaptability of this approach provides feasible functionalities in the final products with a better homogeneity of chemical components in a specific solution. Even though this method has several advantages, it also reflects some limitations such as usage of toxic reagents, solvents during synthesis, and the unavoidable introduction of a byproduct. Thereby, their derivatives make them enable for the processing like purification under a time-consuming approach. However, the chemical approaches have blossomed recently due to their impressive features like simplicity, cheap, and easy fabrication of nanomaterials. This approach offers diverse sizes and shapes of nanomaterials synthesis. This method also offers alternate chemicals and conditions. As an example, from this approach synthesis of

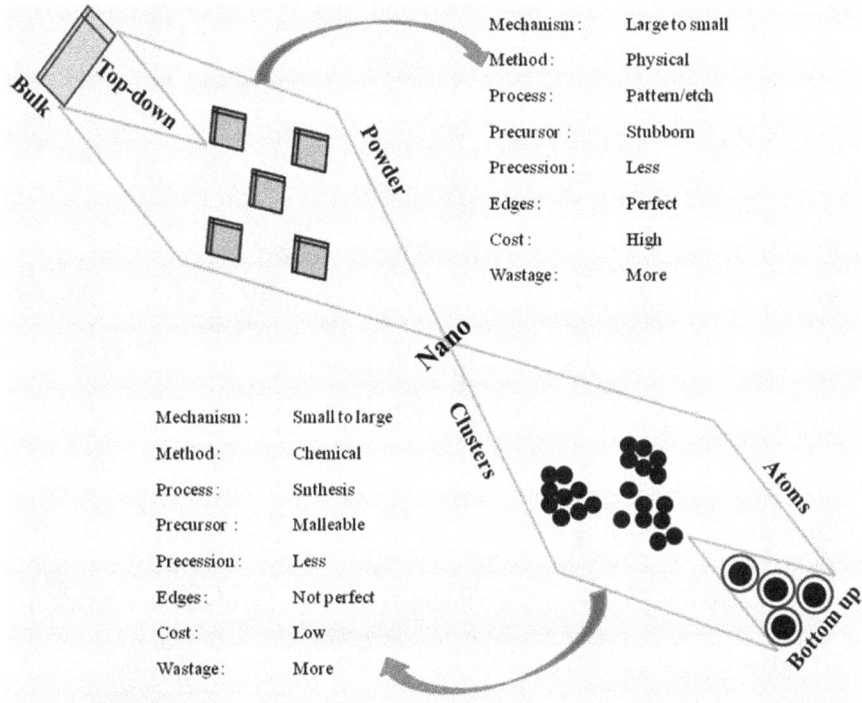

Mechanism :	Large to small
Method :	Physical
Process :	Pattern/etch
Precursor :	Stubborn
Precession :	Less
Edges :	Perfect
Cost :	High
Wastage :	More

Mechanism :	Small to large
Method :	Chemical
Process :	Snthesis
Precursor :	Malleable
Precession :	Less
Edges :	Not perfect
Cost :	Low
Wastage :	More

FIGURE 1.9 Techniques for the conversion of particles at nanoscale.

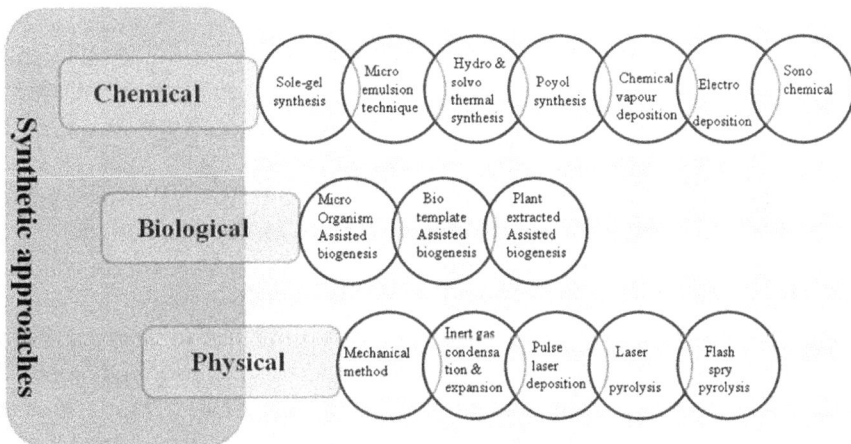

FIGURE 1.10 Schematic illustration of various synthetic approaches.

nanomaterials is feasible at any temperature (low to high) with the suitable doping of foreign material (ion). Adopting this method, a large number of materials can be obtained. This chemical approach is also far better than physical approaches in view of instrumentation. The nanomaterials obtained from this approach could enable self-assembly/patterning for the materials. Specifically, nanomaterials are synthesized through chemical methods in the form of colloids. The obtained colloids consist of two or more phases like solid, liquid, or gas. They can be in the form of particles, plates, or fibers. Nowadays, there are various chemical approaches of nanomaterials synthesis are available, some of them are discussed here below.

1.4.1.1 SOL–GEL

This is a well-known and extensively used colloidal method for the synthesis of nonmaterials, which is controlled by the chemistry of materials. In this method, usually sols are the solid particles in a liquid, while the gels are polymers containing liquid. The formation of sol–gel in this method generally proceeds at low temperature. The low-temperature formation of sol–gel shows their less energy consumption and less pollution. Usually this process starts from metal alkoxide (M-O-M) or metallic inorganic compounds (M-H-M), to allow easier reaction in solution for the production of particles. The existing precursor in a specific solvent further brings into hydrolysis and polycondensation reactions to form xerogel, which consists of invariable inorganic lattice like M-O-M or M-H-M. Thus, sol–gel process can be used to make oxide matrix under the polycondensation reactions with the subatomic pattern in a liquid [41, 42]. Sol–gel synthesis process usually correlates with the hydrolysis precursors condensation followed by polycondensation to form particles. However, gelation can be related to the drying practices from diverse routes. Hence, from the sol–gel process can be easily obtained better controlled size particles distribution and stability for the quantum confined semiconductors like metal and metal oxide. This approach also provides the foundation of a variety of material syntheses including paints, ceramics, cosmetics, detergents, and cells. Sol–gel process synthesized materials can also have noticeable advantages like ultra-small particle size, high specific surface area, extended triple phase boundary, management of composition at the molecular scale with the good homogeneity.

The versatile sol–gel method can also provide synthesis of high-crystalline birnessite (like d-MnO_2-layered manganese oxide). In recent years, synthesis of manganese-oxide-layered birnessite is getting much attention due to their usage as cathode materials for the lithium-ion batteries, as super-capacitors and as water oxidation catalysts. The synthesis of outstanding pioneer oxide-based manganese materials like $LiMn_2O_4$ was achieved using sol–gel method at low temperatures. Usually, highly crystalline birnessites are prepared either from the hydrothermal reforming process or by the high post-treatment of temperature (400–500 °C). From this method, the highly crystalline birnessites can be obtained within 1 h without any additional postprocessing to improve crystallinity. Their perfect crystallinity can be patterned effectively in the presence of Li^+, Na^+, and K^+ with the time varying adjustable crystallite size [43]. Similarly, ZnO nanoparticles can also be synthesized from the sol–gel method for the antimicrobial functions with the varying stirring conditions (500, 1000, 1500, and 2000 rpm) under the controlled particles size, morphology, and thermal stability [44].

1.4.1.2 *HYDRO/SOLVOTHERMAL APPROACH*

The hydro- and solvo-thermal synthesis process of nanomaterials is also one of the popular methods. It has been gathering more attention from scientists and technologists of distinct areas. The hydro- and solvo-thermal techniques can be differentiate only from a small diversity in the usage of the solvent. The word hydrothermal is self-explanatory; "hydro" means water, while term "thermal" represents the heat. In a similar way, the word solvothermal, "solvo" means solvents. According to the principle, the hydrothermal synthesis involves water as a catalyst in a closed stainless steel container at an elevated temperature of above 100 °C, and at a pressure which is typically greater than a few atmospheres. The usually used vessel in this type synthesis is called an "autoclave," which can sustain a high temperature and pressure. To get larger crystal size with better quality of nanostructured materials, the hydrothermal technique is suitable. This is a well-established concept that materials with high vapor pressure close to their melting point or crystalline phases, usually is not stable at their melting point. To synthesize this kind of materials, the hydrothermal process offers a better approach. With this method, variety of oxides-, sulfides-, carbonates-, and tungstates-based nanomaterials have been synthesized with uniform shape and size, by altering the experimental parameters including reaction time, temperature, type of solvent, surfactant type, and precursors in hydrothermal synthesis. In comparison to other available advanced methods, the hydrothermal method offers a lower cost of instrumentation, energy, and precursors. This synthesis process is also environmental friendly as compared to other chemical methods. It does not require any seed or catalyst which is harmful or expensive surfactant/ templates. Therefore, this process is promising for the large-scale synthesis and high-quality crystals at low cost. Similarly, the solvothermal processes also receive wide consideration by utilizing a range of solvents or mixed solvents. This method can be utilized for the "in situ" type of reactions, such as oxidation reduction, hydrolysis, thermolysis, complex formation, and metathesis reactions [45].

1.4.1.3   POLYOL SYNTHESIS

The polyol method is a liquid phase synthesis process; this method is useful for the multivalent alcohols at high temperatures up to its boiling level. For this synthetic process, ethylene glycol is the simplest representative in polyols. It is well established that there are diethylene glycol, triethylene glycol, tetraethylene glycol, and continues up to polyethylene glycol. Further, it also contains more than 2000 ethylene groups with the molecular weight roughly up to 100,000 g/mol. These chemicals are extensively used for the synthesis of nanostructured materials, where the boiling point of the polyols plays the crucial role for the reactions. Specifically, when the number of OH increases then boiling point of the polyol is also increased with an increase in its molecular weight. The newly formed nanomaterial polarity and viscosity are also increased with increasing molecular weight. The solubility of polyol compounds is comparable to water which provides an easy path to use, with the low-cost metal salts, such as nitrates, sulfates, and halides as precursors. The polyol can be considered equivalent to water with their high-boiling solvents. Among these, polyol also has a property of chelating effect; this is one of the major advantages to control the nucleation process of a particle, particle growth, and agglomeration which may arise during the synthesis. The high viscosity of polyols can also provide the better benefits. Usually in this synthetic process, the temperature is around 200–320 °C, without high pressure or autoclave. In this synthetic process, an instant reduction reaction is possible in metal cations toward nanoparticles with simultaneous adequate surface functionalization. Thus, polyol synthesis is the combination of numerous characteristics under a one-pot reaction [46]. For example, by using this synthetic approach ultrathin diameter (2-nm) sized palladium wavy nanowires can be synthesized [47].

1.4.1.4 SONOCHEMICAL APPROACH

A few typical nanomaterials are synthesized using the ultrasound irradiation to achieve the nucleation process; this approach is so-called the sonochemical synthesis method. Usually the irradiation frequencies are in the range of 15 kHz–1 MHz. In this process, creation, growth, and collapse of a bubble in solution liquid are the important parameters for the sonochemistry. As we know, the ultrasound is not audible when it is transmitted in the air. But in the case of a liquid medium due to ultrasonic oscillations, a pressure is formed which acts as the foundation for in-phase expansion and contraction of the dissolved gas bubbles. The expansion and collapse of gas or vapor bubbles under an acoustic field is called acoustic cavitation. This is due to the interaction between sound waves and bubbles in aqueous solution. This process instantly generates extreme temperatures within the cavitation bubbles. Therefore, this results in the development of extremely reactive radicals. As an example, the homolysis of water molecules creates the high temperature within the bubbles generates H and HO radicals. However, available surface-active solutes perform the reactions due to the formation of secondary reducing radicals, usually such process being utilized to reduce metal ions. The different shapes and size distributions of gold nanoparticles have been synthesized through the sonochemical process.

1.4.1.5 MICROEMULSION METHOD

The formation of emulsion is nothing but the stirring or mechanical agitation of two immiscible liquids. In the range of up to a few millimeters to 100 nm that is generally turbid in appearance. Similarly, an additional category of immiscible liquids is called microemulsions that are transparent in the range of approximately 1–100 nm. Usually they are optically clear fluids and thermodynamically stable. In the microemulsions process, size and shapes are stabilized by surface-stabilizing active agents, this is called surfactants. The surfactants are having unique physical properties. The most important aspect of the microemulsion technique for the formation of cavities is that it leads to better synthesis of nanomaterials with the superior advantages such as biocompatibility and biodegradability.

1.4.1.6   CHEMICAL VAPOR DEPOSITION

The CVD is one of the widely used chemical methods to fabricate highly pure advance performing materials. CVD is a hybrid process for the coating of various inorganic and organic materials by using vapor phase of different chemicals. The chemical reactions and thermal decomposition of gas phase species occurs at an elevated temperature in the temperature range 500–1000°C. Generally reactants crack themselves to form dissimilar products at a fixed high temperature. These intermediate products diffuse to the surface of the substrate and undergoes in various surface chemical reactions at a suitable site. Subsequently, nucleation and film growth are formed for the required material on the substrate [48]. In the CVD process, there are five main steps involved: (1) the mixture of reactant gases and diluent inert gases are positioned into the chamber and their flow rate adjusted with the help of mass flow controller. (2) The gas species should be in motion to the surface site. (3) The adsorption of gas species takes place on the surface site. (4) The nanostructured materials are produced due to chemical interactions between the reactants and substrate. (5) The final by-products that are in gaseous/solid particles form are separated by desorption from the evacuated chamber.

CVD process deposited materials have different forms, such as monocrystalline, polycrystalline, amorphous, and epitaxial. Those produced materials could have different elemental compositions, such as silicon, carbon fibre, filament, silica (SiO_2), silicon–germanium (Si–Ge), tungsten (W), silicon carbide (SiC), titanium nitride (Ti_3N_4), distinct high-K dielectrics, and synthetic diamonds [49]. Usually aligned CNTs are synthesized from CVD, with the solution of xylene–ferrocene precursor. Nevertheless, xylene acts as the carbon source, while at the same time ferrocene acts as seed catalyst to supply the iron metal nanoparticles [50]. Further advancement in conventional CVD technique, a few upgraded techniques have been introduced, such as moderate temperature CVD and metal organic CVD (MOCVD) (metal organic precursors). These processes decompose materials at a moderately low temperature of approximately 500 °C. With these techniques, it is possible to deposit nanostructures at just above the room temperature by using plasma or laser beam to activate the vapor phase reactions [51]. Furthermore, the highly flexible and transparent graphene films can be fabricated through CVD and use as transparent conductive electrodes for the organic photovoltaic (OPV) cells. CVD process deposited graphene films are then transferred to transparent substrates to form organic solar cell heterojunctions, such as, polystyrene sulfonate/copper phthalocyanine/

fullerene/bathocuproine/aluminum [52]. The CVD process nanomaterials synthesis schematic is represented in Figure 1.11(a) and (b).

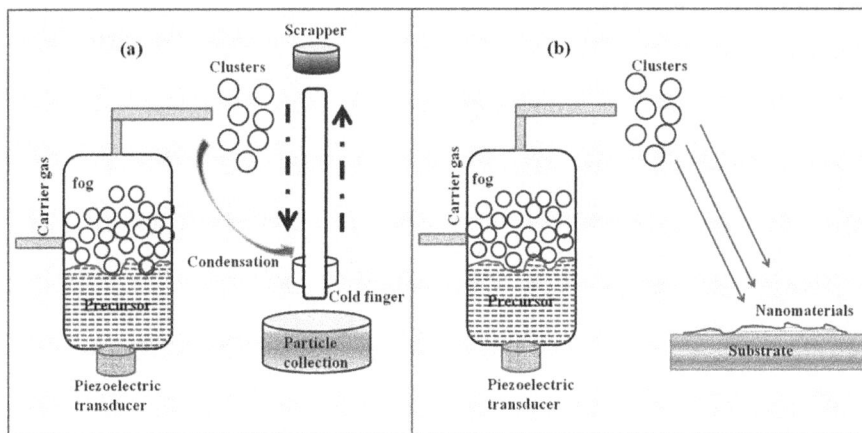

FIGURE 1.11 (a, b) Schematic representation of the nanoparticle and particulate film formation from CVD process.

1.4.1.7 ATOMIC LAYER DEPOSITION

Atomic layer deposition (ALD) has also become a unique deposition technique due to their properties, such as high conformality, high layer quality, and thickness control down to Angstrom level. However, deposition rate is very low in conventional ALD reactors. To get high throughput with the reduce cost, investigators have introduced the spatial ALD in the recent years. The special ALD has remarkable difference to conventional method in terms of precursors that are dosed separately using the purge (or pump step); under the simultaneous and continuous process. This allows the high speed operation rate of ALD. The advance ALD equipment deposition rate can exceed by 1 nm/s. Thus the versatility of this equipment attracted much attention in the running decade to deposit various kinds of thin films on various substrates for different scientific and technical uses.

1.4.2 PHYSICAL APPROACHES

Physical approaches of synthetic processes provide an ecological and environmental pathway to construct clean surface nanostructured materials.

Usually these methods are time and energy consuming. To overcome this problem, they synthesized nonmaterials at high temperatures. Therefore, such processes are governed under the top-down approach. Physical synthetic approaches have major advantages with solvent free uniform monodispersed particles.

1.4.2.1   MECHANICAL METHODS

Mechanical method is one of the key synthetic physical approaches. It is used to synthesize nanostructured materials in the powder form. This approach is further classified into two important subcategories.

1.4.2.1.1   Ball Milling

To make the powder form of nanoparticles, the high-energy ball milling is one of the finest techniques. In this process, numerous types of mills can be used, such as planetary, vibratory, rod, and tumbler. The selection of mills depends on the requirement of nanomaterials; in this process one or more containers can be used to make large quantities of fine particles. The advisable mass ratio of balls for the material is typically in the range 2:1. In this synthetic process, half-filled containers are advisable, because exceeding the quantity can reduce the efficiency of the milling process during collision. To get small grain size, it is better to use larger balls for milling. Usually temperature increases in the range of 100–1100 °C during collision process. From this process, amorphous particle formation is also possible at lower temperatures. For the materials nanoparticle synthesis, liquid can also be used at some stage in the process of milling. In this approach, mostly containers are rotated on their own axis with high speed and sometimes around their central axis also. The rotation in this method is called planetary ball milling. To achieve fine uniform-sized nanoparticles, the speed of rotary motion of the central axis and the container with the time span of milling should be controlled. This method offers synthesized material quantity from a few milligrams to several kilograms. The metallic materials, such as Co, Cr, W, Ni–Ti, Al–Fe, and Ag–Fe, nanoparticles can be prepared using the ball milling process. The nitrogen-doped carbon nanoparticles can also be produced at large scale from this technique. Graphite nanoparticles can also be obtained from this process.

1.4.2.1.2 Melt Mixing

The most extensively used mechanical process is melt mixing. This method is one of the oldest methods used especially for the preparation of polymer composites with nanoparticles in the form of fillers. This synthetic approach involves mechanical mixing of a polymer with modified nanofillers through extrusion or kneading. From this method, one can also obtain required material characteristics in terms of an eco-friendly approach for the commercial and industrial applications [53].

1.4.2.2 INERT GAS TECHNIQUE

For the vapor condensation fabrication process, the most common method is used to synthesize the fine noncrystalline or amorphous alloys with the varying temperature of the substrate as well as reaction conditions. This technique mostly used the synthesis of high-quality pure nanostructured materials, specifically materials containing metals. The stable complex core–shell and three-layer Mn–Bi nanoparticles can also be synthesized from the single step inert-gas condensation method. In the sputtering process, a mixture of Ar and He gas can be used as a carrier gas. The core–shell and three-layer nanostructure can also be achieved under the controlled thermal environment. In this process, there are two forms of particles, such as crystalline Bi core with an amorphous Mn-rich shell and a crystalline Bi annular shell between two amorphous layers with high Mn concentration. These synthesized nanoparticles have been shown significant magnetic hysteresis with a change in bond length between Mn and Bi atoms [54]. There are two key processes in this category.

In the first process, an inert gas (usually argon or helium) is periodically introduced into a vacuum chamber, which contains the vapor of an inorganic material. To vaporize the specimen, an evaporation crucible, a sputtering target, or a laser-ablation target is typically used. Due to boil-off, atoms rapidly collide with the inert gas and lose their energy. Furthermore, the vapor cools quickly to allow supersaturate vapor to form the desired nanoparticles and collected on a finger with the help of liquid nitrogen, in the range of 2–100 nm. Finally, the particles are scrapped and collected for further processing. The whole process proceeds under inert-gas environment. It is also possible to obtain the alloys using the same procedure by utilizing dual metal sources.

In the second process, the evaporated atoms are transferred through a high-pressure helium gas stream that pushes the atoms to a low-pressure chamber with supersonic velocities; this is called inert-gas or free-jet expansion. In this process, sudden cooling and adiabatic expansion occurs of the evaporated atoms and form clusters of a few nanometers diameter. Though this technique is fairly good but it has limitation like agglomeration.

1.4.2.3   PULSE VAPOR DEPOSITION

This process is widely used for the deposition of thin layers of material onto the substrate. Usually their surfaces are in the range of a few nanometers to several micrometers. Therefore, the collection of procedures and technologies are collectively called physical vapor deposition (PVD). This process mainly covers the following technologies;

- pulsed laser deposition (PLD),
- arc deposition from the cathode,
- electron beam PVD, and
- sputtering deposition.

In all the above-described techniques, a vacuum chamber is required with very high vacuum chamber. To propagate the reaction under a vacuum, it is necessary to have free space in the chamber when the atoms are emitted from the target.

1.4.2.3.1   Pulsed Laser Deposition

The pulsed laser deposition method is also known as laser-ablation method, which uses the high-energy laser pulses to evaporate the material from the solid-source target. Hence the use of laser is mostly in the form of pulses, so it is called PLD; however, sometimes the laser may be continuous. The laser-ablation technique provides flexibility toward to construct micro- and nanostructures of polymeric materials. In the PLD process, the high power laser pulses strike to the surfaces of the target. Due to high energy of laser pulses, it leads to the melting and evaporation process and at the end ionization of the material. The ionized evaporated target material particles are deposited onto the specified substrate. In general, excimer laser or Nd:YAG (neodymium-doped yttrium aluminum garnet; $Nd:Y_3Al_5O_{12}$) laser is used for the ablation process. Using PLD process, various kinds of nanostructures can be synthesized, which include oxides, metallic systems, fullerenes, polymers, carbides, nitrides, etc. This technique can also be used for the hybrid system with the combination of PLD and magnetron sputtering (MS) to deposit high-quality thin films. Both processes PLD and MS can be used together for the same target to get an enhancement in deposition rate, such as deposition of titanium dioxide and bismuth-based perovskite oxide Bi_2FeCrO_6 (BFCO) thin films on Si and $LaAlO_3$ substrates [55].

1.4.2.3.2   Electron Beam Vapor Deposition

Electron beam vapor deposition technique is the vacuum-based PVD process for the nanostructured materials. Electron beam system consists of a vacuum unit having an electron beam source and target materials. In this process, usually a charged tungsten filament is used for the electron source, which heats through the passing current to produce the electron beam. In this method, the generated electron beam can be directed onto the target material using the efficient magnets. Hence, in this process the high kinetic energies electrons strike at the target material and heated at a tiny spot and sublimate the materials instantly that is used for the deposition. The key advantage of this technique is the high deposition rate with the possibility of depositing materials in the conducting to insulating states range.

1.4.2.3.3   Sputtering Deposition

Sputtering technique is the PVD method based on vacuum condition; this involves ejection of electrons from a target or by striking the target with high-speed ions. Sputtering deposition technique works on the principle of the transfer of momentum when the atoms of the target are ejected by ion bombardment. In this process, gas atomic weight may be close to the atomic weight of the target materials; this is the most favorable criterion. As an example, deposition of oxide materials from the sputtering, in which only oxygen can be used as the reactive gas. The low melting point materials can also be deposited by sputtering technique; this is one of the key advantages of this method. This technique is mostly employed for the semiconductor industries and antireflective covering of optical lenses.

Typically, the sputtering deposition technique of materials contains the following steps:

1.   Usually Ar-like neutral gases are used to create plasma between the two electrodes due to the collision of electrons with gaseous molecules.

2. The potential difference between the two electrodes accelerated the ions toward the target material.
3. The ions those have suitable energy strike on the target species that leads to ejection of the ionized materials.
4. In the final step, the ejected material can transport and deposit onto the substrate.

1.4.2.3.4   Arc Deposition

The electric arc deposition technique is a method where material evaporates from the cathode target. Due to the effect, location of electric arc proceeds toward the formation of materials, this location is known as cathode spot. This is the small spot at which electric charge strikes and evaporation of the materials takes place from the target. In this process, the arc is moved across the whole region of the target to avoid the burning or holes creation. During this process of deposition, highly moving particles with a huge power can be noticed. This method frequently has been used for the deposition of metallic, ceramic, and composite materials for the hard layers (tools for cutting, drilling, etc.), that enhances their effective usage with extending service life.

1.4.2.4   LASER PYROLYSIS

As compared to physical vapor deposition methods, the laser pyrolysis provides more confined deposition in a small area with a quick heating and faster quenching of particle growth (in a few milliseconds). Additionally, this technique also offers the possibilities of narrow size nanoparticles distri-bution in the range 5–60 nm. In this process, reaction takes place between the laser beam and the molecular flow of gaseous/or vapors phase reactants and their condensable products are generated at the interface. The governing criterion of this process is either the precursor or the reactant that is able to absorb the energy which is supplied through the resonant vibrational mode of infrared CO_2 laser radiations. For this method, other chemicals, such as ammonia (NH_3), sulfur hexafluoride (SF_6), and ethylene gas (C_2H_4), can be used. Along with a large number of oxide nonmaterials with the binary and ternary compositions, such as TiO_2, SiO_2, Al_2O_3, Fe_2O_3, Si, SiC, Si_3N_4, MoS_2, and Si/C/N, Si/Ti/C can also be prepared by laser pyrolysis.

The advancement in this technique offers the direct laser writing as a novel process toward the fabrication of various nanostructures with low cost, high efficiency, and flexible design ability. The direct laser writing method leads to in-situ growth and patterning facility with high design ability which enables growing materials at the desired locations. As an example, the hybrid MoS_2/carbon materials can be synthesized by using electrocatalysts under the hydrogen evolution reactions [56].

1.4.2.5   FLASH SPRAY PYROLYSIS

This technique is based on the flame of aerosol; this is a one-step incineration method where the precursor is in the liquid form having a high combustion enthalpy (more than 50% of total energy of combustion), this is usually kept in an organic solvent. For the nanoparticles formation, this approach favors the liquid precursor by following one of the routes: the droplet-to-particle route or gas-to-particle route. To get nanomaterials from this method following mentioned sequential steps should be followed:

1. Formation of metal vapors from precursors either using evaporation/ decomposition,
2. Nucleation stage as a result of supersaturation,
3. The growth is due to coalescence and sintering
4. The structural growth may be due to aggregation (via chemical bonds) or agglomeration (via physical interactions) of particles.

This technique is usually employed to construct complex and functional nanostructured materials. This approach is widely used to design and fabricate various oxide and nonoxide ceramic nanomaterials, such as CeO_2, SiO_2, Al_2O_3, MgO, CeCu, $BaCO_3$, CaO, $CaCO_3$, and $CaSO_4$ [57].

1.4.3   BIOLOGICAL SYNTHETIC APPROACHES

Several chemical processes occur in the naturally living cells which have yet to be fully understood. Their atomic/ionic/molecular interactions and diffusions are not controlled thermodynamically but are determined kinetically. Nature in this way provides one of the fine paths to synthesize novel materials. In the previous studies, it was well accepted that most of the natural biological systems follow the bio-lab rule under the eco-friendly circumstances. Such eco-friendly materials processing is known as green synthesis. From this process, various pure metals, metal-oxide nanoparticles, and composites with the special structural features can be synthesized; this is also called biomimetic approach. Usually biological approach offers a few special features than chemical methods such as green synthesis process, economic, and energy saving. A large number of metals, alloys, semiconductors, insulating materials with their composites have been synthesized from the biosynthetic process. Using the green synthesis process, several biological materials with bacteria, fungi, and plant extracts have also been synthesized. Predominately, the existing phytochemicals in biological systems are responsible for the formation of metal/metal-oxide nanostructured materials. Though they contribute in the synthesis process, it is not easy to drive the required mechanism. As per available evidences, it is concluded that nanomaterials are formed due to phytochemicals, while, in metal oxide formation inclusion of oxygen either from the atmosphere or from the degrading phytochemicals [58]. The biological materials are classified in the following categories based on the type of biological materials or the organism:

- microorganism-assisted biogenesis,
- biotemplates-assisted biogenesis,
- plant extracts-assisted biogenesis.

1.4.3.1 *MICROORGANISM-ASSISTED BIOGENESIS*

Organisms like bacteria, fungi, or yeasts are observed only from the microscope, usually they are called microorganisms. Predominately, they are classified into two categories on the basis of utility. For example, in the formation of bread, curd, cheese, vaccines, and alcohols usually bacteria are used. In the formation of these substances, some bacteria are dangerous; these are mainly responsible for spoiling foodstuffs or laying the foundation for so many diseases. This could be understood as; the microorganism's interactions with the metals through their cell walls under the unstable process of cells complexity during the synthesis. Such biological systems interaction with bacteria depend on the temperature of the environment, pH of the reaction and system pressure which can alter materials shape, size, and possessing under the high catalytic activity [59]. In the microorganism-assisted process, initially microorganisms hold of the target ions from their surroundings and subsequently release the metal ions into their corresponding element metal through enzymes. This process formation can be classified on the basis of location of formed nanomaterial as intracellular (transfer ions into the microbial cell) and extracellular (catch the metal ions on the surface of the cells and reduce) synthesis in the presence of suitable enzymes [60]. Commonly, bacteria are used in the nanomaterials synthesis, such as gold, platinum, palladium, silver, cadmium sulfide, and cadmium telluride, because to a certain extent mycosynthesis can provide an easy path for the synthesis of stable nanomaterials. Under a superior bioaccumulation capability of fungi with the downstream processing, these are considered to be important strategies toward efficient culture design and economic fabrication of nanoparticles. However, fungi can have greater tolerance capacity as well as binding potential with metal salts with the superior yield of nanomaterials as compared to microorganisms. For the large-scale production, yeasts have also been extensively used for the fabrication of nanoparticles from the extracellular processes under the basic downstream processing. However, viruses can also be utilized for the synthesis of nanowires with functional components for diverse applications, such as PV devices, electrodes for battery, and supercapacitors. Though microorganisms have been used for effective synthesis but it is a time-consuming process with less productivity of nanoparticles. The complex steps involvement in microbial sampling, isolation, culturing, and maintenance are the major problems in such synthesis processes [61].

1.4.3.2   BIOTEMPLATES-ASSISTED BIOGENESIS

Several biological systems have self-assembled tendency to form hierarchical structures which is called biotemplates that exhibit specific dimensions and arrangement of nanostructured materials. The biotemplates are widely used to produce nanoparticles with the arrays, superlattices, or hierarchical structures of various inorganic materials. A lot of biotemplates are available, such as bacteria, textiles/paper, hair cells, insect wings, spider silk, wool, wood, onion, eggshell membranes, DNA, viruses, and diatoms. Though a large number of templates are available, only a few of them have shown technical feasibility and economic value. The DNA, S-layers (i.e., surface layers of cell walls of some bacteria) and some membranes have also possessed biotemplates due to their ordered arrangement in nature along with some constituent groups. Therefore, a number of periodic active positions are created for the anchoring of nanomaterials due to the interactions between templates and materials. In case of proteins biotemplates, numerous binding sites can easily attach to a metal ion and further reduce. The proteins' biotemplates can also provide outstanding scaffolds for the template-driven nanostructured materials with special shape and size arrangement [62]. Such materials are classified into two types on the basis of their structure/shape and filamentous proteins. The ferritin and bovine serum protein are the most commonly used spherical proteins, while, collagen, silks, wool, elastins, actins, keratins, myosins, and flagellins are the widely used filamentous proteins for the synthesis of diverse structured nanomaterials [63].

1.4.3.3   PLANT-ASSISTED BIOGENESIS

Plant-assisted biogenesis process has found a significant attention for extra-cellular synthesis of nanostructured materials, due to their nature of sample and effective alternative scaffolds. In this process, various kinds of live plants can be used, such as plant biomass and biomolecules for the extraction. This process provides a novel research which is named as phytonanotechnology. This also provides us a new possibility toward eco-friendly, simple, fast, stable, and economical synthesis process of nanostructured materials. This method uses the universal solvent (water) as a medium for reduction, biocompatibility, and scalability. The mechanism of the plant-mediated synthesis process elucidates the reduction with the formation of specific shapes/structures and the components that are involved [64].

1.5   IMPORTANCE OF NANOMATERIALS

The major importance of nanomaterials is due to their small size of the building blocks with the high density of interfaces, such as surfaces, grain, and phase boundaries containing various kinds of useful defects such as pores [65]. These additional novel physical and chemical properties can improve the functionalities of nanomaterials substantially. These novel nanostructured materials can have distinct physical and chemical properties, such as optical absorption and fluorescence, melting point, catalytic activity, magnetism, and electric and thermal conductivity, than the corresponding same bulk materials. In the broader way, the nanostructure-induced effects can be distinguished in two types:

- The size effect at the quantum size level, where the normal bulk elec-tronic structure is replaced by a series of discrete electronic levels.
- The surface- or interface-induced effect is important due to a large number of particles increase in the specific surface of the systems.

Usually particle size effect is considered to describe physical properties, while the role of chemical processing is related to the explanation of the surface/or interface-induced effect eminent property, for the particular connection with heterogeneous catalysis. The experimental evidence has been demonstrated the quantum size effect for the small particles produced from different methods. However, the surface-induced effect has been evidenced from the measurement of thermodynamic properties, such as vapor pressure, specific heat, thermal conductivity, and melting point of small metallic particles. Both physical- and chemical-processed nanostructured materials size effects have also been separated on the basis of their optical properties in the case of metal-cluster composites. For the extremely small nanoparticles (<10 nm), semiconductor, metal particles in glass composites, and semiconductor/polymer composites have demonstrated the interesting quantum effects with the nonlinear electrical and optical properties in terms of optoelectronics applications [66]. Nanostructured materials have also shown beautiful color alternation with their unique properties. The gold nanoparticles are one of the most suitable examples owing to their historical art, medicine, and recent applications in optoelectronics, sensors, and photovoltaics. Due to particle size effect makes their unique combinational properties in a range of medical diagnostic and therapeutic applications. The gold nanoparticles morphology (size and shape) and their surface/colloidal properties play a vital role for the various applications [67].

Thus, in terms of technical applications the ultrahigh surface effect, ultrahigh volume effect, and quantum size effect can have a strong influence on the material performance. Likewise, the ultrahigh surface effect can increase surface atoms, the ultrahigh volume effect can be correlated to more availability of electron due to particle size reduction, while the quantum size effect is related with the increase in energy; therefore, quasi-discrete energy of the electron orbital around the Fermi energy level can exist as a consequence of particle reduction.

1.6 NANOSCIENCE SIGNIFICANCE

There is a large number of information available on websites related to nanomaterials news as well as scientific journals belonging to the area of nanotechnology. This technology is developed to meet the global energy needs with clean solutions provide abundant clean water globally, increase the health and longevity of human life, maximize the productivity of agriculture, making powerful information technology available everywhere, and even enable for the development of space travelling. The nanotechnology covers the whole scientific and engineering field and is not limited to a single product or group of products. Therefore, there are numerous areas of nanotechnology associated with many applications. Due to increasing understanding of the relationship between shape, size, and their physiochemical and biological properties, nanomaterials can be considered as futuristic material for the diverse technologies. The recent potential areas of nanomaterials applications are transport, manufacturing, biomedicine, sensors, environmental management, information and communications technology, materials, textiles, equipment, cosmetics, skin care, and defence [68]. Here some of the emerging potential features of nanomaterials are listed in Table 1.4.

TABLE 1.4   Nanomaterial Characteristics and Importance

Characteristic	Importance
Size	Key defining criteria for a nanomaterial
Shape	Carbon nanosheets with a flat geodesic (hexagonal) structure to improved performance in epoxy composites versus carbon fibre.
Surface charge	Surface charge can impact adhesion to surfaces and agglomeration characteristics. Nanoparticles are often coated or "capped" with agents like polymers (PEG) or surfactants to manage the surface charge.
Surface area	This is defining the surface-to-weight ratio for nanomaterials is huge. As an example, 1 g of an 8-nm-diameter nanoparticle has a surface area of 32 m^2
Surface porosity	Several nanomaterials show characters of zeolite-type porous surfaces. These engineered surfaces are designed for maximum absorption of a specific coating or to accommodate other molecules with a specific size.
Composition	This deals with the nanomaterials stoichiometry which is being achieved. The purity of nanomaterials, impact of different catalysts used in the synthesis, and the presence of possible contaminants need to be assessed along with possible coatings that may have been applied.
Structure	This is defining the nanomaterials heterogeneity and providing information concerning crystal structure and the required grain boundaries.

1.7   OPTOELECTRONICS APPLICATIONS OF NANOMATERIALS

Nanostructured materials have a wide range of applications in the field of electronics, optoelectronics, fuel cells, batteries, agriculture, food industry, and medicines. It is evident that nanomaterials split their conventional counterparts because of their superior chemical, physical, and mechanical properties and of their exceptional formability. Herein, we are specifically concern to a few key optoelectronics application of nanomaterials.

1.7.1 PHOTOVOLTAIC CELLS

Photovoltaic cell is one of the highly demanding technologies in this modern era. Basically, photovoltaic solar cells are the devices that provide a direct conversion of solar radiation into electrical energy (photoelectric effect). The photovoltaic energy (or renewable energy) conversion has become more prominent due to pollutant-free source of electrical energy. Essentially, there are three types (or generations) of solar cells that have attracted much attention: first, based on poly- and monocrystalline-silicon wafers (first generation), second is the thin films (second generation) based cells, and third-generation solar cells, including multilayer or "tandem," organic, dye-sensitized, quantum dot (QD), and perovskites solar cells [69, 70]. However, since the innovation of the photovoltaic mostly are silicon-based solar cells covering around 85% of commercial solar cells and remaining 15% contribution are from the thin film solar cells (second generation) [71]. The conversion efficiency of solar cells was first established by the Shockley–Queisser, considering around 31% theoretical upper limits for a massive semiconductor with 1.1 eV energy gap, such as silicon [72]. While in the case of a tandem device containing several semiconductor absorbers or with hot carrier solar cells (e.g., a few third-generation solar cells), the maximum thermodynamic limit for solar photon conversion to free energy without solar concentration can reach 66% [73]. This highlights the area of photovoltaic science and technology that has been considered as a potential to produce low cost third-generation solar cells. In order to do this, we discussed progress (few important) in nano-based solar cells made by different techniques.

1.7.1.1 SILICON AND THIN FILMS SOLAR CELLS

Usually silicon solar cells consist of mono or polycrystalline material p–n junctions and their wafers are approximately 0.3 mm thick, as shown in Figure 1.12 [74]. At present, the highest efficiencies of such devices are around 25.3% for single crystal and 21.9% for multicrystalline silicon [75]. However, thin-films solar cells, such as (CdTe), copper indium gallium diselenide (CIGS), and amorphous thin-film silicon [76, 77] efficiencies are around 21.2%. However, these silicon and thin-film solar cells are not based on nanotechnology, nanomaterials can be used in a combination to improve their light management and energy harvesting ability to make them thinner and cheaper [78]. The high-transparency antireflective coatings of the silica nanoparticles in a polymeric matrix can improve the peak power output of the solar module [79]. In the prior innovations [80], it has been demonstrated that nanotechnology approach can improve light absorption in silicon solar cells by deposition of silicon nanoparticles surface layer. This can decreased the light loss by reflection of nanoparticles that absorb part of the reflected light by the adjacent particles. Similarly, it has been established that anti-reflective solar cell surface with a nanostructured inverted pyramid array shape can help to avoid the use of an extra antireflecting coating [81]. To make electric contact for solar cells, aluminum and silver [82, 83] nanoparticles embedded paste composite can be used. This could enhance the high contact durability and lower the ohmic resistance to improve cell efficiency. However, all back contact solar cells are manufactured with low-cost hybrid emitters by using a thin silicon dioxide dielectric layer (20–40 Å) between the emitter single crystalline silicon substrate (backside surface) and doped nanopolycrystalline silicon and doped Si nanoparticles paste [84, 85].

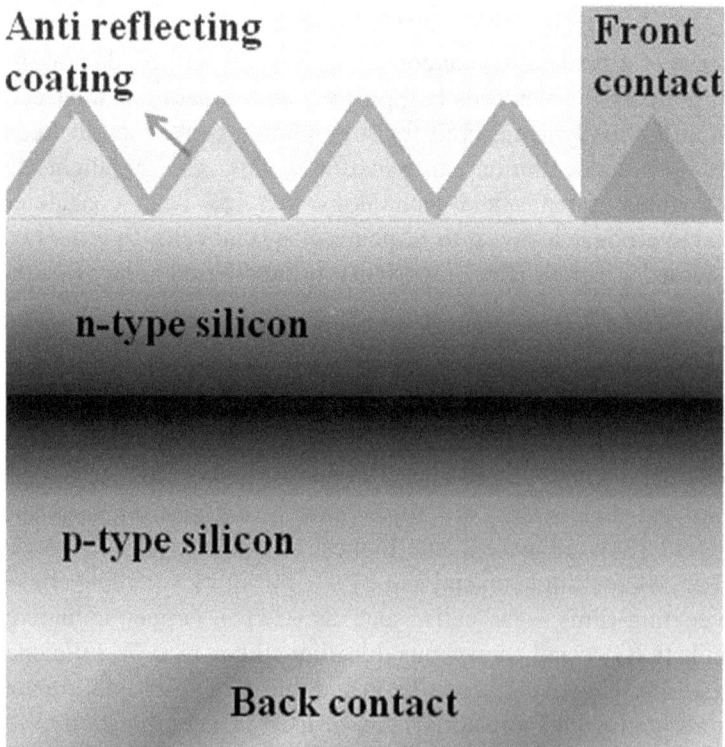

FIGURE 1.12   Schematic of silicon-based solar cell.

1.7.1.2 MULTIJUNCTION SOLAR CELLS

Multijunction photovoltaics are also known as tandem solar cells, where a combination of semiconductor materials are involved having different band gaps (and multiple p–n junctions) for the better absorbance of solar spectrum; the scheme of multijunction solar cell is given in Figure 1.13 [86]. As per National Renewable Energy Laboratory (NREL) 2018 efficiency chart, multijunction solar cells have the highest efficiencies between 27.8% and 46%. The multijunction solar cells fabrication cost are more than others because they require high order material purity for the crystals growth. Since they have been shown to be more robust under the outer space radiation and therefore mostly are used for space applications [87]. However, some reports were on tandem solar cells with nanowires as active component where nanowires are made of a plurality of light absorbing segments with decreasing band gap to allow sequential light harvesting. Mainly GaAsP or GaInP has been used for the top segment of the nanowires and GaInAs or InAsP for the bottom segment [88]. Their energy range arises due to material combinations of the energy span harvesting between 0.4 eV (InAs) and 2.24 eV (GaInP). Besides, their maximum spacing between adjacent nanowires is shorter than the wavelength of light that the solar cell structure is able to absorb, and the presence of nanostructure of a sequence of quasi-continuous absorbing layers are for the incoming photons. Therefore, an array of GaAs, AlGaAs, InP vertical nanowires, or alloys thereof absorbed the light in the first spectral range are deposited on a bulk solar cell of Si or CIGS, while, in the second spectral range density of the nanowires can be selected in predetermined portion of the incident photons. These materials may pass through the region between the nanowires and at the end absorbed by the bulk material [89]. Besides, it has also been reported that a photovoltaic device having an upper cell of Cd or Si may absorb a first range of wavelengths of light and a bottom cell that can absorb a second range of wavelengths of light in crystalline or polycrystalline nanowires heterojunction solar cells [90].

1.7.1.3 ORGANIC SOLAR CELLS

Organic solar cells can be constituted from the conductive organic polymers or small organic molecules for light absorption and charge transport processes. Such types of solar cells are currently running with the maximum efficiency in laboratory around 11.5% [91]. Though this type of solar cell achieved the highest efficiency yet it is lower than the widely used Si solar cells and

CIGS. However, they can offer potential light weight, mechanically flexible, easy to process, and low cost solar cells [92]. The organic solar cells are n-type materials with embedded organic nanowires, such as polythiophene, poly(phenylenevinylene), and perylene tetracarboxylic diimide. Owing to their improved charge movement and collecting characteristics, these solar cells offer better reduction in the recombination of holes and electrons, thereby overall increasing the efficiency of the solar cell [93]. Additionally their core–shell nanoparticles, such as Ag–SiO_2, Ag–TiO_2, Au–TiO_2 [94] and TiO_2, ZnO, WO_3, and MoO_3, can be combined with fullerenes, carbon nanotubes, and nanorods to produce better electron transport layers [95, 96]. In the bulk heterojunction solar cells, it has been reported that the nano-ink-processed fullerene derivatives electron acceptor material can achieve 3% higher power conversion efficiency [97].

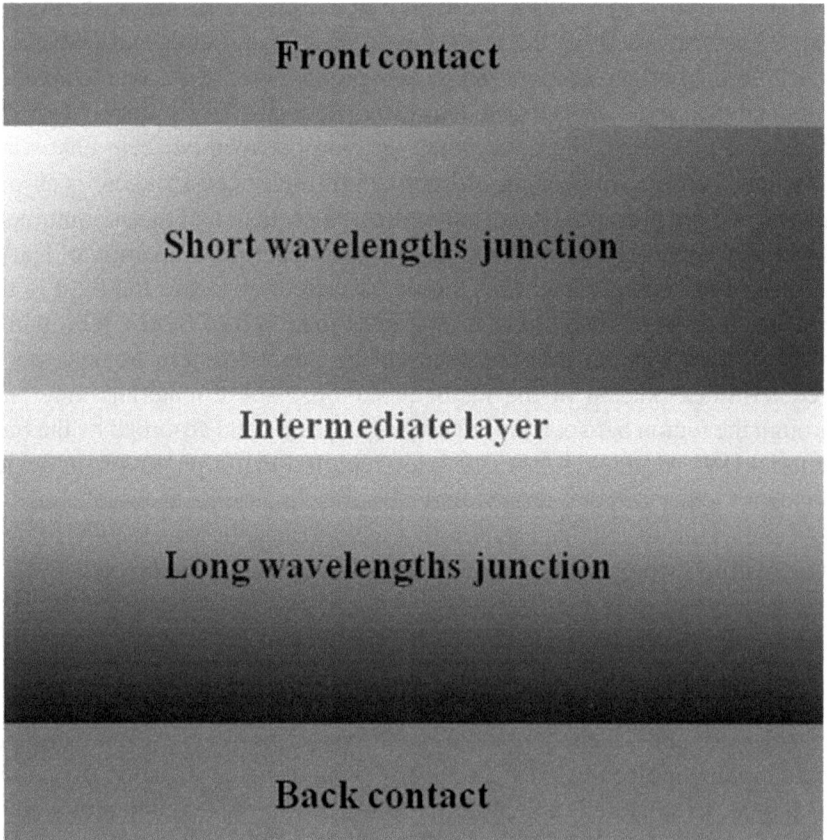

FIGURE 1.13   Schematic of the multijunction solar cell

1.7.1.4   DYE-SENSITIZED SOLAR CELLS (DSSCS)

The DSSCs solar cells have been extensively explored due to their potential features. DSSCs solar cells were first proposed by O'Regan and Gratzel in 1991 [98]. The key advantages of these solar cells are the relatively high conversion efficiency (11.9%) with their low fabrication cost and nontoxic feature. It is beneficial to fabricate DSSCs that can use various materials with the possibility to produce flexible and optically transparent solar cells [91,99]. This kind of solar cell basically consists of a photoanode, which is made from a conductive transparent substrate covering a wide band gap n-type nanostructured semiconductor. Usually, TiO_2 can be modified with a monolayer of an organic dye, and a schematic is represented in Figure 1.14. Photoanode and an electrocatalyst supported on the counterelectrode are immersed in an electrolytic solution for a redox system. Typically this kind of solution contains iodide/triiodide couple, specifically in case of solar photons are expose the photoanode can be absorbed by the organic dye. As a consequence, the electrons are excited from the highest-occupied molecular orbital to the lowest-unoccupied molecular orbital of the dye. The excited electrons are injected into the conduction band of the nanocrystalline oxide and transport toward the substrate. The existing organic dye is restored by the electron donation from the redox coupling process, while iodide regenerates the reduction of triiodide at the counter electrode; thereby, the circuit is completed through electron migration from the external load [74,100]. The nanostructured TiO_2 metal oxide layer on the conductive glass surface is the backbone of DSSCs. This is owing to their dual functional behavior; as photoactive centers at which dye is adsorbed and as pathways for transport of the photoexcited electrons from dye to the back contact [101]. Thus the key importance of using a nanostructured material is that it provides a very high surface area which can accommodate a huge load of dye molecules results in a better light-harvesting efficiency. In DSSCs, usually a 10-μm thick film comprised of a three-dimensional network of interconnected nanocrystalline semiconductor, typically 15–20 nm in diameter is synthesized from the sol–gel method at the temperature 230 °C [70]. The TiO_2 nanotubes vertical alignment can be obtained from the electrochemical Ti anodization, after thermal annealing at the temperature between 400 and 500 °C, in order to achieve high electron mobility anatase phase [102]. The DSSCs solar cells can also offer low-temperature alternatives, which are important for plastic-flexible substrates. Some evidences are also available that to use viscous dispersion liquid of anatase crystalline titanium dioxide nanoparticles; in that metal oxide proportion relative to the solvent is higher than 50% in weight. The nanoparticles can be synthetized from the liquid of gas phase methods with their subsequent dispersion in water containing small amounts of C3 to C10 linear or branched alcohol, such as 1-propanol, tertpentyl alcohol, and cyclohexanol, and at the end mix the solution at the temperature around 15 °C. The well-dispersed viscous solution can be deposited onto the transparent substrate by using different methods, such as screen printing, metal masking, gravure printing, doctor blade coating, squeegee coating, or spray coating, followed by the low-temperature heat treatment between 120 and 150 °C. At the end, a mesoporous film with nanosized pores can be obtained [103].

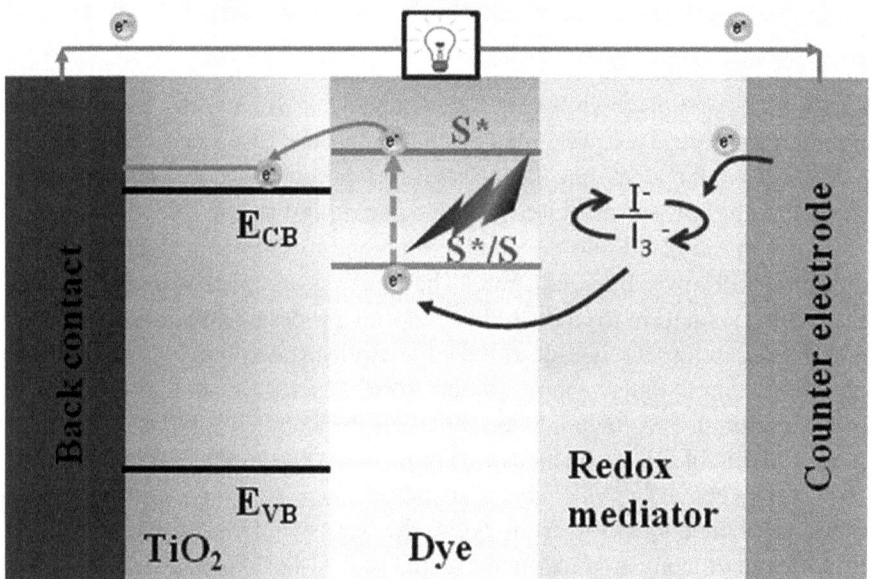

FIGURE 1.14   Schematic of the dye-sensitized solar cell.

There is another approach to make core–shell of TiO_2–SnO_2 nanostructures. The use of SnO_2 can provide 100-times more electron mobility as compared to TiO_2 with the improved charge transfer from the excited dye to SnO_2 conduction band. Because the energy of the SnO_2 conduction band is lower than the corresponding band of the TiO_2. Therefore, charge injection from the dye is easier due to larger energy drop, this leads to that SnO_2 can be used in conjunction with smaller band-gap dyes. Further to suppress the back reaction of electrons from SnO_2 conduction band to the electrolytic solution, a wide band gap with very thin insulating oxide dielectric layer shell of NbO_3, Al_2O_3, MgO, SiO_2, or TiO_2 is formed between the SnO_2 and photosensitizer. Such dielectric layer inhibits the charge recombination and enables a significant enhancement in the photovoltaic conversion efficiency. Therefore, overall carrier collection efficiency can be improved due to one-dimensional core–shell nanostructures, like nanowires and nanotubes instead of nanoparticles [104]. Furthermore, to improve metal oxide electron collection efficiency, a dense electron transport layer or multilayer having thickness between 20 and 700 nm is usually deposited onto the back contact (transparent conductor substrate). This additional dense layer can be made of a single semiconductor (such as Si and Ge) to a compound semiconductor like metal oxides, sulfides, and selenides or compounds, which has a perovskite structure. Such compound semiconductor formation increases the surface area of the nanostructured electron transport layer of TiO_2; this can be realized using electroplating in a titanium tetrachloride solution [105]. DSSCs can also be made with a porous reflecting element or photonic crystals to enhance the power conversion efficiency of the solar cell device through the increasing selective optical absorption of the electrode. In the photonic crystal, alternating layers are comprised with different refraction indexes, which could localize efficiently from the incident light within the absorbing layer. Each layer of the multilayer photonic crystal can be made from the nanoparticles and their porosity allows the electrolytes and the absorbing compounds to flow through the multilayer. It is important to ensure the good electrical contact with remaining components of the cell and make sure that there is no detrimental effect on the charge transport [106]. Through the use of nanostructures materials it has also been extended to the electrocatalyst supported on the counter electrode. Hence these are providing different options of materials like metal nanoparticles (platinum, gold, silver, copper, and aluminum), a lower cost carbon-based compound (graphite, fullerene, carbon nanotube, and graphene) or an electroconductive polymer (polythiophene and polyaniline) [104–106]. Thus, the plurality of DSSCs solar cells can be connected to various types of modules. With this advantage, photocurrent could be controlled by the less efficient photoelectrode and photovoltages additive, then finally achieve much higher photovoltage than for a single module [107, 108].

1.7.1.5 QUANTUM DOTS-SENSITIZED SOLAR CELLS (QDSSC)

Quantum dots (QDs) are the nanocrystals of semiconducting materials which is less than 10 nm accompanied with unique optical properties owing to their band gap, absorption, and emission spectra, which can be tuned from the varying particle size [69]. The QDs can provide an alternative to molecular dyes for the quantum DSSC. Such solar cell structure, operation principles, and materials could be similarly accepted as the photoactive material, as represented schematically in Figure 1.15 [100, 109]. QDs can offer the possibility of absorbing the high-energy photons for the generation of multiple charge carriers or capturing hot electrons before their thermalization, which can boost the operational efficiency of QDSSC to a theoretical external QF which is >100% [110]. Yet the best research QDSSC efficiency reported by NREL is around 13.4%, which is higher than the corresponding DSSC. Therefore, they can be considered as promising materials for the development of high-efficiency solar cells [75]. The most extensively employed method to synthetize QD is the hot injection process. This synthesis process can be used for the preparation of many different colloidal nanocrystals with tuneable size, shape, and surface passivation. In this synthesis process, one of the precursors at room temperature is injected into a hot solution (200–300 °C) of another precursor; this gives an instantaneous formation of nuclei. With the decrease in the temperature formation of new nuclei is inhibited, therefore, it only allows the slow growth of the tiny nanoparticles initially formed. This nucleation and growth stages are separated, and thus a high degree of monodispersity can be achieved [100,111]. There are several compositions of different materials, such as Si, GaAs, InAs, PbS, PbSe, CdSe, CdTe, CuInGeSe, CuInGeS, CuZnGeSe, and CuZnGeS, have been tested for the QDs [112]. In this order, the boron-doped silicon core–shell nanocrystals with enhanced electroconductivity [113] can also be synthesized from the hot injection method and it was verified for the QDSSC. The PbS and PbSe QDs have also been tested for the infrared (IR) absorbing material (for wavelengths higher than 700 nm) in photodetectors, IR-to-visible upconversion devices (night vision) and IR solar cells, which are synthesized from the hot injection method and multiinjection method [114, 115]. Furthermore, all kinds of inorganic perovskite QDs have a chemical formula of $CsPb(Cl_aBr_{1-a-b}I_b)_3$, here $0 \leq a \leq 1$, $0 \leq b \leq 1$, encapsulated in SiO_2 or polymer as a protective shell in QDSSC and light-emitting devices [116]. The QD polymer composite where a barrier coating protects the QD from moisture or oxygen in order to secure its high efficiency and color purity has been investigated [117]. It is recently established a quantum well solar cell design with a PN or PIN junction containing materials of different energy band gap within the active region [118]. In such solar cells, a wider energy gap barrier layer is used to suppress carrier injection across the junction. This leads to the increase in the power output and maximize the photocurrent of semiconductor solar cells. In this order, it is also reported that the replacement of the ligand molecules bonded to the QDs (e.g., oleic acid) with a molecular chain length of about 2 and 3 nm, and found that the shorter molecules can be used to increase electrical conductivity [119].

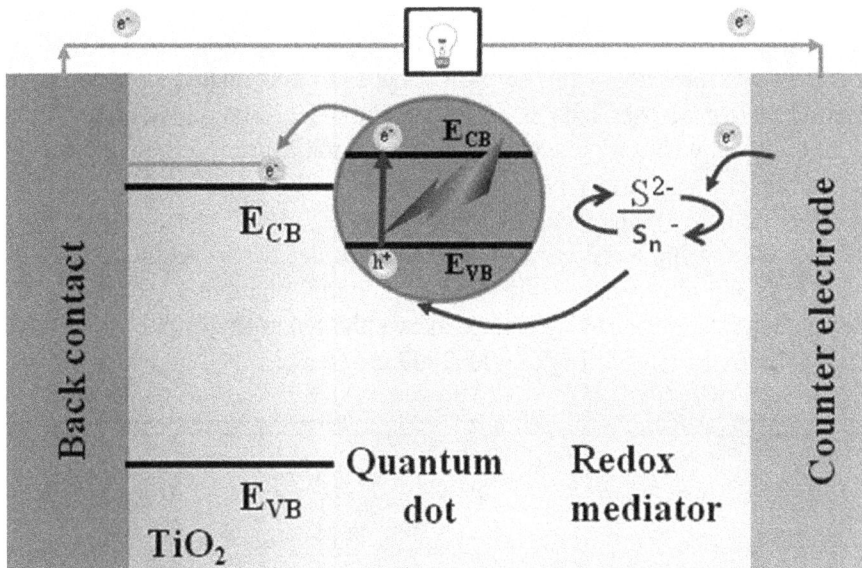

FIGURE 1.15 Scheme of the quantum dot solar cell.

1.7.2 *CARBON NANOTUBES FOR OPTOELECTRONICS*

Semiconducting carbon nanotubes (CNTs) have a direct band gap which incorporates in a variety of optoelectronic devices, such as light detectors, light emitters, and transparent conductors [120–122]. It has been demonstrated that the van Hove singularities in the one-dimensional density of states and strongly bound excitons make CNTs interesting candidates for optoelectronics; the corresponding schematic is illustrated in Figure 1.16 [123]. Their binding energy exciton depends on the diameter of CNTs [124, 125] and dielectric constant of the surrounding environment [125]. While in the prior experiments, it has been demonstrated that CNT optical excitations may become conductive in dispersed aqueous solutions containing individual CNTs coated with surfactants [123, 126]. In general, the photoexcitation generates excitons in the second sub-band of CNTs followed by radiative decay to the first sub-band (see Figure 1.16) [123]. As a consequence, a two-dimensional plot of photoluminescence as a function of excitation and emission energies peaks with the unique identify of CNT chirality can be achieved [123]. It has also been well established that the fluorescence QF of dispersed CNTs is in the order of 10^{-3}–10^{-4} with an effective radiative lifetime of 1–10 ns at room temperature [123, 127]. Further advancements in CNTs have been demonstrated that with as-grown suspended CNTs QF up to 10% can be achieved [128]. Their lower QF can be correlated to the multiple nonradiative processes such as exciton–exciton annihilation [129], the presence of low-energy dark excitons cannot relax to the ground state radiatively and efficient nonradiative, phonon-mediated decay of excitons [130, 131]. A wide range of QF values can be achieved as reported in various literatures by considering the sensitivity of the radiative decay rate to the quality of the CNTs and the nature of the surrounding environment. CNTs diameter-dependent excitonic binding energy allows the tunability of photoresponse, especially in the near-infrared (900–2000 nm) range.

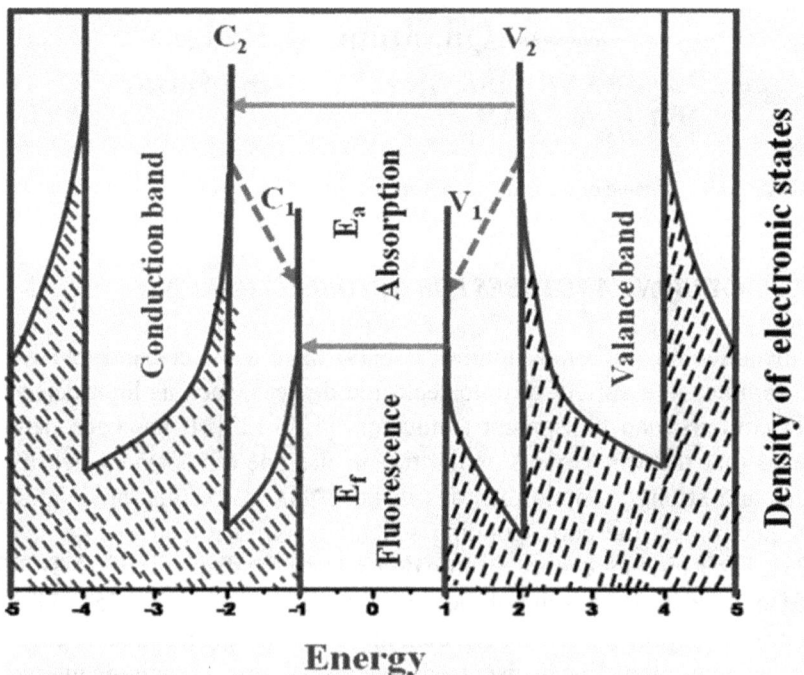

FIGURE 1.16 Schematic of the van Hove singularities in the one-dimensional electronic density of states and typical optical transitions in CNTs.

Semiconducting CNTs are also prominent candidates for the photocurrent and electroluminescent devices. For example, CNTs can be utilized as the basis of light-emitting transistors and photodetectors [132]. Firstly, it was recognized in a study of electroluminescence of ambipolar CNT–field-effect transistors (FETs), which revealed a number of interesting observations: (1) Unlike conventional p–n junctions, electroluminescence in ambipolar CNT FETs does not require extrinsic doping; (2) The maximum electroluminescence efficiency is observed in the off state; (3) Polarization of the emitted light is parallel to the CNT axis [133]. In this sequence it has been demonstrated that the position of light emission from the CNT channel can be tuned through the controlling recombination site via biasing conditions [134]. But the achieved emitted light power (~100 pW) and QF (10^{-6}) were rather low for typical light-emitting applications [135]. The observed spectral broadening was also in a small channel device due to hot carrier recombination mediated by optical/zone boundary phonons. Furthermore, there are several attempts have been made to improve the emission power per unit area by utilizing large arrays of electrolyte gated aligned CNTs [136] and top-gated aligned 99% semiconducting CNTs [137] in ambipolar device operation. Though these devices still showed the decreased QF (10^{-9}) as compared to individual CNT FETs [136]. These devices show an increase in the electroluminescence intensity for the random CNT TFTs [138]; however, the red shift and spectral broadening can be achieved due to exciton transfer from large band gap CNTs (narrow diameter) to small band gap CNTs (large diameter) in heterogeneous mixture of as-grown CNTs. In this order, the exciton energy transfer mechanism has been confirmed through a spatially and spectrally resolved photoluminescence experiment with a crossed junction of CNTs from different chiralities [139]. For the electroluminescence spectrum, spectral red-shifting from aligned bundles of monodisperse CNTs has also followed the same mechanism [137]. The electroluminescence can also be achieved for the unipolar CNT FETs under the impact of the excitation processes from hot carriers [140–143]. Under the high electric fields, CNTs electrons can gain sufficient kinetic energy for the electronic excitations across the band gap upon scattering. This excitation process is at least three times more efficient than the CNTs as compared to their conventional bulk semiconductors [121, 140]. Such enhancement of excitation can also be induced by high local electric fields from inhomogeneities, such as defects, trapped charges and CNT–metal contacts [144]. The localized unipolar CNT electroluminescence can be obtained in artificially constructed regions of high electric field in the channel. Another approach that is based on the fabrication of CNT FETs also demonstrated that a portion of the CNT is consistent with the dielectric and the remainder can be suspended over a trench etched in the channel (Figure 1.5c) [140]. The abrupt discontinuity in their dielectric constant leads the band bending that accelerates electrons [145]. The QF of such devices has been achieved 1000× higher than that from ambipolar CNT FETs. Similarly other efficient approach to achieve sufficiently large electric fields due to the impact of excitation is reported by creating p–n junctions between electrostatically doped p-type and n-type regions in the same CNTs [123,146]. Thus the impact of excitation processes in CNTs is not govern by the same selection rules like optical excitations, thereby it can lead to the creation of excitons as well as free electron–hole pairs [141]. As an example, the electrically driven thermal light emission corresponding to inter-band transitions in comparison to featureless blackbody radiation has also been reported in the suspended quasi-metallic CNTs [147]. In the optoelectronics applications, the semiconducting CNTs have been extensively studied as photodetectors [120, 121]. The photodetection can be viewed as the inverse process of electroluminescence where optically generated excitons are separated into free electrons and holes to produce a photocurrent under an applied field [148] or an open-circuit photovoltage in an asymmetric field configuration [149]. However, in beginning the photoconductivity measurements with the individual CNT ambipolar FETs had showed internal QF up to 10% with expected resonances at optical energies corresponding to CNT excitonic states [148]. Under the sufficiently large fields, the separate excitons can also be generated locally from the asymmetric Schottky contacts, p–n junctions, and local charge defects [149–151]. The electrostatically gated p–n junction in a CNT can also efficiently generate the multiple electron–hole pairs as per absorbed photon through a process similar to the impact excitation process in electroluminescence [152]. Therefore, the mechanism of photodetection in individual CNT devices is mainly based on generation and dissociation of excitons, whereas the photoconductivity in CNT networks is often dominated by the thermal effects of nonradiative recombination. More specifically, the bolometric increases in CNT resistance due to the increased local temperature is appeared to be a viable method of photodetection [153].

Yet the highest QF for light emission in CNTs could not exceed beyond 10^{-3}; hence, this limits the scope of practical applications for CNTs in light-emitting optoelectronic applications. Considering this key issue, researchers have attempted to incorporate CNTs into alternative technologies, such as light-emitting diodes based on conjugated polymers to get enhanced electroluminescence efficiencies [154]. The CNT thin films provide good flexibility such as optically transparent and highly conductive; therefore, they can be considered as an alternative to indium tin oxide (ITO) in organic light-emitting diodes (OLEDs) and organic TFTs [155]. Particularly, CNTs hold promise to overcoming the limitations of ITO for large-area flexible electronics like brittleness and patterning-related issues with the desirable properties such as high optical transparency and low sheet resistance [155]. With the 99% pure metallic CNTs, it has also been reported the enhancement in sheet resistance is more than five times as compared to as-grown CNTs [156]. Along with sorting by diameter of CNTs, it can also produce conductive films with the tunable optical transmittance [156]. CNT thin-film electrodes can also provide three times lower contact resistance than commonly used Au electrodes for the organic semiconductor TFTs [157,158]. Moreover, CNT-based transparent conductors can also have useful application for the photovoltaics as well as components of nonlinear optics [159]. As an example, the dispersed CNTs can be useful for the suitable absorbers up to 40% attenuation efficiency and passive mode-lockers for the femtosecond lasers [160]. The advantage of CNTs also offers the facile fabrication with the wide tunable wavelengths with various diameters of CNT.

1.7.3 GRAPHENE FOR OPTOELECTRONICS

Graphene can have several potential utilities, which are well-suited for the optoelectronic applications [161]. For example, graphene linear dispersion with zero band gap leads to the possibility to be used for the tunable optical excitations, a schematic of which is shown in Figure 1.17 [162]. This gives an optical absorption spectrum that can be featureless over wavelengths ranging from 300 to 2500 nm [145]. The single-layer graphene can absorb 2.3% of the incident light with the minimum reflection (<0.1%) in this wavelength range [163]. The transmittance spectrum of the graphene can also provide $(1-\pi\alpha \sim 97.7\%)$ the effective fine structure constant (α), which depends on the dielectric constant of the environment [163]. Their absorbance peak around at 250 nm reflects a saddle-point singularity in the Brillouin zone of graphene [164, 165]. Usually, graphene could not exhibit the luminescent but its chemical derivatives can, like graphene oxide exhibits photoluminescence over a broad range [166, 167]. Such light emission has been speculated to occur in islands of sp^2 carbon within graphene oxide or at the oxygen-induced defect sites. Furthermore, electroluminescence from pristine graphene has also well explored [168], this light emission mechanism has been underlined different from graphene oxide. Particularly, optical phonon-assisted radiative recombination carriers provide the light emission in graphene (likewise to metallic CNTs) [168].

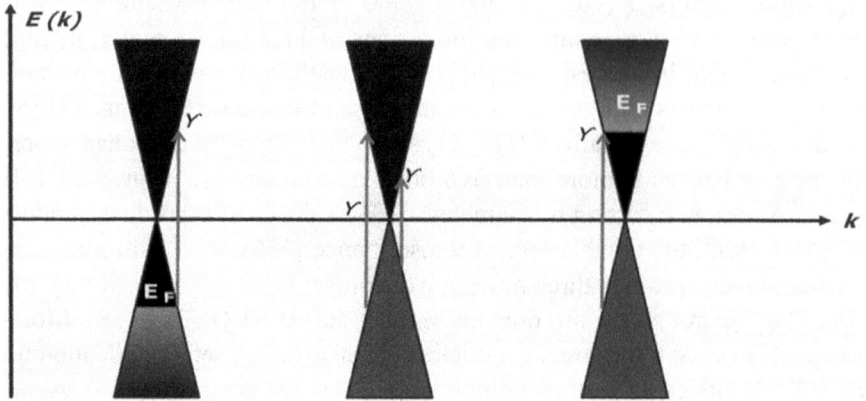

FIGURE 1.17 Schematic of the linear dispersion of graphene near the wave vector (K) point for the tunable photoresponse in unbiased graphene. Here light and dark colors are representing empty valence band or filled conduction band.

However, graphene itself may have limited potential as a light-emitting material, but it is effective when use as a transparent conductor in flexible OLEDs, organic TFTs, and OPVs. Graphene-based transparent conductors' performance has been achieved close to ITO with the superior mechanical flexibility [161]. The minimum finite conductivity of undoped graphene (4–8 e^2/h) offers sheet resistance (R_s) 4–7 kΩ in a high-quality single-layer graphene. This implies the necessity of the chemical doping from different methods to graphene-based transparent conductors. In comparative performances, matrices of the CNT thin films and graphene transparent conductors with ITO have demonstrated that CNT thin films sheet resistance R_s is slightly lower than ITO at the same frequency [16], while the graphene transparency can be above 90% of ITO. It has also been predicted that transparency of graphene can be further enhanced by innovation on the higher quality larger area graphene.

Graphene can also be used in photodetectors where optically generated electron–hole pairs can be separated through an externally applied electric field. The unique linear dispersion of graphene provides a uniform photoresponse in the THz to the ultraviolet range [161], while the CNTs and conventional bulk semiconductors are much lower than this range. Further, with the high mobility of graphene has also been reported to yield ultra-fast photodetection up to 40 GHz [169], their operational speed can be as high as 500 GHz under the intrinsic photoresponse. An internally built-in electric field from the metal–graphene contacts can also provide an efficient way to achieve photodetection; their photocurrent can be up to 15%–30% with external photocurrent yield at 6 mA/W [170]. Graphene-based photoresponse has also been examined in a p–n junction creation via top-gate architecture on graphene FETs [171,172]. This device exploits hot carriers in graphene under the nonlocal transport which contributes to the photoresponse in addition to the photovoltaic effect.

Another optical application of grapheme-related materials is in bioimaging. Specifically, photoluminescence property of the nanoscale graphene oxide can be used for the live cell imaging in the visible and near infrared range [173]. To the vivo applications, graphene can be used to address the issues surrounding the potential toxicity. Though a recent development demonstrates that encapsulation of graphene in the biocompatible block copolymer, as a consequence, pluronic can reduce the toxicity and inflammation in the lungs of mice [174]. Therefore, graphene biocompatible dispersion optoelectronics property can provide an additional opportunity to fabricate biomedical applicable devices including imaging contrast agents and drug delivery.

1.7.4 TRANSITION METAL DICHALCOGENIDES

Transition metal dichalcogenides (TMDCs) are another class of the important 2D materials. TMDCs are generally comprised of three atom layers; the central layer is composed of a lower transition metal and upper and lower layers are made up of chalcogenide atoms, these two layers typically consist of sulfur, selenium, and tellurium [175]. An individual TMDC mono-layer can have different phases in contrast to graphene. The most crucial monolayers are the so-called 2H and 1T, those are having ABA and ABC-like stacking of the lattice atoms; here capital and lower letters denote chalcogen and metal atoms, respectively [175].

Though group VIB TMDCs have received a lot of attention of interest among researchers in the electronic, optoelectronic, and spintronic fields [176]. The several TMDCs of VIB group are the semiconductors having a band gap in the eV range. This makes them useful for an efficient integration of the ultra-thin FET devices. This can be considered as the prime motivation about TMDCs. The TMDCs intrinsic semiconductor characteristic is responsible for promoting a good on/off ratio in FET devices, which is easily outperforming graphene with its semimetallic band structure. While, chalcogenides containing molybdenum and tungsten are the naturally layered materials with a 2H stacking. In these materials, the band structure varies as a function of the number of layers. However, in the bulk form most of the TMDCs have an indirect gap which can be characterized by a suitable technique.

In recent years, another versatile method has been introduced for the study of the conduction band of the samples. That is called time-resolved angle-resolved photoemission spectroscopy (ARPES) which involves pump and probe experiments, the schematic of which ARPES is represented in Figure 1.18. In this experiment, two photons with a femtosecond delay of each other are sent to the sample. The source of photon is called "pump" for the function to excite the electrons into the conduction band, while, the other source is called "probe" which contributes to extraction of the photoelectrons. This technique can be used to reveal the dynamics of the charge carriers, and is capable to visualize the electron relaxation path in excited states along with momentum resolution [177]. Note that the spin-resolved ARPES and time-resolved ARPES can be used for the bulk crystals [97] or for the epitaxially grown MoS_2 [178,179]. Due to their large area, there are common problems with low signal and long acquisition times. In order to overcome these problems, several modifications have been introduced such as using time-resolved sources for EF-PEEM [100] and TOF PEEM [180,181].

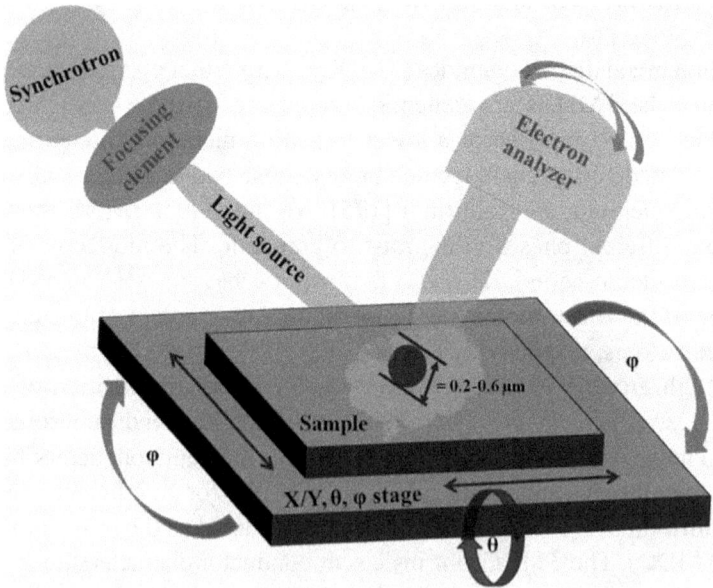

FIGURE 1.18 Schematics for the spatially localized ARPES with synchrotron light source. The synchronous light is focused on a spot of few hundreds of nanometers. The specimen can be visualized from the scanning in (X/Y) real space and photoelectrons can be collected by the energy analyzer.

The recent study on most common TMDC ReS_2 has an unusual anisotropy in-plane. Therefore, ReS structure can be considered as a distorted 1T crystal structure, while in contrast, VIB group is considered as their 2H TMDCs structure. This leads an additional valence electron for the formation of Re chains along the axis of the crystal. Their low crystal symmetry results in the high-order anisotropic optical, vibrational, and electron transport properties; therefore, adds an additional interest for the applications in sensor and electronic devices. These materials are in very early stage of their study. Therefore, studies of spatially localized ARPES are essential not only for probing layers with different thickness [182] but also for investigating small bulk pieces [183,184]. However, the direct/indirect band gap transition of the ReS_2 has not been clearly established and further studies are needed to fully understand such material [185].

1.8 SHORTCOMINGS OF NANOMATERIALS

Although nanomaterials have numerous potential advantages but they are not free from the certain limitations, such as instability of the particles, specifically retaining the active metal nanoparticles is highly challenging due to their rapid kinetics. In order to retain anodize particles; they can be encapsulated in some other matrix. The thermodynamically metastable nanomaterials lie in the region of high-energy local minima. Therefore, they are prone to attack and undergo transformation. These include poor corrosion resistance, high solubility, and phase change in nanomaterials. This can create the deterioration in materials, therefore, retaining the original structure becomes challenging. Further, the fine metallic particles could act as strong explosives due to their high surface area directly contacts with oxygen. Nanomaterials exothermic combustion makes them explosive. The nanoparticles' highly reactive property also allows us to inherently interact with the impurities. Furthermore, the encapsulation of nanoparticles is necessary during synthesis in a solution (chemical route). The nanoparticles stabilization occurs owing to a nonreactive species overcome the reactive nano-entities. Thereby, a major difficulty arises in the pure nanoparticles synthesis due to secondary impurities, which makes it difficult to get a highly pure form of the synthesized nanomaterials. The formation of oxides, nitrides, etc., can also get aggravated from the impure environment/ surrounding during the nanoparticles synthesis. Hence ensuring the high purity in nanoparticles is the challenge and hard to overcome. Moreover, usually nanomaterials are considered biologically harmful when they become transparent to the cell-dermis. Such toxicity in biological nanomaterials is also predominantly appeared due to their high surface area and enhanced surface activity. Nanomaterials are also the cause of irritation, which indicates about their carcinogenic behavior. If nanoparticle inhaled then due to their low mass entraps inside the lungs and there is no way for them to be expelled out of body. Nanoparticles interaction with liver/blood has also been proved to be harmful (though this aspect is still being debated on). Thus it is extremely hard to safely retain nanoparticles once they are synthesized in a solution. This makes difficulty in synthesis, isolation, and application. Therefore, nanomaterials have to be encapsulated in a bigger and stable molecule/material. Because free nanoparticles are hard to be utilized in isolation, thereby, they have to be interacted for projected use through the secondary exposure. The grain growth is inherently present in nanomateirals during the processing. The smaller grains provide bigger and stable grains at high temperatures with the processing time.

The nanomaterials recycling and disposal is also one of the major issues for their optimization. Since there is no hard-and-fast safe disposal policies evolved for nanomaterials, which require further attention in the near future. The nanomaterials' toxicity is also still under questions and investigations and their experimental results are not available. Hence the uncertainty associated with these materials affects nanomaterials is still to be assessed to develop their disposal policies.

1.9 CONCLUSIONS

In the conclusive remarks, in this chapter we have briefly introduced nano-materials including nanoscience, nanomaterials, and interpretation of the nanotechnology. A complete segment (Section 1.2) has been devoted for the naturally occurring nanomaterials including their various forms, such as natural erosion and volcanic activity, colloids, clays, natural mineralized materials as well as nanoscience utility in the biological world, including their historical background. The nanomaterials dimensionality is also an important parameter for their classification and their specific use. Therefore, a complete segment/subsegments has been devoted to define zero-, one-, two- and three-dimensional nanomaterials including their interpretations for the nanostructures, nanowires, nanotubes, nanolayers (or nanocoatings), and nanoporous materials. Since strategies of the synthesis of nanomaterials is also a crucial part; therefore, a segment/subsegments is devoted to various synthesis process including chemical approaches (such as, sol–gel, hydro/solvothermal, polyol, sonochemical, microemulsion, chemical vapor deposition, atomic layer deposition), physical approaches (such as; mechnatical, ball milling, melt mixing, inert gas technique, pulse vapor deposition, laser pyrolysis, flash spray pyrolysis) and biological synthetic approaches (e.g., microorganism assists biogenesis, biotemplates-assists biogenesis, plant-assists biogenesis). Moreover, it is also customary to provide a comprehensive description on the importance of nanomaterials and the significance of nano-science; therefore, these two segments are also incorporated in this chapter. The most important thing in the area of nanoscience and nanotechnology is to developed nanomaterials applicability. Though nanomaterials can have enormous utility in various scientific and technological fields, herein our aim confined on their optoelectronic applications, more specifically one the emerging area, that is, photovoltaic cells. Therefore, a complete segment/subsegments has been provided on different kinds of solar cells (including silicon and thin films, multijunctions, organic dye-sensitized, quantum-dots-sensitized solar cells). In this segment, optoelectronics applicability of the carbon nanotubes, graphene, and transition-metal dichanlcogenide have also been discussed. Since every kind of materials has advantages and disadvantages, therefore it is worth to include a description on shortcoming of nanomaterials, which has been provided in the last section of the chapter. Hence on the basis of different physiochemical properties of nanomaterials, it is significant to say that these materials can have potential for different optoelectronics utility, more specifically for the solar cells application.

ACKNOWLEDGEMENT

The author, Abhay Kumar Singh, is thankful to the University of Johannesburg, Department of Mechanical Engineering Science (APK), Faculty of Engineering and the Built Environment (FEBE), for the support under the PDRF program. Prof. TC Jen would like to acknowledge the support from NRF and URC funding from the University of Johannesburg.

CONFLICT INTEREST

It is disclose that there is no conflict interest between the authors.

KEYWORDS

- **nanomaterials**
- **optoelectronics**
- **nanoscience**
- **nanotechnology**
- **nanoforms**

REFERENCES

1. Tjong, S.C. Nanocrystalline Materials: Their Synthesis–Structure–Property Relationships and Applications, Elsevier, Boston, 2006.
2. Gleiter, H. Nanostructured materials: basic concepts and microstructure. *Acta Mater.* 2000, 48, 1–29.
3. Tiwari, J.N.; Tiwari, R.N.; Kim, K.S. Zero-dimensional, one-dimensional, two-dimensional and three-dimensional nanostructured materials for advanced electrochemical energy devices. *Prog. Mater. Sci.* 2012, 57, 724–803.
4. NIST, Manufacturing and Biomanufacturing: Materials Advances and Critical Processes, 2011.
5. A synthetic biology roadmap for the UK (TSB, 2012).
6. Arulmani, S.; Anandan, S.; Ashok Kumar, M. Introduction to Advanced Nanomaterials, Elsevier. 2018.
7. Zi, J.; Yu, X.; Li, Y.; Hu, X.; Xu, C.; Wang, X.; Liu, X.; Fu, R. Coloration strategies in peacock feathers. *Proceedings of the National Academy of Sciences* 2003, 100, 12576–12578.
8. Morgan, A. B. Polymer–clay nanocomposites: design and application of multi-functional materials. *Mater. Matters* 2007, 2, 20–23.
9. Neinhuis, C.; Barthlott, W. Characterization and distribution of water-repellent, self-cleaning plant surfaces. *Ann. Bot.* 1997, 79, 667–677.
10. Freestone, I. Meeks, N.; Sax, M.; Higgitt, C. The Lycurgus Cup—A Roman nanotechnology. *Gold Bull.* 2007, 40, 270–277.
11. Sanchez, C.; Beatriz, J.; Belleville, P.; Popall, M. Applications of hybrid organic–inorganic nanocomposites. *J. Mater. Chem.* 2005, 15, 3559–3592.
12. Feynman, R. P. There's plenty of room at the bottom, presentation at the American Physical Society Conference, Caltech, CA, USA, 1959.
13. Timp, G. et al., "The Nano-transistor," www.electrochem.org/dl/ma/200/pdfs/1233.pdf.
14. Drexler, K. E. Engines of Creation, The Coming Era of Nanotechnology, Anchor Press, Doubleday, New York, 1986.
15. Drexler, K. E. Nanosystems: Molecular Machinery, Manufacturing and Computation, John Wiley & Sons, New York, 1992.

16. Bashir, S.; Liu, J.. Overviews of Synthesis of Nanomaterials, Advanced Nanomaterials and Their Applications in Renewable Energy, Elsevier Science, 2015, 51–115.
17. Brichkin, S.B.; Razumov, V.F. Colloidal quantum dots: synthesis, properties, and applications. *Russ. Chem. Rev.* 2016, 85 1297–1312.
18. Xu, X.; Ray, R.; Gu, Y.; Ploehn, H.J.; Gearheart, L.; Raker, K.; Scrivens, W.A. Electrophoretic analysis and purification of fluorescent single-walled carbon nanotube fragments. *J. Am. Chem. Soc.* 2004, 126, 12736–12737.
19. Wang, R.; Lu, K.Q.; Tang, Z.R.; Xu, Y.J. Recent progress in carbon quantum dots: synthesis, properties, and applications in photocatalysis. *Mater. Chem.* 2017, 5, 3717–3734.
20. Mai, L.; Tian, X.; Xu, X.; Chang, L.; Xu, L. Nanowire electrodes for electrochemical energy storage devices. *Chem. Rev.* 2014, 114, 11828–11862.
21. Wei, Q.; Xiong, F.; Tan, S.; Huang, L.; Lan, E.H.; Dunn, B.; Mai, L. Porous one dimensional nanomaterials: design, fabrication and applications in electrochemical energy storage. *Adv. Mater.* 29 (2017) 1602300.
22. Ji, L.; Meduri, P.; Agubra, V.; Xiao, X.; Alcoutlabi, M. Graphene-based nanocomposites for energy storage. *Adv. Energy Mater.* 2016, 6, 1502159.
23. H. Li, Y. Li, A. Aljarb, Y. Shi, L.J. Li, Epitaxial growth of two-dimensional layered transition-metal dichalcogenides: growth mechanism, controllability, and scalability. *Chem. Rev.* 2018, 118, 6134–6150.
24. Li, H.N.; Shi, Y.M.; Chiu, M.H.; Li, L.J. Emerging energy applications of two-dimensional layered transition metal dichalcogenides. *Nano Energy* 2015, 18 293–305.
25. Shi, Y.M.; Kim, K.K.; Reina, A.; Hofmann, M.; Li, L.J.; Kong, J. Work function engineering of graphene electrode via chemical doping. *ACS Nano* 2010 4 2689–2694.
26. Dean, C.R.; Young, A.F.; Meric, I.; Lee, C.; Wang, L.; Sorgenfrei, S.; Watanabe, K.; Taniguchi, T. Kim, P.; Shepard, K.L.; Hone, J. Boron nitride substrates for high-quality graphene electronics. *Nat. Nanotechnol.* 2010, 5, 722–726.
27. Radisavljevic, B.; Radenovic, A.; Brivio, J.; Giacometti, V.; Kis, A. Single-layer MoS_2 transistors. *Nat. Nanotechnol.* 2011, 6 147–150.
28. https://cen.acs.org/articles/95/i22/2D-materials-beyond-graphene.html.
29. Anasori, B.; Lukatskaya, M.R.; Gogotsi, Y. 2D metal carbides and nitrides (MXenes) for energy storage. *Nat. Rev. Mater.* 2017, 2, 16098.
30. Tai, G.; Hu, T.; Zhou, Y.; Wang, X.; Kong, J.; Zeng, T.; You, Y.; Wang, Q. Synthesis of atomically thin boron films on copper foils. *Angew. Chem. Int. Ed.* 2015, 54 15473–15477.
31. Shi, Y.; Zhang, H.; Chang, W.H.; Shin, H.S.; Li, L.J. Synthesis and structure of two-dimensional transition metal dichalcogenides. *MRS Bull.* 2015, 40, 566–576.
32. Bhimanapati, G.R.; Glavin, N.R.; Robinson, J.A. Chapter 3 2D boron nitride: synthesis and applications. *Semiconduct. Semimet.* 2016, 95, 101–147.
33. Cai, S.L.; Zhang, W.G.; Zuckermann, R.N.; Li, Z.T.; Zhao, X.; Liu, Y. The organic flatland-recent advances in synthetic 2D organic layers. *Adv. Mater.* 2015, 27, 5762–5770.
34. Bojdys, M.J. 2D or not 2D-layered functional (C, N) materials "beyond silicon and graphene," *Macromol. Chem. Phys.* 2016, 217, 232–241.
35. Cao, X.; Yin, Z.; Zhang, H. Three-dimensional graphene materials: preparation, structures, and application in supercapacitors. *Energy Environ. Sci.* 2014, 7, 1850–1865.
36. Mao, S.; Lu, G.; Chen, J. Three-dimensional graphene-based composites for energy applications. *Nanoscale* 2015, 7, 6924–6943.
37. Chen, L.; Luque, R.; Li, Y. The controllable design of tunable nanostructures inside metal-organic frameworks. *Chem. Soc. Rev.* 2017, 46, 4614–4630.

38. Wang, R.; Jin, D.; Zhang, Y.; Wang, S.; Lang, J.; Yan, X.; Zhang, L. Engineering metal-organic framework derived 3D nanostructures for high-performance hybrid supercapacitors. *Mater. Chem. A* 2017, 5, 292–302.

39. Liu, R.; Duay, J.; Lee, S.B. Heterogeneous nanostructured electrode materials for electrochemical energy storage. *Chem. Commun.* 2011, 47, 1384–1404.

40. Wu, Z.; Yang, S.; Wu, W. Shape control of inorganic nanoparticles from solution. *Nanoscale* 2016, 8, 1237–1259.

41. Tadic, M.; Panjan, M.; Markovic, D.; Milosevic, I.; Spasojevic, V. Unusual magnetic properties of NiO nanoparticles embedded in a silica matrix. *J. Alloy. Compd.* 2011, 509 7134–7138.

42. Bognor-Legare, V.; Cassagnau, P. In situ synthesis of organic-inorganic hybrids or nanocomposites from sol–gel chemistry in molten polymers, *Prog. Polym. Sci.* 2014, 39, 1473–1497.

43. Ziller, S.; Von Bulow, J.F.; Dahl, S.; Linden, M. A fast solegel synthesis leading to highly crystalline birnessites under non-hydrothermal conditions. *Dalton Trans.* 2017, 46, 4582–4588.

44. Khan, M.F.; Ansari, A.H.; Hameedullah, M.; Ahmad, E.; Husain, F.M.; Zia, Q.; Baig, U.; Zaheer, M.R.; Alam, M.M.; Khan, A.M.; AlOthman, Z.A.; Ahmad, I.; Ashraf, G.M.; Aliev, G. Sol–gel synthesis of thorn-like ZnO nanoparticles endorsing mechanical stirring effect and their antimicrobial activities: potential role as nano-antibiotics. *Sci. Rep.* 2016, 6, 27689–12.

45. Demazeau, G. Solvothermal reactions: an original route for the synthesis of novel materials. *J. Mater. Sci.* 2018, 43, 2104–2114.

46. Dong, H.; Chen, Y.C.; Feldmann, C. Polyol synthesis of nanoparticles: status and options regarding metals, oxides, chalcogenides, and non-metal elements. *Green Chem.* 2015, 17, 4107–4132.

47. Wang, Y.; Choi, S.; Zhao, X.; Xie, S.; Peng, H. C.; Chi, M.; Huang, C.Z.; Xia, Y. Polyol synthesis of ultrathin Pd nanowires via attachment-based growth and their enhanced activity towards formic acid oxidation. *Adv. Funct. Mater.* 2013, 24, 131–139.

48. Espinos, J.P.; Fernandez, A.; Caballero, A.; Jimenez, V.M.; Sanchez-Lopez, J.C.; Contreras, L.; Leinen, D.; Gonzalez-Elipe, A.R. Ion-beam-induced CVD: an alternative method of thin film preparation. *Chem. Vap. Depos.* 1997, 3, 219–226.

49. Mohapatra, D.; Jain, L.; Rai, P.; Hazra, K.; Samajdar, I.; Misra, D. Development of crystallographic texture and in-grain misorientation in CVD-produced single and polycrystalline diamond. *Chem. Vap. Depos.* 2011, 17, 107–113.

50. Liu, J.; Yuan, Y.; Bashir, S. Functionalization of aligned carbon nanotubes to enhance the performance of fuel cell. *Energies* 2013, 6, 6476–6486.

51. Mazumder, J.; Kar, A. Theory and Application of Laser Chemical Vapor Deposition. Springer, 1995.

52. De Arco, L.G.; Zhang, Y.; Schlenker, C.W.; Ryu, K.; Thompson, M.E.; Zhou, C.W. Continuous, highly flexible, and transparent graphene films by chemical vapour deposition for organic photovoltaics. *ACS Nano* 2010, 4, 2865–2873.

53. Lin, B.; Sundararaj, U.; Potschke, P. Melt mixing of polycarbonate with multi-walled carbon nanotubes in miniature mixers. *Macromol. Mater. Eng.* 2006, 291, 227–238.

54. Mukherjee, P.; Balamurugan, B.; Shield, J.E.; Sellmyer, D.J. Direct gas-phase formation of complex core-shell and three-layer Mn–Bi nanoparticles. *RSC Adv.* 2016, 6, 92765–92770.

55. Benetti, D.; Nouar, R.; Nechache, R.; Pepin, H.; Sarkissian, A.; Rosei, F.; MacLeod, J.M. Combined magnetron sputtering and pulsed laser deposition of TiO_2 and BFCO thin films. *Sci. Rep.* 2017, 7, 2503–5.

56. Deng, H.; Zhang, C.; Xie, Y.; Tumlin, T.; Giri, L.; Karna, S.P.; Lin, J. Laser induced MoS_2/carbon hybrids for hydrogen evolution reaction catalysts. *J. Mater. Chem. A* 2016, 4, 6824–6830.

57. Wang, F.; Chemseddine, A.; Abdi, F.F.; Van de Krol, R.; Berglund, S.P. Spray pyrolysis of CuBi2O4 photocathodes: improved solution chemistry for highly homogeneous thin films. *J. Mater. Chem. A* 2017, 5, 12838–12847

58. Jeevanandam, J.; Chan, Y.S.; Danquah, M.K. Biosynthesis of metal and metal oxide nanoparticles. *ChemBioEng Rev.* 2016, 3, 55–67.

59. Bhattacharya, R.; Mukherjee, P. Biological properties of "naked" metal nanoparticles. *Adv. Drug Deliv. Rev.* 2008, 60, 1289–1306.

60. Zhang, X.; Yan, S.; Tyagi, R.D.; Surampalli, R.Y. Synthesis of nanoparticles by microorganisms and their application in enhancing microbiological reaction rates. *Chemosphere* 2011, 82, 489–494.

61. Singh, P.; Kim, Y.J.; Zhang, D.; Yang, D.C. Biological synthesis of nanoparticles from plants and microorganisms. *Trends Biotechnol.* 2016, 34 588–599.

62. Dickerson, M.B.; Sandhage, K.H.; Naik, R.R. Protein- and peptide-directed syntheses of inorganic materials. *Chem. Rev.* 2008, 108, 4935–4978.

63. Huang, J.; Lin, L.; Sun, D.; Chen, H.; Yang, D.; Li, Q. Bio-inspired synthesis of metal nanomaterials and applications. *Chem. Soc. Rev.* 2015, 44, 6330–6374.

64. Mittal, A.K.; Chisti, Y.; Banerjee, U.C. Synthesis of metallic nanoparticles using plant extracts. *Biotechnol. Adv.* 2013, 31, 346–356.

65. Dresselhaus, M.S.; Chen, G.; Tang, M.Y.; Yang, R.; Lee, H.; Wang, D.; Ren, Z.; leurial, J.P.; Gogna, P. New directions for low-dimensional thermoelectric materials. *Adv. Mater.* 2007, 19, 1043–1053.

66. Rabouw, F.T.; Donega, C.M. Excited-state dynamics in colloidal semiconductor nanocrystals. *Top. Curr. Chem.* 2016, 58, 374.

67. Dreaden, E. C.; Alkilany, A.M.; Huang, X.; Murphy, C.J.; El-Sayed, M.A. The golden age: gold nanoparticles for biomedicine. *Chem. Soc. Rev.* 2012, 41, 2740–2779.

68. Bhuvaneshwari, B.; Palani, G. S.; Iye, N.R. Bio-inspired and bio-mimetic materials an overview. *J. Struc. Eng.* 2011, 38, 1–10.

69. Kamat, P.V. Quantum dot solar cells. The next big thing in photovoltaics. *J. Phys. Chem. Lett.* 2013, 4, 908–918.

70. Gratzel, M. Conversion of sunlight to electric power by nanocrystalline dye-sensitized solar cells. *J. Photochem. Photobiol. A: Chem.* 2004, 164, 3–14.

71. Solar Tech USA. 2017. hwww.solartech-usa.comi

72. Shockley, W.; Queisser, H.J. Detailed balance limit of efficiency of p–n junction solar cells. *J. Appl. Phys.* 1961, 32, 510 519.

73. Konstantatos, G.; Sargent, E.H. (Eds.), Colloidal Quantum Dot Optoelectronics and Photovoltaics, Cambridge University Press, Cambridge, United Kingdom, 2013.

74. El Chaar, L.; Lamont, L.A.; El Zein, N. Review of photovoltaic technologies. *Renew. Sust. Energy Rev.* 2011, 15, 2165–2175.

75. National Renewable Energy Laboratory (NREL) 2017.

76. Green, M.A. Thin-film solar cells: review of materials, technologies and commercial status. *J. Mater. Sci. Mater. Electron.* 2007, 18, 15–19.

77. Becker, C.; Amkreutz, D.; Sontheimer, T.; Preidel, V.; Lockau, D.; Haschke, J. et al. Polycrystalline silicon thin-film solar cells: status and perspectives. *Sol. Energy Mater. Sol. Cells.* 2013, 119, 112–123.

78. Ding, K.; Kirchartz, T.; Bittkau, K.; Lambertz, A.; Smirnov, V.; Hupkes, J. et al. Photovoltaics: nanoengineered materials and their functionality in solar cells, nanotechnology for energy sustainability, Wiley-VCH Verlag GmbH & Co. KGaA 2017, pp. 181–206.

79. Maghsoodi, S.; Brophy, B.L.; Colson, T.E.; Gonsalves, P.R.; Abrams, Z.R. High gain durable anti-reflective coating. 2017, US9399720.

80. Jin, Y.; LI, Q.; Fan, S. Method for Making Solar Cells. 2016, US20160225936.

81. Kuan, C.; Lee, M.; Su, W. Solar cell with surface staged type antireflective layer. 2016, US20160225924.

82. Zhu, P. All-aluminum back surface field aluminum paste for crystalline silicon solar cell and preparation method thereof. 2017, US20170148936.

83. Hardin, B.E.; Connor, S.T.; Groves, J.R.; Peters, C.H. Silver-bismuth Noncontact Metallization Pastes for Silicon Solar Cells. 2017, US20170062632.

84. Loscutoff, P.; Rim, S. Hybrid emitter all back contact solar cell. 2016, US20160204288.

85. Kapur, P.; Deshpande, A.; Rana, V. V.; Moslehi, M.M.; Seutter, S.M. Fabrication methods for back contact solar cells. 2015, US20150171230.

86. Zeitouny, J.; Katz, E.A.; Dollet, A.; Vossier, A. Band gap engineering of multi-junction solar cells: effects of series resistances and solar concentration. *Sci. Rep.* 2017, 7, 1766.

87. Dimroth, F. High-efficiency solar cells from IIIV compound semiconductors. *Phys. Status Solidi* 2006, 3, 373–379.

88. Samuelson, L.; Magnusson, M.; Capasso, F. Nanowire-based solar cell structure. 2017, US20170155008.

89. Bjork, M. Ohlsson, J.; Samuelson, L.; Sauar, E.; Aberg, I. Dual layer photovoltaic device. WO/2016/066630, 2016.

90. Fogel, K.E.; Hekmatshoartabari, B.; Sadana, D.K.; Shahidi, G.G.; Shahrjerdi, D. Heterostructure germanium tandem junction solar cell. 2016, US20160284926.

91. National Renewable Energy Laboratory (NREL), (2017).

92. Xu, T.; Yu, L.; How to design low bandgap polymers for highly efficient organic solar cells. *Mater. Today.* 2014, 17, 11–15.

93. Wang,J.; Wang, H.; Prakoso, A. B.; Togonal, A. S.; Hong, L.; Jiang, C.; Rusli. High efficiency silicon nanowire/organic hybrid solar cells with two-step surface treatment. *Nanoscale*, 2015, 7, 4559–4565.

94. Lee, H.; Lee, J.; Lee, J.; Jang, S.; Choi, D.; Bang, J. Organic Solar Cell and Manufacturing Method Therefore. 2017, US20170040543.

95. Peet, J.H.; Lungenschmied, C. Photovoltaic cells. 2016, US20160329510.

96. Sato, H.; Kobayashi, K. Photoelectric conversion element and solar cell. 2016, US20160133392.

97. Wong, W.L.; Leung, L.T.; Zhang, L.; Chow, C.M.; Zhu, H.; Lai, L.F. et al., Fullerene derivatives and their applications in organic photovoltaics. 2017, 9637384.

98. O'Regan, B.; Gratzel, M. A low-cost, high-efficiency solar cell based on dye-sensitized colloidal TiO_2 films. *Nature* 1991, 353, 737–740.

99. F. Christian, Edith, Selly, D. Adityawarman, A. Indarto, Application of nanotechnologies in the energy sector: a brief and short review, *Front. Energy* 2013, 7, 618.

100. Gratzel, M.; Dye-sensitized solar cells. *J. Photochem. Photobiol. C Photochem. Rev.* 2003, 4, 145–153.

101. Janani, M.; Nair, S.V.; Nair, A.S. Photovoltaics: role of nanotechnology in dye-sensitized solar cells, nanotechnology for energy sustainability, Wiley VCH Verlag GmbH & Co. KGaA, Weinheim, Germany. 2017, pp. 101–132.

102. Roy, P.; Kim, D.; Lee, K.; Spiecker, E.; Schmuki, P. TiO_2 nanotubes and their application in dye-sensitized solar cells. *Nanoscale* 2010, 2, 45–59.

103. Kogure, H.; Aoyama, K.; Kishimoto, S.; Ikegami, M. Viscous dispersion liquid and method for producing same, porous semiconductor electrode substrate, and dye-sensitized solar cell. 2017, US20170140880A1.

104. Lee, J.; Nishimura, K.Y.; Pan, W.; Vail, S.A. Dye-sensitized solar cell, tandem dye-sensitized solar cell, and composite nanostructure. 2015, EP2843674.

105. Horiuchi, T.; Yashiro, T.; Segawa, H.; Uchida, S. Dye-sensitized solar cell. 2016, 20160276609.

106. Miguez, H.; Colodrero, S. Solar to electric energy conversion device. 2016, 20160155573.

107. Lee, J.G.; Lee, S.J.; Joung, I.S. Series-type dye-sensitized solar cell module and method of manufacturing the same. 2016, 20160225535.

108. Chien, H. Solar cell and method for manufacturing the same. 2015, 20150179354.

109. Rennig, A.; Hammermann, M.; Eickemeyer, F. Solar cell structure. 2017, 9584065.

110. Solar Tech USA. 2017 (hwww.solartech-usa.comi)

111. De Mello Donega, C.; Liljeroth, P.; Vanmaekelbergh, D. Physicochemical evaluation of the hot-injection method, a synthesis route for monodisperse nanocrystals. *Small* 2015, 1, 1152–1162.

112. Kubo, S.; Nakayama, T.; Ninomiya, H.; Muramoto, K.; Fujita, K. Quantum dot solar cell. 2017, 20170213924.

113. Jun, S.A.; Won, Y.; Jang, H.S.; Jang, E.J. Semiconductor nanocrystals and processes for synthesizing the same. 2015, 20150218442.

114. So, F.; Kim, Y.; Do; Lee, W.; Jae, Pradhan, K.; Bhabendra. Monodisperse IR-absorbing nanoparticles and related methods and devices. 2017, WO/2017/039774.

115. Wassvik, O.; Bergstrom, H.; Wallander, M.; Craven-bartle, T. Solar panel converter layer. 2016, 20160276501.

116. Wang, H.; Zhang, X.; Lin, S.; Tang, A.; Liu, R.; Tsai, T. et al., Quantum dot composite material and manufacturing method and application thereof. 2017, EP3192846.

117. Kikuchi, T. Jang, E.; Kang, H.A.; Won, N.; Cho, O.; Koh, H.D. Barrier coating compositions, composites prepared therefrom, and quantum dot polymer composite articles including the same. 2016, 20160160060.

118. Welser, R.E.; Sood, A.K. Quantum well waveguide solar cells and methods of constructing the same. 2016, 20160027940.

119. Ono, M.; Kikuchi, M.; Tanaka, A.; Suzuki, M.; Kanemitsu, Y. Semiconductor film, solar cell, light-emitting diode, thin film transistor, and electronic device. 2015, 20150295035.

120. Avouris, P.; Chen, Z. Perebeinos, V. Carbon-based electronics. *Nat. Nano* 2007, 2, 605–615.

121. Avouris, P.; Freitag, M.; Perebeinos, V. Carbon nanotube photonics and optoelectronics. *Nat. Photon* 2008, 2, 341–350.

122. Liangbing, H.; Jianfeng, L.; Jun, L.; George, G.; Tobin, M. Flexible organic light-emitting diodes with transparent carbon nanotube electrodes: problems and solutions. *Nanotechnology* 2010, 21, 155202.

123. Bachilo, S. M.; Strano, M. S.; Kittrell, C.; Hauge, R. H.; Smalley, R. E.; Weisman, R. B. Structure-assigned optical spectra of single-walled carbon nanotubes. *Science* 2002, 298, 2361–2366.

124. Dukovic, G.; Wang, F.; Song, D.; Sfeir, M. Y.; Heinz, T. F.; Brus, L. E. Structural dependence of excitonic optical transitions and band-gap energies in carbon nanotubes. *Nano Lett.* 2005, 5, 2314–2318.
125. Perebeinos, V.; Tersoff, J.; Avouris, P. Scaling of excitons in carbon nanotubes. *Phys. Rev. Lett.* 2004, 92, 257402.
126. O'Connell, M. J.; Bachilo, S. M.; Huffman, C. B.; Moore, V. C.; Strano, M. S.; Haroz, E. H.; Rialon, K. L.; Boul, P. J.; Noon, W. H.; Kittrell, C.; Ma, J.; Hauge, R. H.; Weisman, R. B.; Smalley, R. E. Band gap fluorescence from individual single-walled carbon nanotubes. *Science* 2002, 297, 593–596.
127. Lebedkin, S.; Hennrich, F.; Skipa, T.; Kappes, M. M. J. Near-infrared photoluminescence of single-walled carbon nanotubes prepared by the laser vaporization method. *Phys. Chem. B* 2003, 107, 1949–1956.
128. Lefebvre, J.; Austing, D. G.; Bond, J.; Finnie, P. photoluminescence imaging of suspended single-walled carbon nanotubes. *Nano Lett.* 2006, 6, 1603–1608.
129. Wang, F.; Dukovic, G.; Brus, L. E.; Heinz, T. F. Time-resolved fluorescence of carbon nanotubes and its implication for radiative lifetimes. *Phys. Rev. Lett.* 2004, 92, 177401.
130. Walsh, A.G.; Vamivakas, A. N.; Yin, Y.; Cronin, S. B.; Unlu, M. S.; Goldberga, B. B.; Swan, A. K. Scaling of exciton binding energy with external dielectric function in carbon nanotubes. *Physica E* 2007, 40, 2375–2379.
131. Perebeinos, V.; Avouris, P. Phonon and electronic nonradiative decay mechanisms of excitons in carbon nanotubes. *Phys. Rev. Lett.* 2008, 101, 057401.
132. Engel, M.; Small, J. P.; Steiner, M.; Freitag, M.; Green, A. A.; Hersam, M. C.; Avouris, P. Thin film nanotube transistors based on self-assembled, aligned, semiconducting carbon nanotube arrays. *ACS Nano* 2008, 2, 2445–2452.
133. Misewich, J. A.; Martel, R.; Avouris, P.; Tsang, J. C.; Heinze, S.; Tersoff, J. Electrically induced optical emission from a carbon nanotube FET. *Science* 2003, 300, 783–786.
134. Freitag, M.; Chen, J.; Tersoff, J.; Tsang, J. C.; Fu, Q.; Liu, J.; Avouris, P. Mobile Ambipolar domain in carbon-nanotube infrared emitters. *Phys. Rev. Lett.* 2004, 93, 076803.
135. Freitag, M.; Perebeinos, V.; Chen, J.; Stein, A.; Tsang, J. C.; Misewich, J. A.; Martel, R.; Avouris, P. Hot carrier electroluminescence from a single carbon nanotube. *Nano Lett.* 2004, 4, 1063–1066.
136. Zaumseil, J.; Ho, X.; Guest, J. R.; Wiederrecht, G. P.; Rogers, J. A. Electroluminescence from electrolyte-gated carbon nanotube field-effect transistors. *ACS Nano* 2009, 3, 2225–2234.
137. Engel, M.; Small, J. P.; Steiner, M.; Freitag, M.; Green, A. A.; Hersam, M. C.; Avouris, P. Thin film nanotube transistors based on self-assembled, aligned, semiconducting carbon nanotube arrays. *ACS Nano* 2008, 2, 2445–2452.
138. Adam, E.; Aguirre, C. M.; Marty, L.; St-Antoine, B. C.; Meunier, F.; Desjardins, P.; Ménard, D.; Martel, R. *Nano Lett.* 2008, 8, 2351–2355.
139. Qian, H.; Georgi, C.; Anderson, N.; Green, A. A.; Hersam, M. C.; Novotny, L.; Hartschuh, A. Exciton energy transfer in pairs of single-walled carbon nanotubes. *Nano Lett.* 2008, 8, 1363–1367.
140. Chen, J.; Perebeinos, V.; Freitag, M.; Tsang, J.; Fu, Q.; Liu, J.; Avouris, P. *Science* 2005, 310, 1171–1174.
141. Perebeinos, V.; Avouris, P. *Phys. Rev. B* 2006, 74, 121410.
142. Koswatta, S. O.; Perebeinos, V.; Lundstrom, M. S.; Avouris, P. Computational study of exciton generation in suspended carbon nanotube transistors. *Nano Lett.* 2008, 8, 1596–1601.

143. Perebeinos, V.; Avouris, P. Exciton ionization, Franz–Keldysh, and stark effects in carbon nanotubes. *Nano Lett.* 2007, 7, 609–613.

144. Freitag, M.; Tsang, J. C.; Kirtley, J.; Carlsen, A.; Chen, J.; Troeman, A.; Hilgenkamp, H.; Avouris, P. Electrically excited, localized infrared emission from single carbon nanotubes. *Nano Lett.* 2006, 6, 1425–1433.

145. Ilani, S.; Donev, L. A. K.; Kindermann, M.; McEuen, P. L. Measurement of the quantum capacitance of interacting electrons in carbon nanotubes. *Nat. Phys.* 2006, 2, 687–691.

146. Grosso, G.;Graves, J.; Hammack, A. T.; High, A. A.; Butov, L. V.; Hanson, M.; Gossard, A. C. Excitonic switches operating at around 100 K. *Nat. Photon.* 2009, 3, 577–580.

147. Mann, D.; Kato, Y. K.; Kinkhabwala, A.; Pop, E.; Cao, J.; Wang, X.; Zhang, L.; Wang, Q.; Guo, J.; Dai, H. Electrically driven thermal light emission from individual single-walled carbon nanotubes. *Nat. Nano.* 2007, 2, 33–38.

148. Freitag, M.; Martin, Y.; Misewich, J. A.; Martel, R.; Avouris, P. Photoconductivity of single carbon nanotubes. *Nano Lett.* 2003, 3, 1067–1071.

149. J. U. Lee, Photovoltaic effect in ideal carbon nanotube diodes. *Appl. Phys. Lett.*, 2005, 87, 073101.

150. Ahn, Y. H.; Tsen, A. W.; Kim, B.; Park, Y. W.; Park, J. Photocurrent Imaging of p–n junctions in ambipolar carbon nanotube transistors. *Nano Lett.* 2007, 7, 3320–3323.

151. Balasubramanian, K.; Burghard, M.; Kern, K.; Scolari, M.; Mews, A. Photocurrent imaging of charge transport barriers in carbon nanotube devices. *Nano Lett.* 2005, 5, 507–510

152. Gabor, N. M.; Zhong, Z.; Bosnick, K.; Park, J.; McEuen, P. L. Extremely efficient multiple electron–hole pair generation in carbon nanotube photodiodes. *Science* 2009, 325, 1367–1371.

153. Itkis, M. E.; Borondics, F.; Yu, A.; Haddon, R. C. Bolometric infrared photoresponse of suspended single-walled carbon nanotube films. *Science* 2006, 312, 413–416.

154. Lee, K. W.; Lee, S. P.; Choi, H.; Mo, K. H.; Jang, J. W.; Kweon, H.; Lee, C. E. Enhanced electroluminescence in polymer-nanotube composites. *Appl. Phys. Lett.* 2007, 91, 023110.

155. Kumar, A.; Zhou, C. The race to replace tin-doped indium oxide: which material will win?. *ACS Nano* 2010, 4, 11–14.

156. Green, A. A.; Hersam, M. C. Colored semitransparent conductive coatings consisting of monodisperse metallic single-walled carbon nanotubes. *Nano Lett.* 2008, 8, 1417–1422.

157. Southard, A.; Sangwan, V.; Cheng, J.; Williams, E. D.; Fuhrer, M. S. Solution-processed single walled carbon nanotube electrodes for organic thin-film transistors. *Org. Electron.* 2009, 10, 1556–1561.

158. Sangwan, V. K.; Southard, A.; Moore, T. L.; Ballarotto, V. W.; Hines, D. R.; Fuhrer, M. S.; Williams, E. D. Transfer printing approach to all-carbon nanoelectronics. *Microelectronic Engineering* 2011, 88, 3150–3154.

159. Liu, X.; Si, J.; Chang, B.; Xu, G.; Yang, Q.; Pan, Z.; Xie, S.; Ye, P.; Fan, J.; Wan, M. Third-order optical nonlinearity of the carbon nanotubes. *Appl. Phys. Lett.* 1999, 74, 164–166.

160. Sakakibara, Y.; Tatsuura, S.; Kataura, H.; Tokumoto, M.; Achiba, Y. Near-infrared saturable absorption of single-wall carbon nanotubes prepared by laser ablation method. *Jpn. J. Appl. Phys.* 2003, 42, L494.

161. Bonaccorso, F. Sun, Z.; Hasan, T.; Ferrari, A. C. Graphene photonics and optoelectronics. *Nat. Photon.* 2010, 4, 611–622.

162. Peres, N. M. R. The transport properties of graphene: An introduction. *Rev. Mod. Phys.* 2010, 82, 2673–2700.

163. Nair, R. R.; Blake, P.; Grigorenko, A. N.; Novoselov, K. S.; Booth, T. J.; Stauber, T. Peres, N. M. R.; Geim, A. K.. Fine structure constant defines visual transparency of graphene. *Science* 2008, 320, 1308.

164. Bae, S.; Kim, H.; Lee, Y.; Xu, X.; Park, J.S.; Zheng, Y.; Balakrishnan, J.; Lei, T.; Ri Kim, H. Song, Y. I.; Kim, Y.-J.; Kim, K. S.; Ozyilmaz, B.; Ahn, J.-H.; Hong, B. H.; Iijima, S. Roll-to-roll production of 30-inch graphene films for transparent electrodes. *Nat. Nano.* 2010, 5, 574–578.

165. Mak, K. F.; Shan, J.; Heinz, T. F. Seeing many-body effects in single- and few-layer graphene: observation of two-dimensional saddle-point excitons. *Phys. Rev. Lett.*, 2011, 106, 046401.

166. Eda, G.; Lin, Y.-Y.; Mattevi, C.; Yamaguchi, H.; Chen, H.-A.; Chen, I. S.; Chen, C.-W.; Chhowalla, M. Blue photoluminescence from chemically derived graphene oxide. *Adv. Mater.*, 2010, 22, 505–509.

167. Gokus, T.; Nair, R. R.; Bonet, A.; Böhmler, M.; Lombardo, A.; Novoselov, K. S.; Geim, A. K. Ferrari, A. C.; Hartschuh, A. Making graphene luminescent by oxygen plasma treatment. *ACS Nano*, 2009, 3, 3963–3968.

168. Essig, S.; Marquardt, C. W.; Vijayaraghavan, A.; Ganzhorn, M.; Dehm, S.; Hennrich, F.; Ou, F.; Green, A. A.; Sciascia, C.; Bonaccorso, F.; Bohnen, K. P.; Löhneysen, H. V.; Kappes, M. M.; Ajayan, P. M.; Hersam, M. C.; Ferrari, A. C.; Krupke, R. Phonon-Assisted electroluminescence from metallic carbon nanotubes and graphene. *Nano Lett.*, 2010, 10, 1589–1594.

169. Xia, F.; Mueller, T.; Lin, Y.-M.; Valdes-Garcia, A.; Avouris, P. Ultrafast graphene photodetector. *Nat. Nano.* 2009, 4, 839–843.

170. Mueller, T.; Xia, F.; Avouris, P. Graphene photodetectors for high-speed optical communications. *Nat. Photon.* 2010, 4, 297–301.

171. Gabor, N. M.; Song, J. C. W.; Ma, Q.; Nair, N. L.; Taychatanapat, T.; Watanabe, K.; Taniguchi, T.; Levitov, L. S.; Jarillo-Herrero, P. Hot carrier-assisted intrinsic photoresponse in graphene. *Science* 2011, 334, 648–652.

172. Sun, D.; Aivazian, G.; Jones, A. M.; Ross, J. S.; Yao, W.; Cobden, D.; Xu, X. Ultrafast hot-carrier-dominated photocurrent in graphene. *Nat. Nano.* 2012, 7, 114–118.

173. Sun, X.; Liu, Z.; Welsher, K.; Robinson, J.; Goodwin, A.; Zaric, S.; Dai, H. Nano-graphene oxide for cellular imaging and drug delivery. *Nano Research* 2008, 1, 203–212.

174. Duch, M. C.; Budinger, G. R. S.; Liang, Y. T.; Soberanes, S.; Urich, D.; Chiarella, S. E.; Campochiaro, L. A.; Gonzalez, A.; Chandel, N. S.; Hersam, M. C.; Mutlu, G. M. Minimizing oxidation and stable nanoscale dispersion improves the biocompatibility of graphene in the lung. *Nano Lett.* 2011, 11, 5201–5207.

175. Chhowalla, M.; Shin, H.S.; Eda, G.; Li, L.-J.; Loh, K.P.; Zhang, H. The chemistry of two-dimensional layered transition metal dichalcogenide nanosheets. *Nat. Chem.* 2013, 5, 263.

176. Wang, Q.H.; Kalantar-Zadeh, K.; Kis, A.; Coleman, J.N.; Strano, M.S. Electronics and optoelectronics of two-dimensional transition metal dichalcogenides. *Nat. Nanotechnol.* 2012, 7, 699–712.

177. Wallauer, R.; Reimann, J.; Armbrust, N.; Güdde, J.; Höfer, U. Intervalley scattering in MoS$_2$ imaged by two-photon photoemission with a high-harmonic probe. *Appl. Phys. Lett.* 2016, *109,* 162102.

178. Grubišic Cabo, A.; Miwa, J.A.; Grønborg, S.S.; Riley, J.M.; Johannsen, J.C.; Cacho, C.; Alexander, O.; Čhapman, R.T.; Springate, E.; Grioni, M.; et al. Observation of ultrafast free carrier dynamics in single layer MoS_2. *Nano Lett.* 2015, *15*, 5883–5887.

179. Ulstrup, S.; Cabo, A.G.; Miwa, J.A.; Riley, J.M.; Grønborg, S.S.; Johannsen, J.C.; Cacho, C.; Alexander, O.; Chapman, R.T.; Springate, E.; et al. Ultrafast band structure control of a two-dimensional heterostructure. *ACS Nano* 2016, *10*, 6315–6322.

180. Spiecker, H.; Schmidt, O.; Ziethen, C.; Menke, D.; Kleineberg, U.; Ahuja, R.C.; Merkel, M.; Heinzmann, U.; Schönhense, G. Time-of-flight photoelectron emission microscopy TOF-PEEM: First results. *Nucl. Instrum. Methods Phys. Res.* 1998, *406*, 499–506.

181. Chernov, S.V.; Medjanik, K.; Tusche, C.; Kutnyakhov, D.; Nepijko, S.A.; Oelsner, A.; Braun, J.; Minár, J.; Borek, S.; Ebert, H.; et al. Anomalous d-like surface resonances on Mo(110) analyzed by time-of-flight momentum microscopy. *Ultramicroscopy* 2015, *159*, 453–463.

182. Gehlmann, M.; Aguilera, I.; Bihlmayer, G.; Nemšák, S.; Nagler, P.; Gospodarič, P.; Zamborlini, G.; Eschbach, M.; Feyer, V.; Kronast, F.; et al. Direct observation of the band gap transition in atomically thin ReS_2. *Nano Lett.* 2017, *17*, 5187–5192.

183. Hart, L.S.; Webb, J.L.; Dale, S.; Bending, S.J.; Mucha-Kruczynski, M.; Wolverson, D.; Chen, C.; Avila, J.; Asensio, M.C. Electronic bandstructure and van der Waals coupling of $ReSe_2$ revealed by high-resolution angle-resolved photoemission spectroscopy. *Sci. Rep.* 2017, *7*, 5145.

184. Webb, J.L.; Hart, L.S.; Wolverson, D.; Chen, C.; Avila, J.; Asensio, M.C. Electronic band structure of ReS_2 by high-resolution angle-resolved photoemission spectroscopy. *Phys. Rev. B* 2017, *96*, 115205.

185. Cattelan, M.; Fox, N. A. A Perspective on the Application of Spatially Resolved ARPES for 2D Materials. Nanomaterials 2018, 8, 284–26.

186. Bayan, S., Das, U. & Mohanta, D. Peacock feather supported self assembled ZnO nanostructures for tuning photonic properties. Eur. Phys. J. D . 2010. 61, 463–468 (2011). https://doi.org/10.1140/epjd/e2010-10457-7.

CHAPTER 2

0D Nanomaterials and Their Optoelectronic Applications

ZIAUL RAZA KHAN

Department of Physics, College of Science, University of Hail, Hail 2440, Kingdom of Saudi Arabia. E-mail: zr.khan@uoh.edu.sa

ABSTRACT

Semiconductor nanomaterials one of the most important components of advanced optoelectronics devices. Size-dependent properties of semiconductor nanomaterials have provided broad range of modification in device sizes, power consumption, and efficiency enhancement for the advancement of optoelectronic devices. Zero-dimensional semiconductor nanomaterials (0D nanomaterials) can be a promising candidate due to its unique physical, electronic, and surface properties to meet this objective in the advancement of optoelectronic devices. The biggest challenge of the 21st century is a huge amount of energy demand and supply. Energy harvesting devices, such as solar cells, low power consumption, and efficient lightning devices, light emitting laser diode, laser controlled devices, and energy stored devices based on 0D nanomaterials will be played a major role to resolve this toughest demand of energy in future life. Therefore, 0D nanomaterials have been attracted intense attention of material researchers to develop various types of 0D nanomaterials with suitable, optical, electronic, and electrical properties for manufacturing of advanced optoelectronic device applications. This chapter describes the classification of nanomaterials and growth techniques discussed in brief of nanomaterials. Afterward, this chapter focuses on the development of 0D nanomaterials by various techniques and demonstrates the potential applications of 0D nanomaterials for optoelectronic applications in recent trends.

2.1 INTRODUCTION

History of nanomaterials is thousands of years old. In ancient times, people used nanomaterials such as Au, Ag, and C nanoparticles in various cuisines and medicines but that time nano word and size-effect phenomena were not introduced. People just know about the practical advantages of nanomaterial in food and traditional medicines. In the 21st century, people started to grow clusters of various materials and observed interesting deviation of physical properties with the variation in cluster size. The turning moment for the material researchers was 29 December, 1959. It was a great day, when quantum chromodynamics man, physicist Richard Feynman, surprised to the material researchers of whole world in his plenary talk in the annual meeting of American Physical Society, at the California Institute of Technology (Caltech). He said that "There is plenty of room at the bottom" and predicted big package in small thing in coming era [1]. The Richard Feynman prediction ignited the idea of nanomaterial in materials researchers mind and they started to grow nanomaterial and investigated its physical properties. Recently, a huge revolution was seen in the science and technology from earth to space science owing to the size-dependent material properties. Low-dimensional materials showed the unique physical properties as compared to the bulk counterpart such as 1000 times lighter and 1000 times stronger than the steel which may be useful in spacecraft technology. Therefore, the devices based on the nanomaterials are miniaturize, highly efficient, and low power consumption. In the present era, people are facing tough challenges for green energy production to meet demand of the world. Now most of the energy sources are depended on the fossil fuels to run daily life needs. This chapter is mainly focused on the development and applications of 0D nanomaterials. These materials have attracted much attention due to its excellent physical properties and have potential applications in various fields, such as energy harvesting, storage, smart electronics devices, gas sensing, light sensing, low power consumption light devices, catalyst, pollution detection, and laser controlled devices.

The term nanotechnology term, as a discipline, was first introduced in 1974 by Taniguchi [2]. He reported processing, separation, consolidation, and deformation of materials by single atom or molecules. Drexler gave the idea about applications of low-dimensional materials in nanosize devices in 1980 [3]. The word "nano" means "nanos" which means dwarfs. At meter scale "nano" is used as a prefix for 10^{-9}. Nanoscience is the study of matter at dimension of nanoscale 10^{-9} m and the fabrication of devices is based on the nanodimensional materials which is known as nanotechnology. Less than 100 nm size materials are called nanomaterials. These materials have high surface-to-volume ratio and due to quantum confinement of carriers, nanomaterials show exceptional electronic, optical, and magnetic properties. This change becomes more and more prominent when size reduces below the Bohr radius of molecules, which is intrinsic property of materials. The range of Bohr radius for the most of the materials falls in between 1 and 10 nm. The size of materials reduces to zero as a result of increase in high surface-to-volume ratio. Surface atoms have lesser number of neighboring atoms as compared to the bulk counterpart and hence have low cohesive energy. Therefore, lowering in melting point occurs in nanomaterials. Reduction in cohesive energy leads to a huge variation in thermal properties of nanomaterials and decreases in ferromagnetism and ferroelectric property on reduction in size less than 100 nm. In addition, high surface-to-volume ratio enhances the presence of high number of dangling bond on the surface, which makes nanomaterials higher reactive and catalytic than the bulk. Owing to the presence of higher number of dangling bonds, nanomaterials show the significant enhancement in sensing, catalytic, and surface functionalization and assembly.

The aim of this chapter is to discuss the growth and optoelectronic applications of 0D nanomaterials. Usually, II–VI (CdTe, CdS, ZnS, ZnSe, ZnO, etc.) III–V (GaN, BN, GaAs, GaP, and InN), and oxide (TiO_2, SnO_2, CuO, etc.) group of semiconductor materials has potential applications in optoelectronic devices. In the 1990s, worldwide many countries released huge amount of money in research area of nanoscience and nanotechnology. Many research groups have started the work on growth of 0D, 1D, 2D, and 3D semiconductor nanomaterials and investigated their physical properties. These materials showed excellent opto-dielectric-nonlinear properties of semiconductor nanostrcutures on size reduction and achieved the objectives to develop low power consumption and smart optoelectronic devices.

2.2 CLASSIFICATION OF LOW-DIMENSIONAL MATERIALS

Two decades ago, a revolution started in the growth area of low-dimensional materials and researchers were developing nanomaterials by various approaches of insulators, semiconductors, and conductors. Therefore, classification of low-dimensional materials on the basis of size and dimensions was required. Nanostructured materials (NSMs) are the basic clusters of atoms with less than 100 nm size in at least one direction. In 1995, first classification of NSMs was introduced by Gleiter and later on enlightened by Skorokhod in 2000. But present classification is not enough to cover all types of nanomaterials such as 0D, 1D, 2D, 3D, and fullerenes, nanotubes, and nanoflowers. Consequently, a further modification was needed to consider all types of NSMs. Pokropivny and Skorokhod proposed a modified classification scheme for NSMs, in which 0D, 1D, 2D, and 3D NSMs are included [6]. The detailed discussion of NSMs based on scheme of Pokropivny et al. is as follows:

2.2.1 0D NANOSTRUCTURED MATERIALS

From the last decades, various methods have been employed to grow 0D nanomaterials (0D NMs) with controlled size along with one direction. In 0D nanostructures, confinement of electrons in their motion in all directions is found. In addition, a reduction in size in all three directions gives rise to the noticeable change in the density of states and 0D nanomaterials density of states which is shown in Figure 2.1(a). Quantum dots (QDs), spheres, and nanoclusters are the best examples of 0D nanomaterials. Recently, 0D nanomaterials have been grown by various research groups for optoelectronic applications, such as energy harvesting, storage, photodetecting, light emitting diodes (LEDs), and solid-state laser. Waldiya et al. reported 0D CdS nanostructure which is shown in Figure 2.2, and they found excellent optoelectronic properties of 0D CdS nanostructures [7–10].

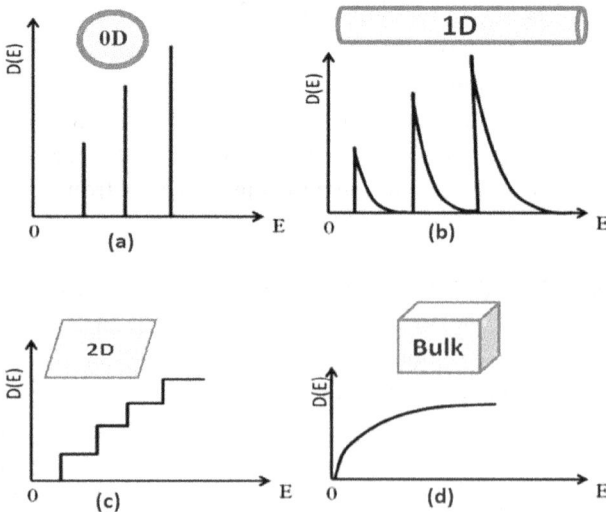

FIGURE 2.1 (a) 0D, (b) 1D, (c) 2D, (d) 3D nanomaterials system density of states (density of states vs. energy) plots.

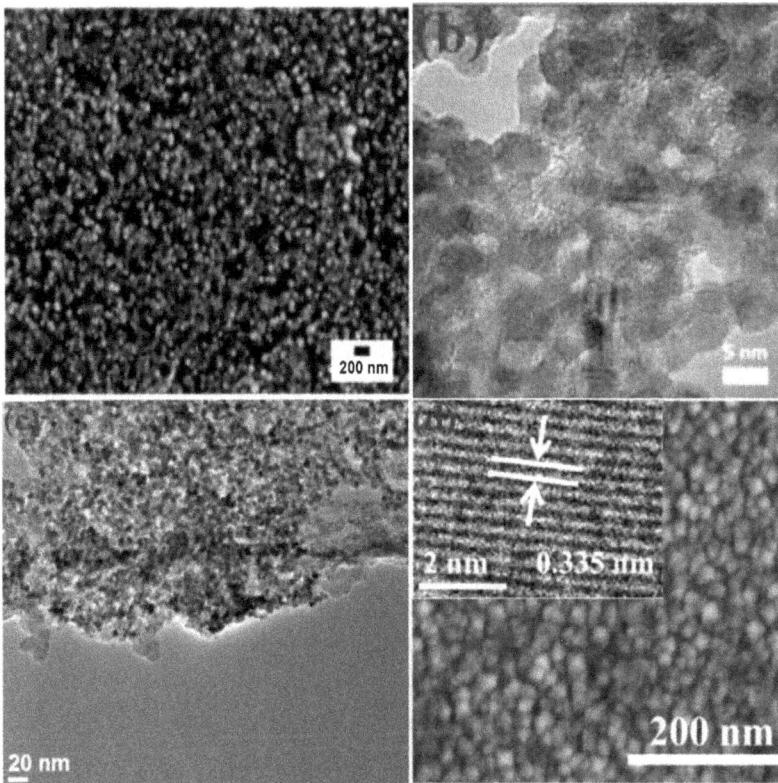

FIGURE 2.2 (a) CdS 0D nanostructure for optoelectronic applications [7], (b) ZnO QDs [8], (c) ZnO quantum dot white LED [9], (d) ZnO/SnO$_2$ QD for solar energy conversion [10].

Sources: (a) Reprinted with permission from Ref. [7]. © 2019 Elsevier. (b) Reprinted with permission from Ref. [8]. © 2019 Elsevier. (c) Reprinted with permission from Ref. [9]. © 2014 Springer Nature. (d) Reprinted with permission from Ref. [10]. © 2019 Elsevier.

2.2.2 1D NANOSTRUCTURED MATERIALS

Recently, 1D nanomaterials have attracted much attention due to their potential applications in wide range of optoelectronics field. In 1D nanomaterials, the electrons are confined in two directions and their free movement in directions is known as 1D NMS. As confinement of carriers in two directions, density of states (DOS) of 1D nanomaterials has changed from continuous to inverse steeped parabola. DOS of 1D nanomaterials is shown in Figure 2.1(b). Nanowires, nanorods, nanofibers, and nanotubes are the best examples of 1D nanomaterials. One-dimensional nanomaterials are playing revolutionary role in the fabrication of low power consumption for smart and miniaturize optoelectronics devices. The discussion about the 1D nanomaterials cannot be completed without the great physicist of nanomaterials, Iijima. He did path-breaking work on carbon (C) 1D nanomaterials and first he developed single-wall carbon nanotube. If you are thinking about the 1D nanomaterials, the field of 1D nanomaterials such as nanotubes has attained a significant attention after the pioneering work by Iijima [11]. Nowadays, many optoelectronic devices, such as solar cells, LEDs, laser, photodetector, light sensor, etc., have been commercialized in market based on 1D nanomaterials. Figure 2.3(a)–(f) shows the 1D nanostructure of CdS reported by Bakhsh et al., where they found size-dependent variation in luminescent properties [12]. Kumar et al. reported the growth of Cu-doped CdS nanowires and their luminescent properties for the applications of LEDs [13]. Enhanced photocatalytic properties of CdS nanowires were reported by the Kim et al. [14].

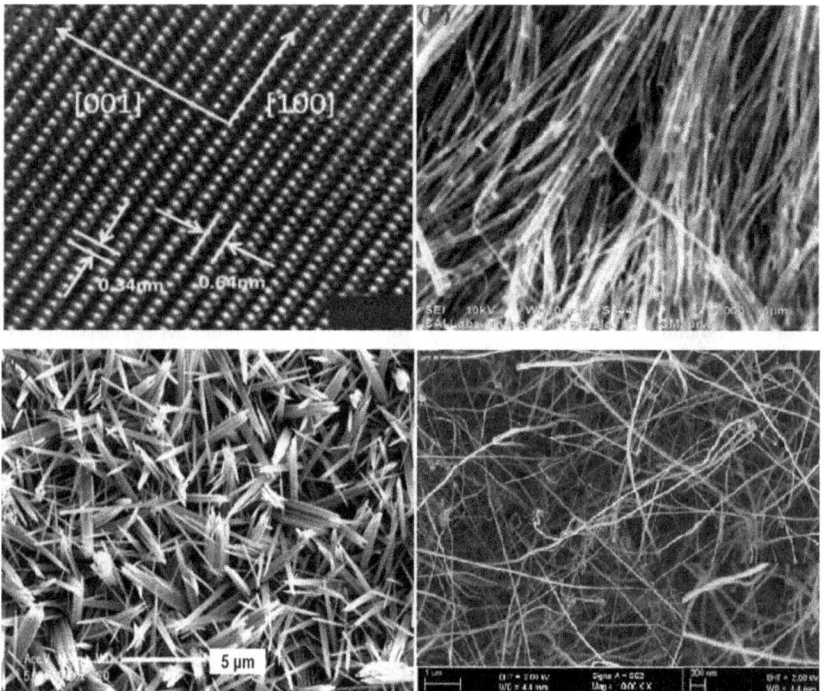

FIGURE 2.3　(a) CdS 1D nanostructure TEM images [12]. (b) Cu–CdS nanowires [13]. (c) CdS nanowires with different growth time [14]. (d) Boron nitride nanotube [15].

Sources: (a) Reprinted with permission from Ref. [12]. © 2016 Elsevier. (b) Reprinted with permission from Ref. [12]. © 2019 Elsevier. (c) Reprinted with permission from Ref. [14]. © 2019 Elsevier. (d) Reprinted with permission from Ref. [15]. © 2015 Elsevier.

2.2.3 2D NANOSTRUCTURED MATERIALS

Two-dimensional nanostructures have electron confinement in one direction and can move freely in two directions. Nanofilms, nanoplates, and branched structures are fallen in the category of 2D nanostructures. DOS of 2D nanomaterials shows the significant variation from continuous to stepped parabola. DOS structure of 2D nanomaterials is shown in Figure 2.3(c) (DOS). From last decades, growth and characterization of 2D nanomaterials are the burning topics among the material researchers because of the key role in advanced devices like solar cells, photocatalyst, sensors, photodetectors, and as seed layers for various materials [16–21]. Figure 2.4 illustrates various 2D nanomaterials [22–24].

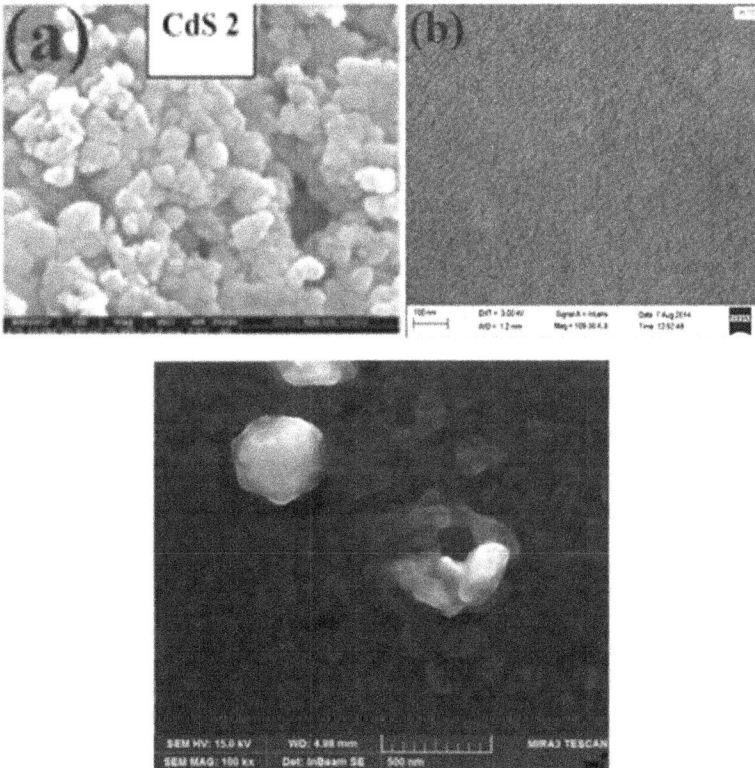

FIGURE 2.4 (a) CdS thin films for optoelectronic applications [22]. (b) CdS thin films [23]. (c) Al–ZnS thin films [24].

Sources: (a) Reprinted with permission from Ref. [22]. © 2018 Elsevier. (b) Reprinted with permission from Ref. [23]. © 2018 Materials Science-Poland. (c) Reprinted with permission from Ref. [24]. © 2019 Elsevier.

2.2.4   3D NANOSTRUCTURED MATERIALS

From last decades, 3D nanostructure materials have been fascinated noticeable research interest among the material research community. These materials have unique physical properties due to high specific surface area comparative to their bulk counterpart. Excellent physical properties of nanomaterials arising owing to quantum confinement effect on the size, shapes, dimensionality, and morphologies are the main reason to potential applications. On the other hand, 3D nanostructures have variety of applications in optoelectronics, such as solar cells, photodetector, photoresistor, phototransistor, and LED. In addition, such materials with high porosity make more efficient for sensing applications due to their high molecular transport.

2.3   METHOD USED FOR DEVELOPMENT OF SEMICONDUCTOR NANOMATERIALS

Plenty of synthesis methods of nanomaterials have been developed and searching of new and cost-effective methods on progress among the material researchers so far. Nanomaterials growth method is basically classified in two main types top-down and bottom-up and schematic illustration of these two approaches is shown in Figure 2.5 [25].

2.3.1   TOP-DOWN

In this approach, bulk particles size reduced to nanodimension particles. Top-down method mainly consisting based on the physical vapor deposition techniques, such as thermal evaporation, sputtering, pulse laser ablation, and mechanical ball milling. Physical vapor deposition methods are commonly used technique to grow device grade 0D, 1D, 2D, and 3D nanostructure materials but disadvantages of physical deposition method is expensive and need high vacuum system. Various top-down methodologies are discussed in the following sections.

2.3.1.1   THERMAL EVAPORATION METHOD

Schematic diagram of thermal evaporation is shown in Figure 2.6 [26]. Thermal evaporation technique is a widely used method to develop various morphology 0D, 1D, 2D, and 3D nanostructures. In this method, bulk materials evaporate in high vacuum chamber and atoms condensate on substrates, and nucleation process initiates and forms atom clusters. We also report the growth and physical properties of CdS nanostructure thin films [27, 28].

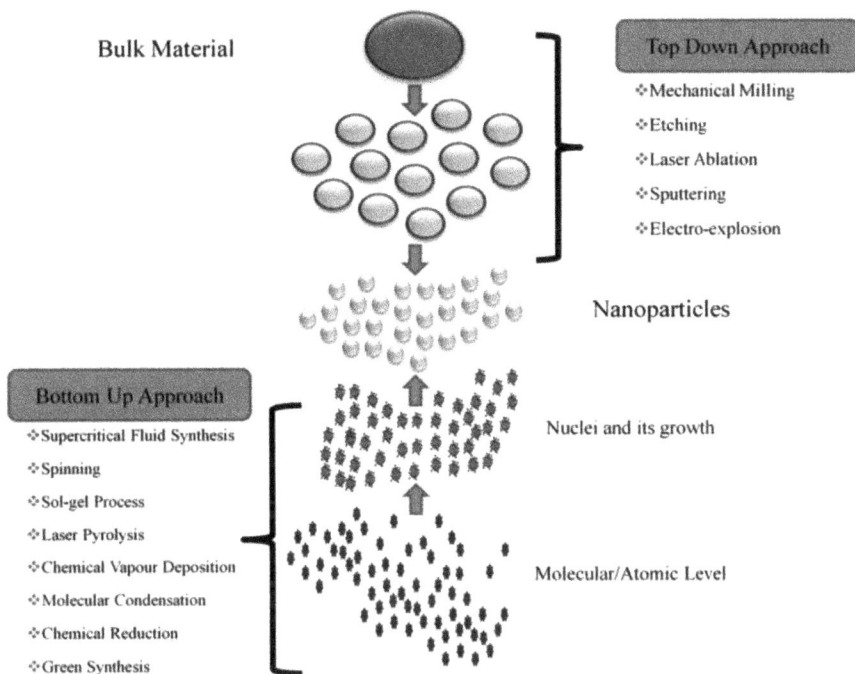

FIGURE 2.5 Schematic illustration of top-down and bottom-up approach of nanostructure materials [25].

Source: Reprinted with permission from Ref. [25]. © 2019. Elsevier.

2.3.1.2 PULSE LASER DEPOSITION

The basic requirement of this method is high power excimer laser [29]. This laser makes the system expensive. But advantages of programmable logic device are that the various nanostructure materials can be grown with high purity and controlled dimensions. This method is the best method to grow metal oxide, nitride, and carbide compound using the gaseous chemical reaction method.

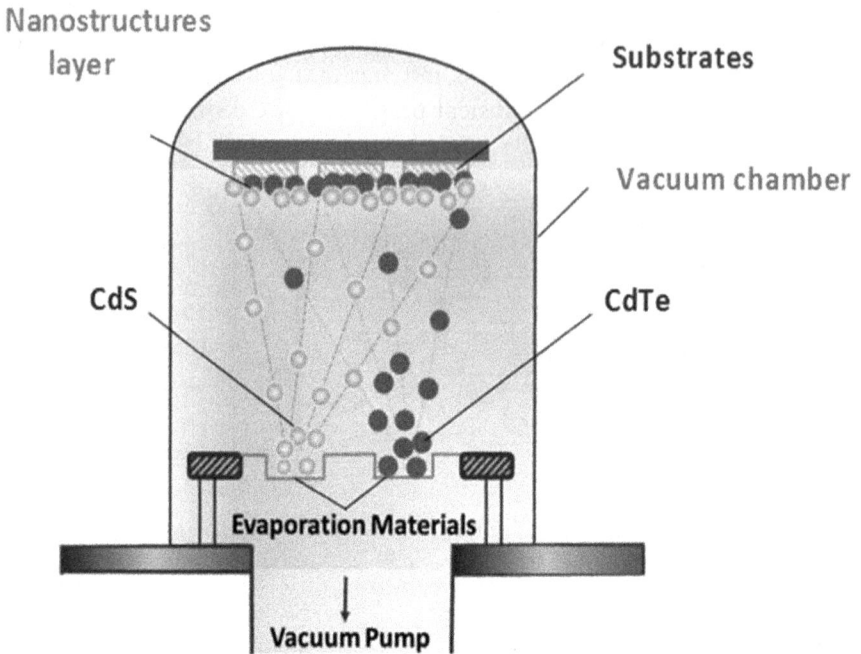

FIGURE 2.6   Thermal evaporation system [26].

Source: Reprinted with permission from Ref. [26]. © 2018 Elsevier.

2.3.1.3   SPUTTERING

Sputtering is a phenomenon where high energy particles are ejected the atoms from target material [30]. Various types of sputtering systems, such as high-target utilization, ion-assisted deposition, ion beam, reactive, high power impulse magnetron, and gas flow sputtering, have been used to grow different types of nanomaterials until now.

2.3.1.4   BALL MILLING

Ball milling system setup is shown in Figure 2.7 [31]. Benjamin and his co-workers developed this method in 1960. In this technique, a material places in the ball mill and high energy collision occurs from the balls. This method is one of the famous techniques to grow various dimension nanomaterials from the bulk materials without any high temperature reaction. In addition, ball milling is also used for mechanical activation and mechanochemistry.

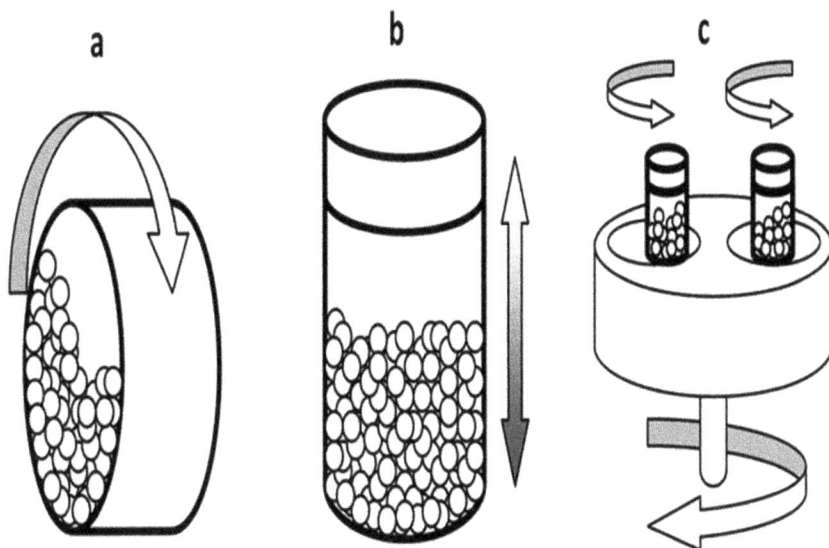

FIGURE 2.7 Schematic illustration of ball milling system [31].

Source: Reprinted with permission from Ref. [31]. © 1999 Royal Society of Chemistry.

2.3.1.5 LITHOGRAPHY PROCESSES

Lithography is one of the promising growth methods of top-down approach to grow self-assembled 0D, 1D, 2D, and 3D nanostructures onto the various types of substrates. Surface patterning only can be achieved by lithography technique which is applicable on various types of materials. This technique contains many steps, first preparation of a hard copy of pattern which is called mask or reticle. Second step is that the transfer of designed mask onto the wafer. The size of patterns or assembly can be altered using de-magnify or magnify via suitable lens. The schematic illustration of lithography process can be seen in [32]. After the development of pattern on the photoresists and then exposed to the suitable wavelength light, the developed area is etched. Various types of lithography techniques are: e-beam lithography, nanoimprint lithography, hot embossing, nanosphere lithography, and colloidal lithography.

2.3.2 BOTTOM-UP

Bottom-up approach of nanomaterials growth consists of the formation of cluster atoms by atoms condensation and as a result of formation of monolayers on the substrates. Various methods are available based on the bottom-up approach, such as spray pyrolysis, sol–gel process, chemical/electrochemical deposition, and chemical vapor deposition.

2.3.2.1   SPRAY PYROLYSIS

Figure 2.8 shows the spray pyrolysis system diagram [33]. Spray pyrolysis method is a sophisticated and cost-effective chemical technique. In this method, nanoparticles are deposited on hot substrates by spray technique and got arrayed. Main phase of spray pyrolysis method is based on the parent solution preparation and atomizer spray rate.

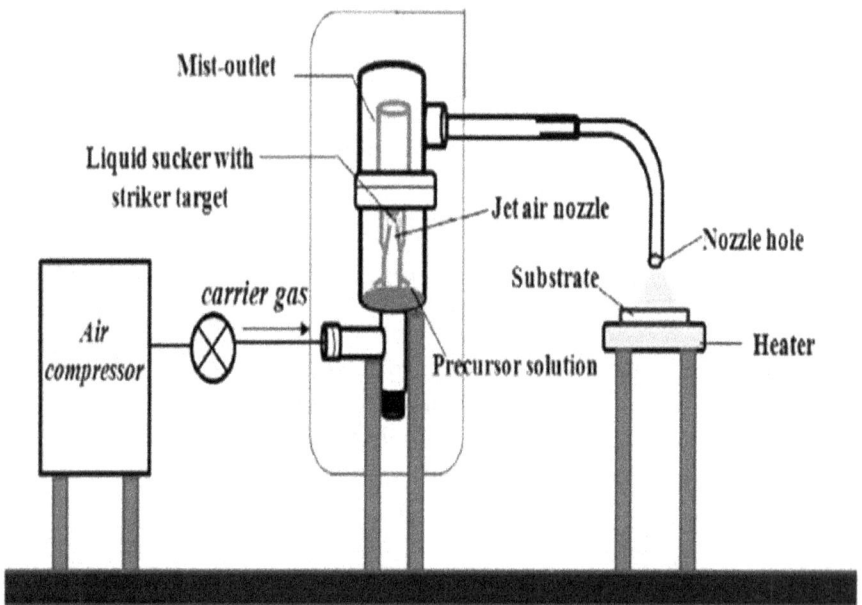

FIGURE 2.8   Schematic sketch of spray pyrolysis setup [33].
Source: Reprinted with permission from Ref. [33]. © 2017 Elsevier.

2.3.2.2   SOL–GEL PROCESS

The schematic diagram of sol–gel process is illustrated in Figure 2.9 [34]. This method is also based on chemical synthesis; here nanoparticles of desired sizes are developed in an integrated gel system. This method is commonly used in the growth of nanomaterials. This method has more controlling parameters to synthesize desired size nanomaterials as compared to the physical vapor deposition methods. On variation of some growth conditions including dopant introduction, heat treatment, and properly choosing some other surfactants, including inverted micelles, polymer matrix architecture based on block copolymers or polymer blends, porous glasses, and ex situ particle capping, it is possible to control the better size distribution and stability control of quantum-confined semiconductors, metal, metal oxide nanoparticles. Disadvantages of sol–gel process are wide range of chemical reactions due to the gel network.

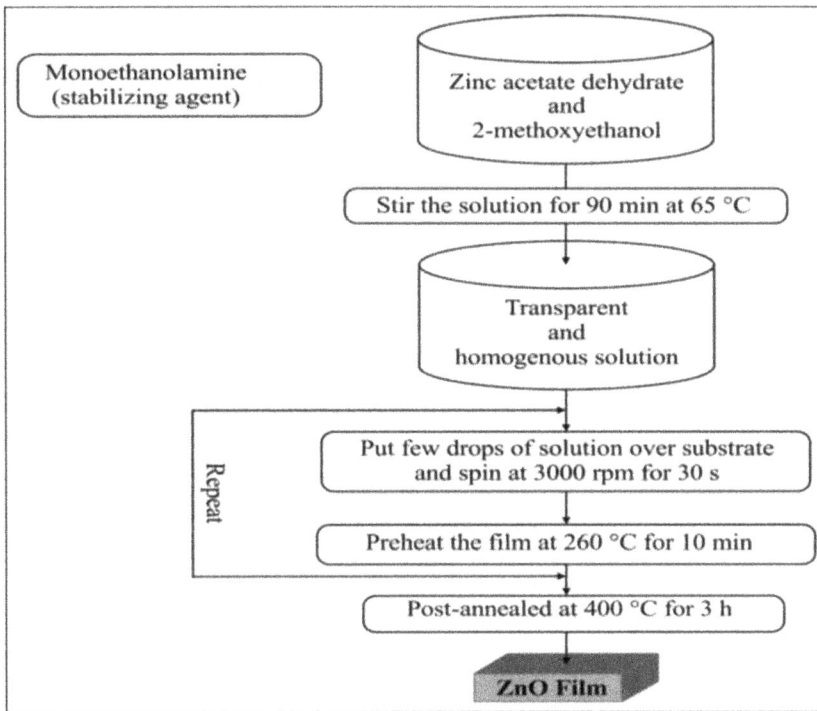

FIGURE 2.9 Schematic sketch of sol–gel process setup [34].

Source: Repinted with permission from Ref. [34]. © 2017 Munirah, Ziaul Raza Khan, Anver Aziz, Mohd. Shahid Khan, M.U. Khandaker .

2.3.2.3 HYDROTHERMAL AND SOLVOTHERMAL GROWTH

Figure 2.10 shows the schematic diagram of hydrothermal method [35]. This method is basically used to grow various nanowire materials and nanostructures. The name of this method hydrothermal is assigned due to the main role of water as a medium. Chemical precursors react and synthesize desired materials in water medium around 120 °C and grow wonderful design nanostructures.

Conceptually, hydrothermal system can be defined as the use of water as reaction medium in a sealed reaction container when the temperature is raised above 100 °C. Solvothermal synthesis is a widely used chemical processing to grow the various types of nanostrcutures; an experimental setup is presented in Figure 2.10. Both chemical routes are very similar. In solvothermal chemical route, this is not always necessary aqueous solution condition. With the variation in temperatures, reaction time, precursors, surfactant type, and precursor type can be grown desired dimension nanostructures.

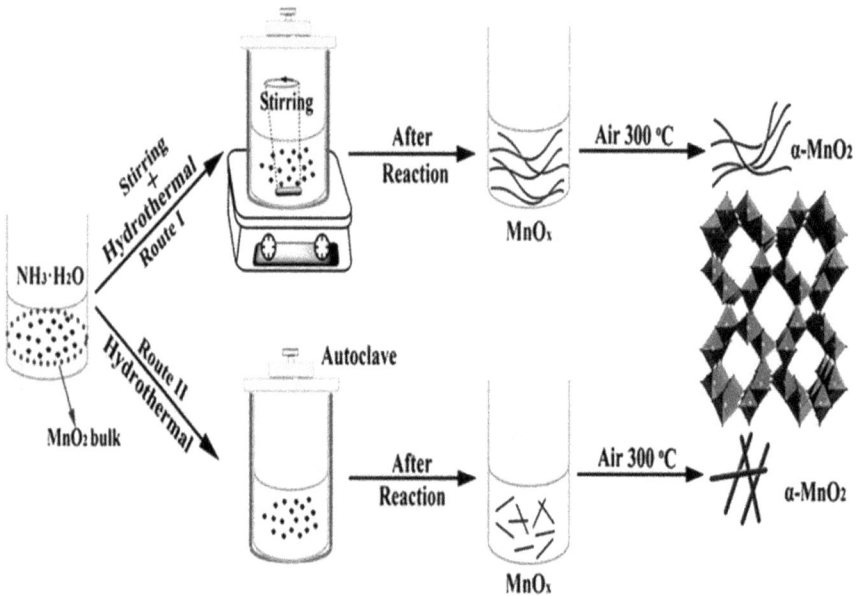

FIGURE 2.10   Schematic sketch of hydrothermal process [35].

Source: Reprinted with permission from Ref. [35]. © 2014 Royal Society of Chemistry.

2.3.2.4   ELECTRODEPOSITION

This method is based on the Faraday's electroplating law [36]. In this process, electrical current played main role to deposit nanostructure materials onto the substrates. Mainly electrodeposition consists of using two-electrode or three-electrode electrodeposition system. In recent times, the electrochemical deposition method offers the good platform to grow various types 0D, 1D, 2D, and 3D nanomaterials. Electrodeposition of the 0D, 1D, 2D, and 3D NSMs were carried out by using pulse electrodeposition (PE) in a three-electrode cell system.

2.3.2.5   CHEMICAL VAPOR DEPOSITION (CVD)

Figure 2.11 shows the schematic diagram of boron nitride nanotube (BNNT) growth mechanism by chemical vapor deposition (CVD) system [37]. In this process, gaseous molecules dissociate using radio frequency (rf) power and hence plasma produced. This plasma condensates and forms clusters and deposits onto the substrates. This method can be deployed to grow all low-dimensional nanoparticles. CVD is basically based on vapor–liquid–solid approach. It consists of several important steps which are discussed as follows:

1. A predefined mixture of reactant and diluent inert gases is placed into the chamber by the mass flow controller at a specified flow rate.
2. The gas species move to the surface site.
3. The reactants get adsorbed on the surface site.
4. The reactants undergo chemical reactions with the substrate to form the nanostructures.
5. The gaseous reaction products are desorbed and evacuated from the chamber.

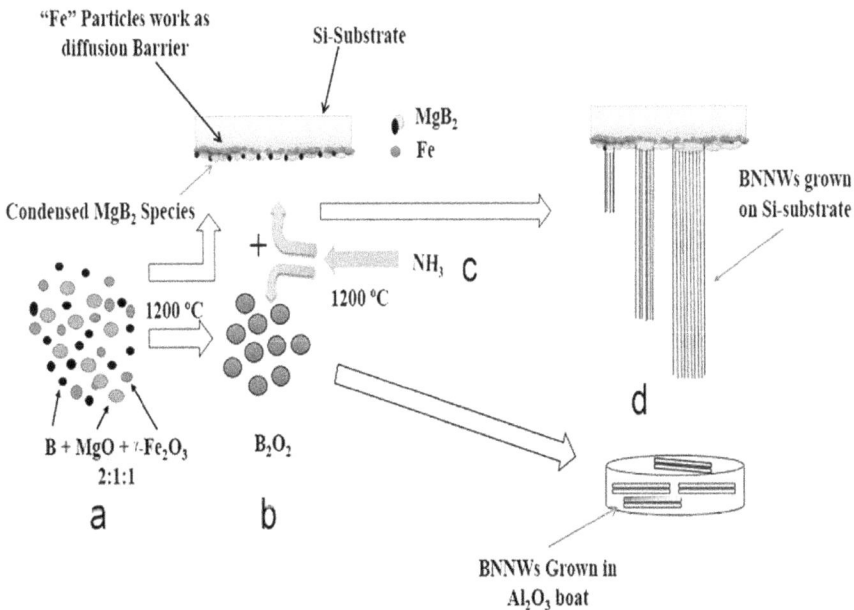

FIGURE 2.11   Schematic illustration of BNNT growth mechanism by CVD method [37].
Source: Reprinted with permission from Ref. [37]. © 2014 Elsevier.

2.4   GROWTH OF 0D NANOMATERIALS

0D nanostructures have been grown by variety of methods, herein first section, we will discuss about the growth of 0D NSMs by top-down approach. Top-down approaches mainly cover the physical deposition system. Thermal vapor deposition system is one of the promising top-down approaches. Thermal evaporation technique is the conventional extensively used methodology to grow 0D nanostructures. Li et al. reported the growth of SnO_2 nanoparticle by thermal evaporation method using Sn powder at 1000 °C [38]. Tawale et al. have developed nanoscale spherical and octahedral shapes of tin oxide (SnO_2) nanoparticles which were synthesized by using direct sublimation process employing thermal evaporation at 1350 °C without any catalyst [39]. Chen et al. reported the fabrication of graphene nanofluid onto the substrates using thermal evaporation method [40]. Lachebi et al. fabricated aluminum nanoparticles via thermal deposition on heated substrates [41]. In the 1990s, new growth techniques based on the pulse laser were added in the queue of physical vapor deposition approach, later on this is known as pulse laser deposition. In the past few years, many researchers have developed 0D nanostructures via pulse laser deposition techinques. Svetlichnyi et al. reported ZnO by pulse laser ablation method [42]. Huotari et al. developed WO_3, SnO_2, and V_2O_5 novel nanostructures for gas-sensing applications [43]. Sputtering is the important method from the category of physical vapor deposition. No year is empty without publication of sputtering techniques since 1939, when F. M. Penning patented this method. Here some recent reports discussed about the growth of 0D nanomaterials. Liu grown Au–Pd alloy nanoparticles on ZnO nanorod arrays [44]. Koo et al. reported the excellent sensing quality of Al-doped ZnO 0D nanoparticles for the sensing of toxic gases [45]. These techniques have some leads in comparison with other physical deposition techniques along with drawbacks. Outcomes of these techniques are highly clean and good quality nanoparticles but drawbacks are controlling parameters which are limited to alter the size and morphology of 0D nanoparticles. Velasquez et al. reported 0D nanoparticles of magnetite nanoparticles via ball milling method [46]. Protesescu et al. reported that wet ball milling of bulk $APbBr_3$ (A=Cs, FA) with solvents and capping ligands yields green luminescent colloidal nanocarbons (NCs) with a high overall yield and optoelectronic quality [47]. Zhou et al. described the synthesis of black phosphorus 0D nanostructures by different ball milling, such as planetary milling, shake milling, and plasma-assisted ball milling [48]. Lithography is the most widely used technique for the synthesis of ordered 0D nanostructures and patterning onto the substrates. It is a low-cost technique among the top-down synthesis approaches of 0D nanostructures. It also played a significant role to developing integrated circuit devices. Recently, some researchers reported on the growth of 0D nanostructures. An array of platinum nanocrystals of controlled size and shape using lithography was developed by Komanicky et al. [49]. Chang et al. developed Ag, Al, and Au nanotriangle arrays and In nanoparticles were developed using nanosphere lithography [50]. Nanosphere lithography is an important technique to employ the development of high quality and ordered 0D nanostructures. Šahbazović et al. reported the synthesis of silicon nanostructures using the nanosphere lithography [51].

Bottom-up approach of synthesis of nanostructures is the natural development of low-dimensional materials. Most of the chemical synthesis of 0D nanostructures falls in the category of bottom-up techniques. Bazta et al. reported the fabrication of high quality n-type ZnO 0D nanostructure films on glass substrate at 450 °C via spray pyrolysis techniques with different doping concentration (0, 2, 5, and 7 at%) of yttrium [52]. Prasad et al. have developed Ce–ZnO 0D nanostructures using spray pyrolysis method with different cerium doping concentrations. They investigated the effect of Ce doping on the microstructural, morphological, optical, and gas sensing properties of 0D nanostructures [53]. Karakaya et al. reported the replacement of tin oxide (SnO_2) or indium transparent electrode (ITO) as a III-group-doped ZnO transparent conducting electrodes via sol–gel process in solar cell devices and flat panel displays due to competitive electrical and optical properties [54]. Rathinamala et al. reported a novel CdS/PS (porous silicon) biosensor with enhanced electrochemical properties using electrochemical method. They studied electrochemical properties by cyclic voltammetry (CV), electrochemical impedance spectroscopy (EIS), and chronoamperometry. They found that 0D nanostructures CdS/PS heterostructure have good electrochemical properties [55]. Liu et al. developed 0D SnO_2 compact nanostructures via sol–gel process and found suitable materials for photoanodes in photoelectrochemical cells for the first time, which simply enabled SnO_2 gel coatings on conducting glass substrates at 500 °C, after annealing coating change into closely packed SnO_2 nanoparticles [56]. Bao-yun et al. synthesized hierarchical hollow spherical CdS nanostructures using simple microwave-assisted hydrothermal techniques by $CdCl_2 \times 2H_2O$ and $Na_2S_2O_3 \times 5H_2O$ as a chemical precursor materials. They investigated structural, morphological, elemental, and high-resolution transmission electron microscopy [57]. Zhang et al.

synthesized CdS micro/nanostructures by facilely solvothermal method using ethylenediamine, ethanolamine, and ethylene glycol as pure and mixed solvents with different S and Cd sources. They found hexagonal wurtzite structures of CdS 0D nanostructures [58]. Cao et al. synthesized gold (Au) 0D nanostructures on the surface of Pt electrode via electrodeposition method. They found noticeable change in the morphology of nanostructures with the increase of electrodeposition amount. They also optimized the applied potential, injection time, and buffer pH [59]. Nisanci et al. developed highly crystalline ZnO 0D nanostructures using a novel electrochemical approach based on uderpotential deposition (UPD) and simultaneous oxidation of Zn atomic layers from an oxygenated aqueous suspension of ZnO [60]. Narin et al. described the growth of 0D nanostructures of ZnO via ultrasonic spray chemical vapor deposition (USCVD) techniques at atmospheric pressure. They investigated the surface morphology, structural and optical properties of nanostructures. 0D ZnO structures showed the strong peak along the (002) plane with polycrystalline in nature [61]. Chen et al. developed large-scale AlN 0D nanostructures with different morphology onto the flexible carbon substrates via chemical vapor deposition without catalyst. They obtained single crystalline and growth along (001) preferential plane. The effect of growth time and temperature on the morphology and field emission characteristics of AlN 0D nanostructures were investigated in details. AlN showed significant enhancement in the field emission characteristics due to the high aspect ratio [62].

2.5   ROLE OF 0D NANOMATERIALS IN OPTOELECTRONIC DEVICES

In the last few decades, 0D nanomaterials have been extensively investigated for the various potential applications. The devices based on light emitting, light sensing, and energy conversion and storage are the essential part of our daily life. As day by day energy consumption demand increases around the world, we need to develop smart devices with high efficiency and low power consumption. So, 0D nanomaterials can played vital role to rectify energy production and consumption problems. 0D semiconducting nanomaterials are suitable contender specially for optoelectronics applications such as photovoltaic, phototransistors, photoresistors, photoconductive camera devices, laser diodes, quantum cascade lasers, light emitting diode, optical fiber communications, etc.

2.5.1 SOLAR CELLS

One of the most important optoelectronics applications of 0D semiconductor nanomaterials is in solar energy conversion. In future, we need to move from fossil fuels to green energy due to high global warming and limited sources of fossil fuels. Now, the main aim of researcher is to develop solar cells with high efficiency and cost effective. Maximum theoretical efficiency of solar cells has been estimated by various models around 29%. However, experimental efficiency of solar cells using various 0D nanostructures observer layers has been reported 22% so far [63]. From the last few decades, Si-based solar cells commercialize in the market but these solar panels have low efficiency and expansive. Recently, researchers are extensively investigating the solar cells with various 0D semiconductor nanostructure materials to enhance the efficiency and make them inexpensive. Solar cells based on II–VI group semiconductor 0D nanostructures material have been reported recently [64–66].

2.5.2 LIGHT EMITTING DIODE

Wide band gap organic and inorganic semiconductor 0D nanostructures have attracted much attention for light emitting diode (LED) devices. LED devices manufacture using 0D semiconductor nanostructures can be efficient and low power consumption devices. LED can replace the thermal filament bulb for lightning in homes and streets. Filament bulbs were consumed 20 times higher power than the LED. Recently, many researchers have fabricated 0D nanostructures-based LEDs with the high efficiency. Several organic and inorganic 0D semiconductor nanomaterials-based LEDs have been reported [67, 68].

2.5.3 PHOTODETECTOR

A photodetector is the device to detect the light of certain wavelength and convert the light signals into an electrical current which is known as photocurrent. When the device is exposed to light with antireflecting window layers, as a result absorption of photon energy equivalent to the band gap of materials causes electron–hole pairs creation occurs in the n depletion region. Intrinsic semiconductor materials such as Si and Ge and binary semiconductor materials II–VI group (CdTe, CdS, ZnSe, ZnS, and ZnO), III–V group (GaAs, InP, GaN) are the key materials to manufacture desired light photodetectors. Shao et al. reported high responsivity of ultraviolet light with the ZnO 0D nanoparticles which is shown in Figure 2.12 [69]. Recently, photodetecting properties of 0D semiconductor nanostructures have investigated extensively by various research groups. Several research groups reported photodetector using 0D nanomaterials with high responsivity [70, 71].

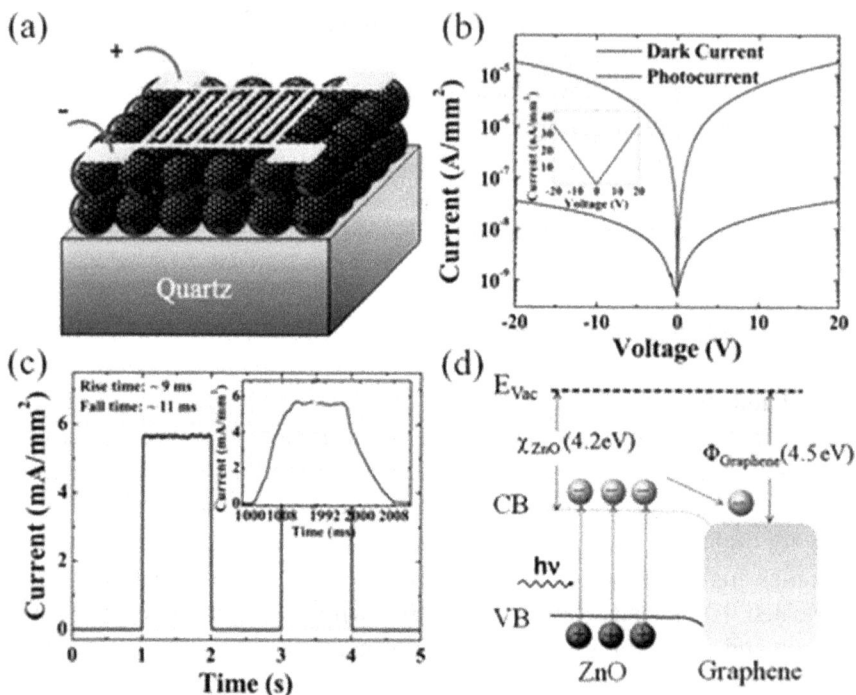

FIGURE 2.12   (a) 0D ZnO UV detector device. (b) Dark and illuminated *I–V* characteristics. (c) Response. (d) Schematic diagram of band structure [69].

Source: Reprinted with permission from Ref. [69]. © 2009 Royal Society of Chemistry.

2.5.4   *PHOTORESISTOR*

Resistance of materials depends on the intensity of light is known as photoresistor or light dependent resistor. Photoresistor is also organic and inorganic photoconducting semiconductor materials based devices. Photoresistor is exposed to light, the material absorbs the radiation and electrons move from the valance band to the conduction band and carrier concentration increases, hence resistance of photoresistor varies with intensity of light. These devices have potential applications in automatic door openings and auto switching. Nowadays, 0D nanostructure materials play a noticeable role to grow efficient with high responsivity photoresistors. Currently, highly efficient photoresistors were reported by various research groups; few are discussed in [72, 73].

2.5.5 PHOTOTRANSISTOR

Phototransistor is one of the most important optoelectronics devices; it converts light energy into electrical energy as shown in Figure 2.13. Phototransistor devices are based on same principle as photoresistors but producing both current and voltage due to the bipolar junction device. Phototransistors are also made up of using semiconductor materials. Phototransistors activate on illumination of light and works in various electronics devices very well. Phototransistor is also a bipolar junction transistor device and made up of photoconducting semiconductor materials. When the base of phototransistor is exposed to the light, photogenerated free carriers produced in the base region. As a result, excess current increases and current flows through the base and emitter and current is converted into the voltage. Phototransistors have variety of potential applications in technological advancement such as smoke detectors, infrared receivers, and CD players, etc. 0D nanostructures photoconducting semiconducting materials are backbone for the fabrication of efficient and highly sensitive phototransistors. Present decades, various research groups have fabricated phototransistors using 0D nanostructures with enhanced performance [74, 75].

2.5.6 QUANTUM CASCADE LASER

Quantum cascade lasers (QCLs) are semiconductor lasers and other most important applications of 0D semiconductor nanostructures materials. QCL emits the wavelength in the mid-IR range from 4 to 10 μm. QCL has potential applications in mid-IR region such as molecular gas analysis and absorption spectroscopy. Nowadays, QCL is the open area of research among material researchers. They are fabricating QCL using 0D nanostructures semiconductor nanomaterials such as QDs and nanoparticles [76–78].

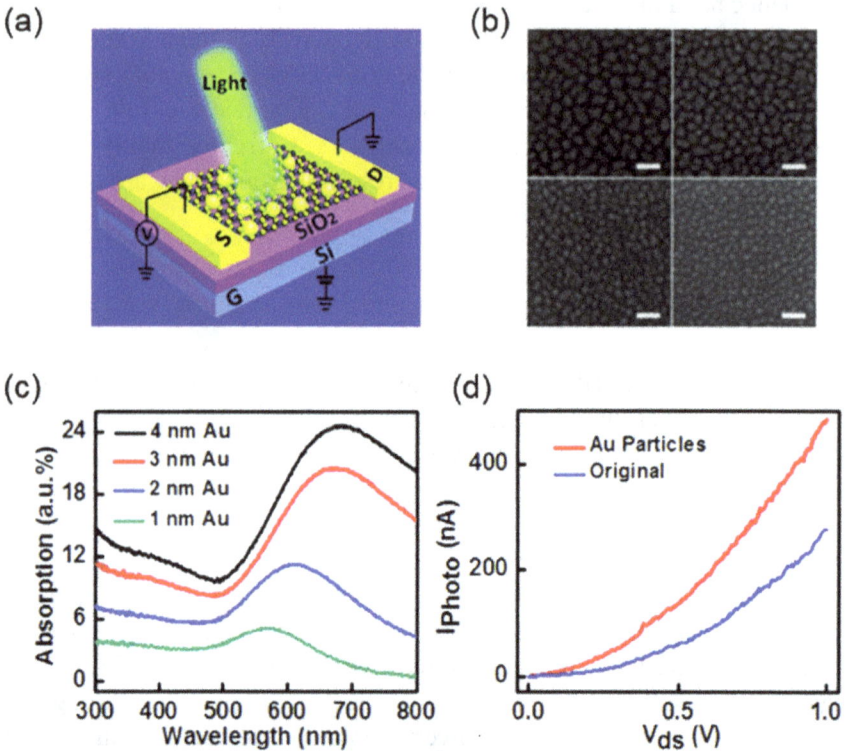

FIGURE 2.13   0D nanostructures based phototransistors device [75].
Source: Reprinted with permission from Ref. [75]. © 2017 Elsevier.

2.5.7   OPTICAL FIBER COMMUNICATIONS

Optical fiber communication technology has revolutionized the information technology industry. It is a widely used method to communicate higher data with high speed and minimum loss in signals. As a result, these leads over the traditional data transfer technology; fiber optic communication technology has potential applications in Ethernet systems, broadband distribution, and general data networking. Optical fiber basically works on total internal reflection, when the incident angle is greater than the critical angle, condition for total internal reflection satisfied. Further, improvement in efficiency and performance in optical fiber communication technology, optical fiber research, and development work have extensively under investigation by various research groups. Recently, semiconductor 0D NSMs have been attracted much attention in fabricating high quality optical fiber. Ding et al. reported CdTe/CdS QDs-based fiber optic for nitric oxide sensing. They found excellent response up to picomolar sensitivity [79]. Currently, Zhou et al. developed a novel ratiometric fluorescence sensor based on CdTe QDs doped hydrogel optical fiber for selective and on-site detection of Fe^{+3} ions in real time. They observed ratiometric sensing based on multi-QDs doped fiber which can be feasibly applied for real-time and on-site analysis of heavy metal ions in aqueous environment [80]. Here, we have discussed the applications and role of 0D nanostructures in several important optoelectronics applications in the above sections. In addition, many other potential optoelectronic applications of 0D NSMs, such as charge coupled imaging devices, transducers, wave guide, optical receivers, or detectors have helped in defense and aviation industry widely. 0D semiconductor NSMs have given new opportunities to design nanosatellites. Finally, 0D NSM-based optoelectronic devices have high efficiency low power consumption devices than traditional devices. Fabrication of optoelectronic devices using 0D NSMs into substrates is still challenging and is an open area of research.

2.6 CONCLUSIONS AND FUTURE PROSPECTS

The main aim of this chapter is to describe the zero-dimensional nanostructures definition, synthesis methods, and applications in optoelectronic devices. Here, we summarily discuss types of nanostructures based on dimensions and growth method employed to develop various shape and dimension NSMs. Zero-dimensional NSMs are the promising candidate for optoelectronics applications such as photovoltaic, photodetector, photo resistor, phototransistor, photodiode, light emitting diode, optical fiber, optical receivers, and charge coupled device. Optoelectronics devices are played a vital role for the advanced technological applications in information technology, defense, space, computing, and laser technology. However, semiconductor 0D nanostructures are a noticeable role in safe and secure information transmission via quantum walk concept. Nowadays, 0D nanostructures have widened scope to improve optoelectronic devices performance, quality, and reliability. Many optoelectronic devices have commercialized in the market based on 0D semiconductors nanostructures. But, optoelectronics technology is still an open area of research to improve the performance of devices and hence this technology has bigger room to research and development work in future for material science researchers. Researchers and corporate houses are looking toward the bright future in optoelectronics technology. Recent progress in optoelectronics technology shows the transformation of human life from conventional era to smart era where all things controlled by one hand means all in one slogan raise.

ACKNOWLEDGMENT

The author is highly grateful to acknowledge the permissions to include the figures in this chapter from various publishers.

CONFLICT OF INTEREST

There is no conflict of interest.

KEYWORDS

- **0D nanomaterials**
- **growth techniques**
- **density of states**
- **types of nanomaterials**
- **role of 0D in optoelectronics**

REFERENCES

1. R. Feynman, There's plenty of room at the bottom. Engineering Science, 23 (1960) (5) 22–36.
2. N. Taniguchi, On the basic concept of nanotechnology. Proceedings of the International Conference on Production Engineering, London, Part II British Society of Precision Engineering, (1974).
3. K.E. Drexler, Nanosystems: Molecular Machinery, Manufacturing, and Computation, MIT PhD thesis, (1991), John Wiley & Sons, Inc., New York. ISBN: 0471575186.
4. H. Gleiter, Acta Materialia 48 (2000) 1.
5. V. Skorokhod, A. Ragulya, I. Uvarova, Physico-chemical kinetics in nanostructured systems. Kyiv: Academperiodica; 2001. p. 180.
6. V. V. Pokropivny, V. V. Skorokhod, Materials Science and Engineering C 27 (2007) 990.

7. M. Waldiya, R. Narasimman, D. Bhagat, D. Vankhade, I. Mukhopadhyay, Materials Chemistry and Physics 226 (2019) 26–33.
8. J. Sowik, M. Miodyńska, B. Bajorowicz, A. Mikolajczyk, W. Lisowski, T. Klimczuk, D. Kaczor, A. Z. Medynska A. Malankowska, Applied Surface Science 464 (2019) 651–663.
9. J. Chen, J. Pan, D. Qingguo, G. Alagappan, W. Lei, Q. Li, J. Xia, Applied Physics A 117 (2014) 589–591.
10. S. Li, J. Pan, H. Li, Y. Liu, W. Ou, J. Wang, C. Song, W. Zhao, Y. Zheng, C. Li, Chemical Engineering Journal 366 (2019) 305–312.
11. S. Iijima, Nature 354 (1991) 56.
12. A. Bakhsh, I. Hussain Gul, A. Maqsood, S. H. Wu, C. H. Chan, Y. C. Chang, Journal of Luminescence 179 (2016) 574–580.
13. V. Kumar, K. Kumar, H. C. Jeon, T. W. Kang, D. Lee, S. Kumar, Journal of Physics and Chemistry of Solids 124 (2019) 1–6.
14. W. Kim, D. Monllor-Satoca, W. S. Chae, M. A. Mahadik, J. S. Jang, Applied Surface Science 463 (2019) 339–347.
15. P. Ahmad, M. U. Khandaker, Y. M. Amin, Z. R. Khan, Journal of Physics and Chemistry of Solids 85 (2015) 226–232.
16. F. Göde, S. Ünlü, Materials Science in Semiconductor Processing 90 (2019) 92–100.
17. O. K. Echendu, U. S. Mbamara, K. B. Okeoma, C. Iroegbu, C. A. Madu, I. C. Ndukwe, I. M. Dharmadasa, Journal of Materials Science: Materials in Electronics 27 (2016) 10180–10191.
18. J. Y. Oh, J. M. Yu, S. R. Chowdhury, T. I. Lee, M. Misra, Electrochimica Acta 298 (2019) 694–703.
19. Y. Lu, Y. Lin, D. Wang, L. Wang, T. Xie, and T. Jiang, Nano Research 4(11) (2011) 1144–1152.
20. P. Srinivasan, B.G. Jeyaprakash, Journal of Alloys and Compounds 768 (2018) 1016–1028.
21. S. Arya, A. Sharma, B. Singh, M. Riyas, P. Bandhoria, M. Aatif, V. Gupta, Optical Materials 79 (2018) 115–119.
22. N. Sathiya Priya, S. S. P. Kamala, V. Anbarasu, S. A. Azhagan, R. Saravanakumar, Materials Letters 220 (2018) 161–164.
23. Z. R. Khan, Munirah, A. Aziz, M. S. Khan, Material Science-Poland 36 (2018) 235–241. Sol-Gel Derived Cds Nanocrystalline Thin Films: Optical and Photoconduction Properties.
24. A. Azmand, H. Kafashan, Journal of Alloys and Compounds 779 (2019) 301–313.
25. P. Khanna, A. Kaur, D. Goya, Journal of Microbiological Methods 163 (2019) 105656.
26. W. L. Syu, Y. H. Lin, A. Paliwal, K. S. Wang, T. Y. Liu, Surface and Coating Technology 350 (2018) 823–830.
27. Z. R. Khan, M. Zulfequar, M. S. Khan, Material Science and Engineering: B 147 (2010) 145–149.
28. M. Shkir, H. Abbas, Z. R. Khan, Journal of Physics and Chemistry of Solids 73 (2012) 1309–1313.
29. https://www.egr.msu.edu/eceshop/testingfacility/laser/.
30. A. Jilani, M. S. Abdel-Wahab, A. H. Hammad, Advance deposition techniques for thin film and coating. IntechOpen DOI: 10.5772/65702.
31. G. Gorrasi, A. Sorrentino, Green Chemistry 5 (2015) 2610–2625.
32. N. Thakur, P. G. Reddy, S. Nandi, M. Yogesh, S. K. Sharma, C. Pradeep, S. Ghosh, K. E. Gonsalves, Journal of Micromechanics and Micro Engineering 27 (2017) 125010.

33. A. Kennedy, V. S. Kumar, K. P. Raj, Journal of Physics and Chemistry of Solids 110 (2017) 100–107.
34. Munirah, Z. R. Khan, A. Aziz, M. S. Khan, M. U. Khandaker, Materials Science-Poland 35(1) (2017) 246–253.
35. K. Lei, L. Cong, X. Fu, F. Cheng, J. Chen, Inorganic Chemistry Frontiers 3 (2016) 928–933.
36. S. Sahoo, K. K. Naik, C. S. Rout, Nanotechnology 26 (2015) 455401(8pp).
37. P. Ahmad, M. U. Khandaker, Z. R. Khan, Y. M. Amin, Ceramics International 40 (2014) 14727–14732.
38. Y. Li, R. Peng, X. Xiu, X. Zheng, X. Zhang, G. Zhai, Superlattices and Microstructures 50 (2011) 511–516.
39. J. S. Tawale, G. Gupta, A. Mohan, A. Kumar, A. K. Srivastava, Sensors and Actuators B 201 (2014) 369–377.
40. P. Chen, S. Harmand, S. Szunerits, R. Boukherrou, International Journal of Thermal Sciences 135 (2019) 445–458.
41. I. Lachebi, A. Fedala, T. Djenizian, T. Hadjersic, M. Kechouane, Surface & Coatings Technology 343 (2018) 160–165.
42. V. Svetlichnyi, A. Shabalinalvan, L. D. Goncharova, A. Nemoykina, Applied Surface Science 372 (2016) 20–29.
43. J. Huotari, V. Kekkonen, J. Puustinen, J. Liimatainen, J. Lappalainen, Procedia Engineering 168 (2016) 1066–1069.
44. Y. Liu, Materials Letters 224 (2018) 26–28.
45. A. Koo, R. Yoo, S. P. Woo, H. S. Lee, W. Lee, Sensors & Actuators: B Chmical 280 (2019) 109–119.
46. A. A. Velasquez, C. C. Marin, J. P. Urquijo, Journal of Nanoparticle Research 20 (2018) 72.
47. L. Protesescu, S. Yakunin, O. Nazarenko, D. N. Dirin, M. V. Kovalenko, Applied Nanomaterials 1 (2018) 1300–1308.
48. F. Zhou, L. Ouyang, M. Zeng, J. Liu, H. Wang, H. Shao, M. Zhu, Journal of Alloys and Compounds 784 (2019) 339–346.
49. V. Komanicky, A. Barbour, M. Lackova, M. Zorko, C. Zhu, M. Pierce, H. You, Nanoscale Research Letters 9 (2014) 336.
50. Y. C. Chang, C. B. Tseng, Plasmonics 8 (2013) 1395–1400.
51. A. M. Šahbazović, M. Novaković, E. Schmidt, I. Gazdić, V. Ðokić, D. Peruško, N. Bibić, C. Ronning, Z. Rakočević, Optical Materials 88 (2019) 508–515.
52. O. Bazta, A. Urbieta, J. Piqueras, P. Fernández, M. Addou, J. J. Calvino, A. B. Hungría, Ceramics International 45 (2019) 6842–6852.
53. M. R. Prasad, M. Haris, M. Sridharan, Sensors and Actuators A 269 (2018) 435–443.
54. S. Karakaya, O. Ozbas, Applied Surface Science 328 (2015) 177–182.
55. I. Rathinamala, N. Jeyakumaran, N. Prithivikumaran, Vacuum 161 (2019) 291–296.
56. B. Liu, L. Wang, Y. Zhu, Y. Xia, W. Huang, Z. Li, Electrochimica Acta 295 (2019) 130–138.
57. H. Bao-yun, J. Zhenzi, H. Jian-feng, Y. Jun, Transactions of Nonferrous Metals Society of China 22 (2012) 89–94.
58. Z. Zhang, Y. Rena, Lu Han, G. Xie, Bo Zhong, Physica E 92 (2017) 30–35.
59. G. Cao, Q. Liu, Y. Huang, W. Li, S. Yao, Electrophoresis 31 (2010) 1055–1062.
60. F. B. Nisanci, T. Öznülüer, Ü. Demir, Electrochimica Acta 108 (2013) 281–287.
61. P. Narin, E. Kutlu, G. Atmaca, A. Atilgan, A. Yildiz, S. B. Lisesivdin, Optik 168 (2018) 86–91.

62. F. Chen, T. Wang, W. Su, Ceramics International 44 (2018) 18686–18692.
63. M. Müller, G. Fischer, B. Bitnar, S. Steckemetz, R. Schiepe, M. Mühlbauer, R. Köhler, P. Richter, C. Kusterer, A. Oehlke, E. Schneiderlöchner, H. Sträter, F. Wolny, M. Wagner, P. Palinginis, D. H. Neuhaus, Energy Procedia 124 (2017) 131–137.
64. M. Leoncini, E. Artegiani, L. Lozzi, M. Barbato, M. Meneghini, G. Meneghesso, M. Cavallini, A. Romeo, Thin Solid Films 672 (2019) 7–13.
65. A. A. Ojo, I. O. Olusola, I. M. Dharmadasa, Materials Chemistry and Physics 196 (2017) 229–236.
66. N. Rajamanickam, S.S. Kanmani, K. Jayakumar, K. Ramachandran, Journal of Photochemistry & Photobiology A: Chemistry 378 (2019) 192–200.
67. G. Xie, C. Jiang, J. Wang, C. Mai, G. Huang, Y. Ma, J. Wang, J. Peng, Y Cao, Organic Electronics 71 (2019) 58–64.
68. X. Cheng, S. W. Xu, Y. M. Lu, S. Han, P. Jiang Cao, F. Jia, Y. X. Zeng, X. K. Liu, W. Y. Xu, W. J. Liu, D. L. Zhu, Journal of Alloys and Compounds 776 (2019) 646–653.
69. D. Shao, M. Yu, H. Sun, T. Hu, J. Lian, S. Sawyer, Nanoscale 5 (2013) 3664–3667.
70. Z. Li, Z. An, Y. Xu, Y. Cheng, Y. Cheng, D. Chen, Q. Feng, S. Xu, J. Zhang, C. Zhang, Y. Hao, Journal of Materials Science 54 (2019) 10335–10345.
71. Y. H. Zhou, H. N. An, C. Gao, Z. Q. Zheng, B. Wang, Materials Letters 237 (2019) 298–302.
72. J. Marquez-Marín, C. G. Torres-Castanedo, G. Torres-Delgado, R. Castanedo-Perez, O. Zelaya-Angel, Superlattices and Microstructures 111 (2017) 1217–1225.
73. J. Li, H. Li, D. Ding, Z. Li, F. Chen, Y. Wang, S. Liu, H. Yao, L. Liu and Y. Shi, Nanomaterials 9 (2019) 505; doi:10.3390/nano9040505.
74. J. Yu, S. W. Shin, K. H. Lee, J. S. Park, S. J. Kang, Journal of Vacuum Science & Technology B 33 (2015) 061211.
75. Y. Sun, X. Yang, H. Zhao, R. Wang, Progress in Natural Science: Materials International 27 (2017) 157–168.
76. J. Faist, M. Beck, T. Aellen, Applied Physics Letters 78 (2001) 147.
77. H. N. Van, A. N. Baranov, Z. Loghmari, L. Cerutti, J. B. Rodriguez, J. Tournet, G. Narcy, G. Boissier, G. Patriarche, M. Bahriz, E. Tournie, R. Teissier, Scientific Reports 8 (2018) 7206
78. N. Zhuo, F. Q. Liu, J. C. Zhang, L. J. Wang, J. Qi Liu, S. Q. Zhai and Z. G. Wang, Nanoscale Research Letters 9 (2014) 144.
79. L. Ding, Y. Ruan, T. Lia, J. Huang, S. C. Warren-Smith, H. Ebendorff-Heidepriem, T. M. Monro, Sensors & Actuators B: Chemical 273 (2018) 9–17.
80. M. Zhou, J. Guo, C. Yang, Sensors and Actuators B 264 (2018) 52–58.

CHAPTER 3

1D Nanomaterials and Their Optoelectronic Applications

MEENAKSHI CHOUDHARY[1], SUDHEESH K. SHUKLA[2,3*], VINOD KUMAR[4], PENNY P. GOVENDER[2*], RUI WANG[3], CHAUDHERY MUSTANSAR HUSSAIN[5], and BINDU MANGLA[6]

[1]*Swiss Institute for Dryland Environmental and Energy Research, Blaustein Institutes for Desert Research, Ben-Gurion University of the Negev, Israel*

[2]*Department of Chemical Science-DFC (formerly Department of Applied Chemistry), University of Johannesburg, 17011, Doornfontein, 2028 Johannesburg, South Africa*

[3]*School of Environmental Science and Engineering, Shandong University, Jinan, Qingdao 266237, China*

[4]*Department of Materials Engineering, Ben-Gurion University of the Negev, Beer-Sheva, 84105, Israel*

[5]*Department of Chemistry and Environmental Science, New Jersey Institute of Technology, Newark, NJ 07102, USA*

[6]*Department of Chemistry, J.C. Bose University of Science and Technology, YMCA, Faridabad, India*

*Corresponding author. E-mails: sudheeshkshukla@gmail.com; pennyg@uj.ac.za

ABSTRACT

Nanotechnology enables to control the size, shape, and crystallographic orientation of nanomaterials at nanoscale range. Among various nanoshapes, one-dimensional (1D) nanostructures have their specific advantages as compared to isotropic nanoparticles. 1D nanomaterials are an ideal system for exploring a large number of novel phenomena at the nanoscale level and investigating the dimensional as well as size-dependent properties for different applications including optoelectronics. In recent years, integration of optical switches or interfaces based on 1D nanomaterials with tailored geometrics has made significant advancement. 1D-based optoelectronic devices can be configured either as a resistor whose conduction could be altered by a charge transfer process or as a field effect transistor. Functionalization of the structural surfaces offers numerous promising opportunities for intensifying optoelectronic competences. This chapter provides a comprehensive appraisal on the state-of-the-art research activity on the synthesis and functionalization of 1D nanomaterials along with their respective optoelectronic applications.

3.1 INTRODUCTION

A material is a chemical mixture of substances that constitutes a particular object, which can be classified on the basis of its geological origin, biological functions, and chemical properties. In principle, materials with a single unit size between 1 and 100 nm (10^{-9} m) can be defined as nanomaterials and their properties can differ from those of the same materials with micron- or millimeter-scale dimensions. During the present era, a remarkable attention has been paid off to the nanomaterials [1–3]. Due to the macroscopic quantum tunneling effect, surface size, and quantum size effect, nanomaterials exhibit promising properties in comparison to their respective bulk counterparts. The unique size-dependent optoelectronic properties of nanomaterials made them promising candidate in the area of biolabeling, functionalization, chemical sensors, and in catalysis [4–6]. Since the discovery of carbon nanotube (CNT) in 1991 [7], 1D nanomaterials have attracted the extensive research interest due to their 2D quantum confinement effect for the construction of nanodevices and interconnects [8]. Moreover, 1D nanomaterials have revealed their capabilities as a building block for different nanoscale devices, such as sensors, waveguides, and lasers [9, 10].

3.1.1 ADVANTAGE OF 1D NANOMATERIALS

π-Conjugated molecules transformed into 1D nanostructure materials have exhibited novel optical, chemical, and electrical properties [11, 12]. The assembly of 1D nanostructured materials includes nanorods, nanotube, nanoribbon, nanowires, and nanobelts which possess huge surface-to-volume ratio (as compared to their respective bulk materials) as well as high crystal-linity of 1D nanomaterials (important features for efficient charge transport) is an important obligation for the fabrication of optoelectronic devices [13]. Regarding fabrication of 1D-based devices, it is essential that entire constituent materials have to be well assembled and aligned at specific position. Fabrication of conventional 1D nanostructures relies on the concept of using multicompo-nent nanomaterials, such as core–shell nanowire, composite and multishaped nanostructures, p–n junction, hollow- and branch-shaped structures. Recent years have witnessed significant progress in 1D-based optoelectronics in order to decipher their charge transport mechanisms (which is highly dependent on effective functionality between two materials as well as their unique structural effects). This chapter provides the more recent and relevant advances on different aspects of synthesis and functionalization of 1D nanomaterials and their optoelectronic applications.

3.2   GROWTH/FABRICATION OF 1D NANOMATERIALS

One-dimensional nanomaterials are more suitable for the construction of optoelectronic devices. In the present era, 1D nanomaterials have achieved a great scientific and technical significance for the development of nanodevices and interconnects. In the past couple of years, a number of attempts have been made for the fabrication of 1D nanomaterials.

3.2.1   VAPOR DEPOSITION GROWTH

Vapor deposition (VD) is a facile method where materials in a vapor state are condensed to form a solid material by condensation, chemical reaction, or conversion. This method is used to alter the mechanical, electrical, thermal, and optical properties and has achieved great success toward fabrication of 1D-based devices. In general, it is difficult to control the monodispersity of the synthesized material but it can be controlled by the degree of saturation [14]. In regard to control the saturation of 1D nanomaterials by VD, a number of techniques have been used, such as using anodized aluminum oxide (AAO) templates, controlling the substrate temperature, and introduction of solid phase reaction [15–17]. In comparison to amorphous nanomaterials, crystalline nanomaterials effectively enhance the performance of the corresponding devices due to their intrinsic properties. Zhao et al. reported an adsorbent-assisted physical vapor deposition (PVD), where they used the silica gel or neutral aluminum oxide as adsorbent in PVD technique to control the degree of saturation. The theme behind this idea is to establish adsorption–desorption equilibrium between organic source and adsorbent [18]. VD method is frequently accepted to grow the highly crystalline nanostructures. By applying this method, any substrate can be directly grown with an avoidable contamination. In this process, a quartz tube is loaded in a temperature-controlled furnace and the substrate area is sublimated at the high temperature region and is followed by carrying the vaporized molecule to the lower temperature region by the inert gases. In the lower temperature region, the nucleation of the crystal is taken place, leading to the formation of highly purity of crystal growth. This method is employed to induce the direct growth of highly crystalline nanomaterials onto the desired substrate. In the similar approach is also employed in PVD, where the deposition of thin film or crystal is taken place by the condensation of vaporized target materials onto the desired substrate. Taking advantages of PVD method, as fabrication of single crystal in vapor phase is free of defect, and provides great potential for the fabrication of 1D nanostructure-based optoelectronic devices for the investigation of optical and electrical properties [19, 20].

3.2.2   HYDROTHERMAL METHOD

Hydrothermal method is another popular approach to grow different dimensions of nanostructured materials. At thematic point of view, this method involves the heterogeneous reaction on aqueous media (<100 °C and <1 bar) and increases reactant solubility under the condition of unusual reactions. This method can be defined as a method of synthesis of single crystal that ultimately depends on the solubility of the minerals in hot water under high pressure. Mother Nature supplies the core ideology of the hydrothermal method, where numerous minerals have formed. At the industrial level, hydrothermal method plays an important role for the concept of hydrometallurgy. The main advantage of hydrothermal method over other types of method for crystal growth is due to its ability to create crystalline phases, which are not stable at their respective melting point [21–24]. A typical hydrothermal process involves with the precursor synthesis and reaction in an autoclave oven at certain controlled temperature. Whereas, using different post-treatment processes, heat treatment, surfactant, and precursor could control the morphology of the crystal. Hydrothermal method enables the homogeneous nucleation and remarkable diffusion and allows the growth of high order of crystallinity at low temperature [25, 26]. Exploiting the advantages of hydrothermal method, a number of 1D nanomaterials have been grown for different electrical and optical applications. In 2006, Tam et al. reported well-aligned growth of ZnO nanorod on p-Si-(001) substrate (Figure 3.1) [27]. While in 2011, Chen and Wu studied the nucleation mechanism of nanorods arrays of ZnO [28]. This method is limited in its use because of difficulty to monitor the time-dependent growth of nanostructured materials.

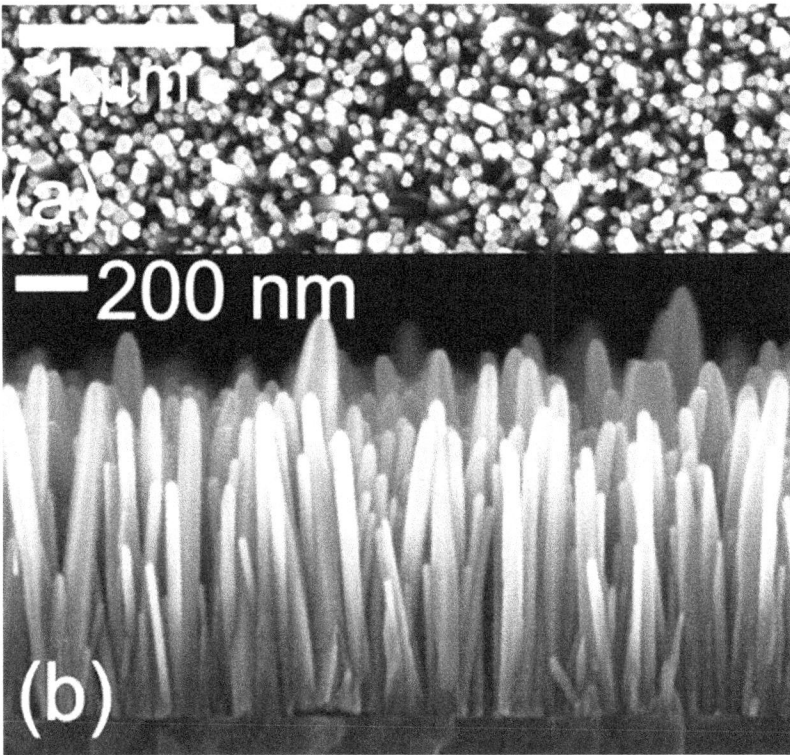

FIGURE 3.1 Typical scanning electron microscopic image of hydrothermally grown ZnO nanorods; (a) top view and (b) side view [27].

Source: Reprinted with permission from Ref. [27]. © 2006 American Chemical Society.

3.2.3 ELECTROSPINNING

Electrospinning is a voltage-driven process, governed by the electrohydrodynamic phenomena. It is a cost-effective and straightforward method to fabricate nanofibers from a polymeric solution or melts. Over the last two decades, a broad range of research has been reported to achieve 1D nanomaterials (in the form of nanotube, nanofibers, and nanobelts) using electrospinning approach. This method is manipulated to produce the fibers structure with polymeric and ceramic composites. This approach involves with applying high voltage to a precursor solution and ejects by the metallic needles. During the initial process of electrospinning, a droplet is formed at the tips of needle, which is deformed by the electrostatic forces (between needle and surface) into a conical shape, which is also known as Taylor cone. When the properties of precursor solution are properly tuned, then a repulsive force is generated between surface charges that produce a stable jet-like mechanism which whips and deforms to produce a fiber at nanoscale dimensions [29–31] (Figure 3.2).

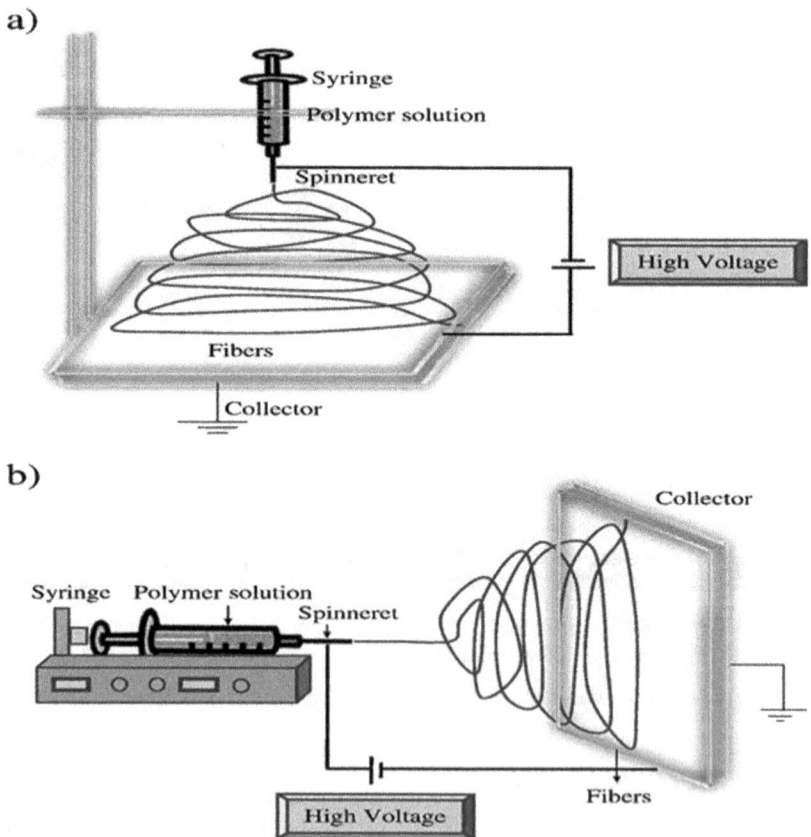

FIGURE 3.2 Schematic presentation of a setup of electrospinning apparatus: (a) typical vertical setup and (b) horizontal setup [31].

Source: Reprinted with permission from Ref. [31]. © 2010 Elsevier.

This method can be used to develop micro-/nanofibers with different compositions. There is an additional postcalcination step for the fabrication of inorganic nanofibers, which is not required for organic nanofibers. Yuh et al. reported the application of electrospinning methods for the fabrication of 1D nanomaterials. Their report highlighted the fabrication of perovskite $BaTiO_3$ nanofibers (after heat treatment at 750 °C for 1 h) with the diameter of 80–190 nm along with the length of 100 μm. This was the first report to develop stand-alone electrospun complex oxide ferroelectric nanofiber [32]. Following this study, in 2006 McCann et al. demonstrated the controlled fabrication of 1D $BaTiO_3$ nanofibers by electrospining. Ribbon-like polycrystalline nanofibers (~75 nm in height and ~200 nm in width with grain sizes <30 nm) were synthesized by electrospining a solution of barium-titanium alkoxide and poly(vinyl pyrrolidone). Preparation of 1D fibers of $BaTiO_3$ (<50 nm in diameter) with ribbon-like morphology was reported by varying the concentration of alkoxide precursor [33].

3.2.4 TEMPLATE-ASSISTED SYNTHESIS

In nanotechnology, template-assisted synthesis of nanostructures with hard templates has become one of the most applied methods for the fabrication of 1D nanomaterials, especially for fabrication of tubular structures with monodisperse diameter. The first use of this technique was reported by Possion in 1970 to synthesize the semiconductor nanowires [34], while in 1994 Martin extended this technique and coined the term template synthesis [35]. Nowadays, different types of common template have been used for this technique, such as porous membrane prepared with silica, AAO, ion-track-etched polymers, and nanochannel glass [36]. The formation of nanostructures in active template-based synthesis resulted from the growth of nuclei that invariably nucleate at the holes and defect of the substrate. The most-used template for the synthesis of 1D nanostructure is the anodic alumina membrane (AAM). In their first study, Li et al. reported the fabrication of ZnO nanotube array (Figure 3.3) by electrochemical deposition using AAM template [37]. Along with all the above remarkable properties of template-assisted synthesis of 1D nanomaterials, there are some disadvantages of this technique too, which are limiting its use for the fabrication of 1D nanomaterials. The disadvantages of these techniques include complex synthesis steps, easy in cracking of the synthesized materials, and poor cohesion.

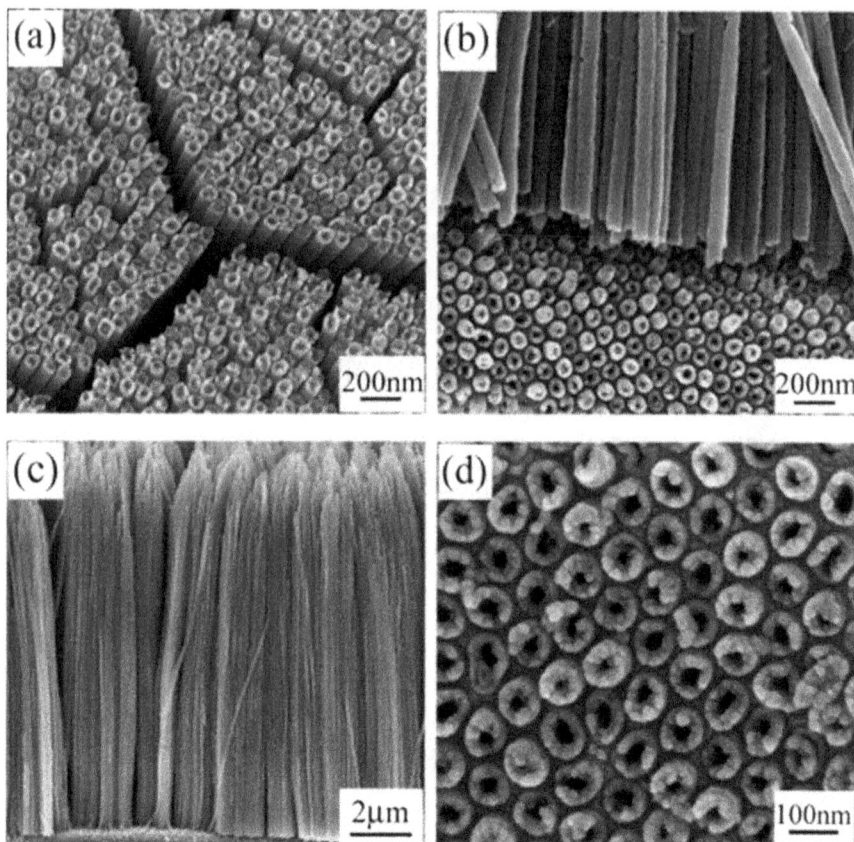

FIGURE 3.3 Scanning electron microscopic image of ZnO nanotube arrays. Grown by direct electrochemical deposition using anodic AAM template: (a) top surface, 3 min, (b, c) cross-section, 5 and 8 min, and (d) magnified image of selected area in (a) [37].

Source: Reprinted with permission from Ref. [37]. © 2006 American Chemical Society.

3.3 PROPERTIES OF 1D NANOMATERIALS

Recent advances in nanomaterials have initiated a number of new routes for the controllable synthesis of different shapes and sizes of 1D nanomaterials, such as nanotube, nanowire, nanofibers, etc., and have attracted attention due to their unique properties and wide range of conceivable applications in the field of electronics, optics, catalysis, and magnetism. For the sake of above-mentioned application area of 1 D nanomaterials, along with novel synthetic strategies, electronics and optical properties (refractive index) of these materials were extensively investigated. The unique properties of these 1D nanomaterials (different from their respective bulk, and source counterpart) are due to the intermolecular integration, such as van der Waals, hydrogen bonding, ionic interaction, and π–π stacking. This section provides the optical properties of 1D nanomaterials, which are further presented under the subsections; optical, electrical, and size-dependent properties [38, 39].

3.3.1 OPTICAL PROPERTIES

Based on the special crystal structural, 1D nanomaterials present the typical optical properties than those of bulk counterparts. In a specific way, the quantum confinement effect introduces many novel photoluminescence (PL) properties to 1D nanomaterials. The particular PL properties of these nano-materials, such as emission enhancement, fluorescence narrowing, defect emission, and switchable emission, played an important role for the develop-ment of new generation of optoelectronic devices. In the solid state, the fluo-rescence efficiency of the organic chromospheres decreases and resulting in concentration quenching. However, this phenomenon is common but it has been observed with an enhanced emission for some particular 1D fluorescent materials. This study reported on the limitation caused by the quenching in order to develop the full color flat panel display and ultrahigh optical memory device [40]. Based on this concept, Lim et al. proposed a multifunc-tional fluorescent 1D molecule for photoswitchable memory device [41]. In 2006, Djurisic and Leung reported the room temperature photoluminescence (RTPL) study of 1D ZnO, which was focused on the ultraviolet (UV)–vis emission or defect emission along some defect emission band, which were observed in the visible region. This observed band is due to the small size of excision Bohr radius of ZnO [42, 43].

Most of the developed devices are based on organic nanomaterials and have the amorphous nature but the crystallinity of the nanomaterials resulted due to the novel luminescent performance [44]. Doping mechanism has also been widely used to tune the fluorescence resonance energy transfer in EL devices. Doping helps to improve the luminescence efficacy and tunes the behavior of emission color. In this approach, the slight variation of the energy acceptor resulted in significant PL color variations. At the particular point of view, that is, white light emission can be achieved by doping blue fluorescent dye with red and green dyes. This concept has been used in several essential applications intended for developing full color displays [45].

3.3.2   ELECTRICAL PROPERTIES

Due to their exception transport properties, significance of 1D nanomaterials is heavily realized for their application in FETs for current transport studies in different device development. Electrical property of 1D nanomaterials is essential for developing the nanoelectronics device fabrication [46], and in this line of research, Xiao et al. fabricated 1D nanomaterial-based FET devices [47]. For using electrical property of 1D nanomaterials, laser has been one of the promising tools for developing devices in present era of research and technology. Smaller molecule exhibited dipolar nature and their functionalization for the device fabrication can be tailored by using dielectrophoretic force. Due to this property, most of the 1D devices have been fabricated under AC or DC electric field.

3.3.3   SIZE-DEPENDENT OPTICAL PROPERTIES

Optical properties of 1D nanomaterials are also affected by the size of materials but their explanation cannot be explained by the quantum confinement theory. This type of property was reported first time in perylene nanocrystals [48] and it was observed that the size of the nanomaterials had crucial role on their respective properties and application. Size of the 1D nanomaterials affects the excitation, emission, and causes the multiple fluorescence spectrum as well as tunes the hard-core properties of the nanomaterial itself. This type of effect is well known and common in organic-materials-based 1D nanomaterials as well as in inorganic semiconductor quantum dots, the reason is the presence of chromophoretic system of organic moieties [49–51].

3.3.4   OPTOELECTRONIC APPLICATION OF 1D NANOMATERIALS

Tremendous properties of 1D nanomaterials attract the interest to be applied in many fields, such as nanoelectronics and optical devices, catalysis and electrocatalysis, chemical/biological sensors, energy, biomedical, and environmental fields. In this section, we are focusing our evaluation especially on the optoelectronic device application of 1D nanomaterials, namely, hetero- and homojunction light emitting diode (LED), laser diode, and photodetector application's point of view. Different 1D nanomaterials (e.g., ZnO and organic materials) are well known and deserved photonic materials in UV region for electronic device development.

LED is a device, which is the source of semiconductor light and emits the light when current flows over. Electrons in the semiconductor devices recombine with the electron holes and releasing energy in the form of photon. LED emits the light by the process of electroluminescence and the color of the light determines by the energy gap of the used semiconductor (Figure 3.4). The application of LED is used in aviation lighting, traffic signals, general lighting, automotive lighting, and in addition LED is also used in video displays.

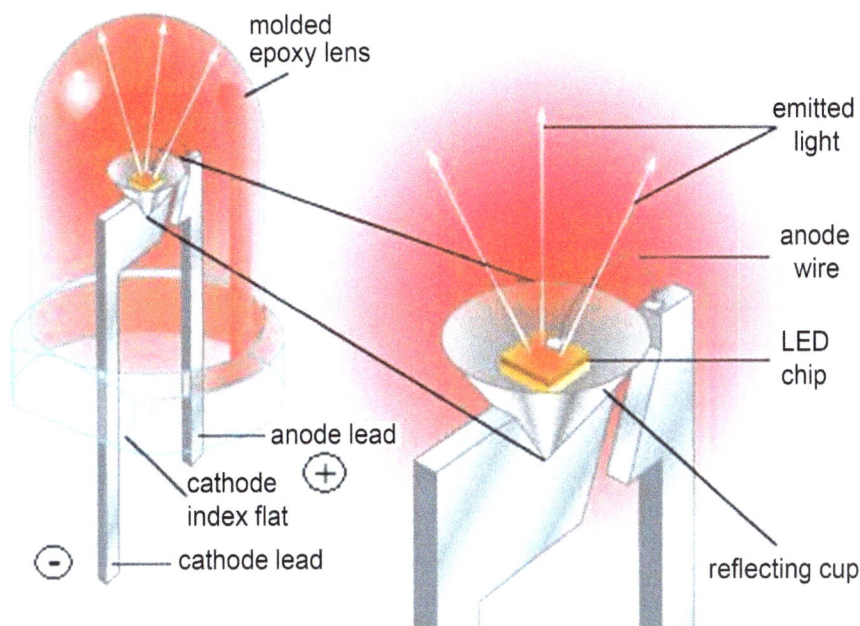

FIGURE 3.4 A schematic presentation of LED setup consists of a semiconductor chip doped with impurities to create p–n junction. In LED, current flows easily from p-side (anode) to n-side (cathode) [52].

Source: Reprinted from Ref. [52]. © 2004 Encyclopedia Britannica.

A number of works were reported about light-emitting device based on organic and inorganic 1D nanomaterials [53, 54]. In the field of device fabrication technology, much attention has been paid on 1D semiconducting organic and metallic nanoparticles for fabricating nanoscale electronic and optoelectronic device due their superb electrical and optoelectrical properties [55–57]. In 2003, Zhou et al. reported on the 1D nanomaterials to study as a model electronic and electro-optical devices. In their study, they synthesized the nanofibers of a doped polyaniline/polyethylene oxide (PANI/PEO) and in their investigations they noticed that the conductivity is due to the diameters of synthesized fibers while the asymmetrical fibers were consistent with the formation of Schottky barriers. In follow-up study by the same group of researchers emphasized the FET behavior of PANI/PEO fabricated devices and observed the surprising saturation channel current at low source–drain voltages, the reported hole mobility in the depletion regime was 1.4×10^{-4} cm^2 V s^{-1} with good air stability [58, 59]. Along with this study, it is also reported by a quite number of reports that electrically bistable devices of 1D based on 1D nanomaterials have shown improved performances. It is well known that organic 1D nanomaterials have also dominated the researcher interest for their application for the fabrication of photovoltaic cells. In organic photovoltaic cell, the active layer basically depends on the bulk heterojunction (BHJ) mechanism, in which the photoactive layer entitled of a continuous amalgam of p-type (electron donor) and n-type (electron acceptor) system. As usual, the excision diffusion length in almost all the photovoltaic cell is about 5–10 nm, the nanodevices control the critical photoactive layer by high power conversion efficiency. The additional promising approach for the fabrication of photovoltaic cell is the implementation of nanowire (NW) in p-type and n-type components to achieve the better performance of the fabricated device. As the diameter of the doped NW can be tuned with the exciton diffusion length, so that the charge separation of the photovoltaic cell could be warranted recombination for the fabricated solar cell. After reviewing different reports, it is concluded that the high absorption coefficient, carrier mobility exciton diffusion length can make a device much competitive and alternate to the BHJ mechanism in the field of photovoltaic cell [60–62].

In addition to the development of single component 1D nanomaterials, multicomponent 1D nanostructured materials (MC1DNM) have also drawn the attention of the researcher for the fabrication of optoelectronic devices. Multicomponent nanomaterials are very promising candidates for the development of high performance optoelectronic device on micro- to nanoscale range. Multicomponent nanomaterials, such as organic–inorganic hybrid nanostructures, p–n junction, and core–shell, have so far been used into photonic devices and nanoscale range. As compared to inorganic NW heterojunction, assembling of multicomponent organic heterostructures is challenging in regard to the fabrication of photonic devices along with understanding the energy transfer phenomenon and charge transfer mechanism [63–65].

Another important application of 1D nanomaterials is laser diode. Laser diode is a semiconductor device, which is similar as LEDs, where a laser beam is created at the diode junctions. Laser diode converts electrical energy into light (Figure 3.5). The doped p–n transition allows electron for the recombination with hole and all these processes govern by the voltage. Laser diodes are the most common lasers, which have a wide range of application for the barcode reader, laser pointer, laser printing, laser scanning, and CD/Blu-ray/CD reading.

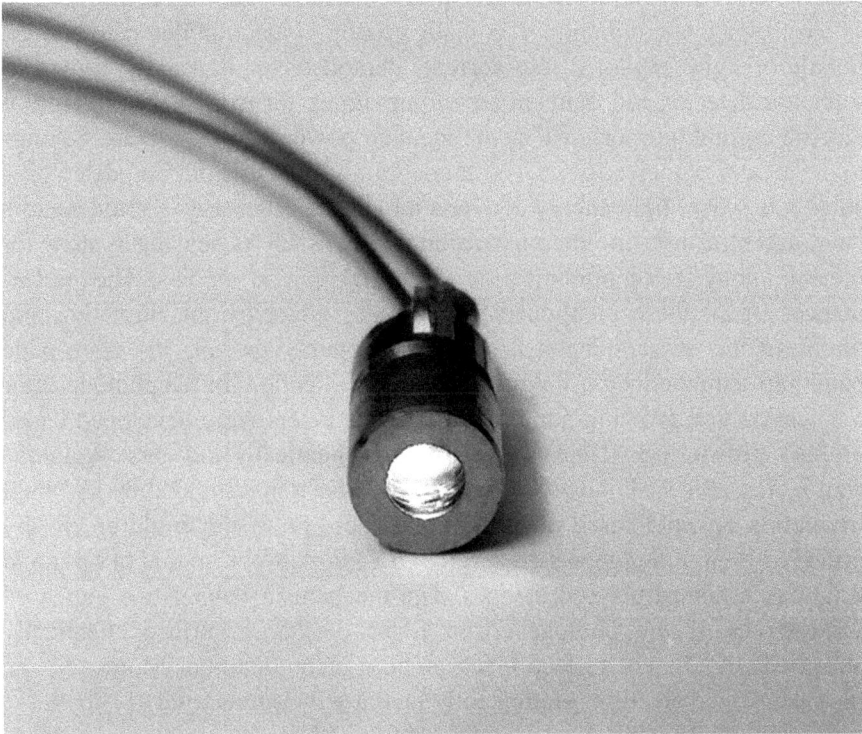

FIGURE 3.5   A systematic presentation of a simple and low-powered metal enclosed laser diode [65].

One-dimensional nanomaterials have also promising applications for the UV lasing action and tremendous works have been reported for the laser diode application based on different types of 1D nanostructured materials (with different shapes and sizes). ZnO is one of the well-known 1D nanomaterials and it has blue optoelectronic applications along with UV lasing action due to the wide band gap. In order to achieve the optical gain for lasing action in an electron–hole plasma method, a high carrier concentration is required. If the exciton binding energy is greater than thermal energy, then the efficient excitonic laser action could be attained at room temperature [66, 67]. The lasing accomplishments have been reported in vapor–liquid–solid (VLS)-based growth of ZnO NW for its evolution in emission spectrum. During the study, it was observed that at low lower excitation, the respective spectrum showed a single but wider emission peak. The spontaneous emission peak was observed at 140 meV below its individual band gap [68].

The third well-known application and widely used device based on 1D nanomaterial is photodetector, which is a sensor of light or is also known as electromagnetic radiation. It is made up of p–n junction that converts the bundle of light (photons) into current. Photodetector is usually known as a photon detector and delivers an output signal, for example, a voltage or electric current, proportional to the incident power, and this device belongs to the area of optoelectronics. Another common example of photodetector is solar cell, where light energy converts into electrical energy. Photodetectors have a central role in emerging technologies, such as sensing system for biotechnology, telecommunication, and atmospheric studies. The performances of all these mentioned technologies affect by the limitation and tuning of the semiconductor based 1D nanomaterials [69, 70]. As a wide band-gap semiconductor, ZnO receives much attention for the photodetector application and growing numbers of photodetectors were developed based on ZnO, grown by different methods such as chemically and so on. A significant increase in the conductivity of ZnO has been investigated by UV light irradiation on ZnO-based devices in air. These processes could be closely observed by introducing a switch (ON/OFF) in-between source of light and ZnO-based nanostructured devices. The mechanism behind this increased conductivity is the photogenerated holes discharge surface chemically absorbed the O_2^- via surface electron–hole recombination. Moreover, the photogenerated electrons significantly increase the conductivity [71].

3.5 CONCLUSION FUTURE PROSPECTIVE

In this chapter, we have presented a review on the synthesis, functionalization, properties, fabrication, and optoelectronic application of 1D nanomaterials. Therefore, we have briefly discussed different synthesis strategies; specific properties of those materials are presented their practical applications in real word scenario. This chapter also highlights that among various nanaoshape, 1D nanostructured materials have specific and sole rewards such as (1) large surface area due to high aspect ratio and (2) specific crystallographic orientation in a way to allocate kinetically favorable direction. In our opinion, there is some point which needs to deserve a special attention for the development of optoelectronic devices-based 1D nanomaterials. First, fabrication method with desired morphology structural orientation is still a key task. Second, there is a need for sufficient performance testing prior to their application to support the long-term stability of 1D nanomaterials. In the light of the literature survey, it is our opinion that the stability of the 1D nanomaterials-based device is improved, which is indeed a requirement for the present scenario of the device and technology development.

KEYWORDS

- **optoelectronics**
- **nanomaterial dimensional**
- **sensing interface**
- **nanotechnology**
- **nanobioelectronics device**
- **sensors**

REFERENCES

1. Goldstein, A. N.; Echer, C. M.; Alivisatos, A. P. Melting in semiconductor nanocrystals. *Science* **1992**, *256*, 1425–1427.
2. Reed, M. A.; Frensley, W. R.; Matyi, R. J.; Randall, J. N.; Seabaugh, A. C. Realization of a three-terminal resonant tunneling device: The bipolar quantum resonant tunneling transistor, *Appl. Phys. Lett.* **1989**, *54*, 1034–1036.
3. Lu, L.; Shen, Y.; Chen, X.; Qian, L.; Lu, K. Ultrahigh strength and high electrical conductivity in copper. *Science* **2004**, *304*, 422–426.
4. Zhao, Y. S.; Fu, H.; Peng, A.; Ma, Y.; Xiao, D.; Yao, J. Low-dimensional nanomaterials based on small organic molecules: Preparation and optoelectronic properties. *Adv. Mater.* **2008**, *20*, 2859–2876.
5. Gu, Y.; Kuskovsky, I. L.; Yin, M.; O'Brien, S.; Neumark, G. F. Quantum confinement in ZnO nanorods. *Appl. Phys. Lett.* **2004**, *85*, 3833–3835.
6. Corma, A. From microporous to mesoporous molecular sieve materials and their use in catalysis. *Chem. Rev.* **1997**, *97*, 2373–2420.
7. Iijima, S. Helical microtubules of graphitic carbon. *Nature* **1991**, *354*, 56–58.

8. Law, M.; Sirbuly, D. J.; Johnson, J. C.; Goldberger, J.; Saykally, R. G.; Yang, P. Nanoribbon waveguides for subwavelength photonics integration. *Science* **2004**, *305*, 1269–1273.

9. Wang, Z. L. Nanobelts, nanowires, and nanodiskettes of semiconducting oxides—From materials to nanodevices. *Adv. Mater.* **2003**, *15*, 432–436.

10. Law, M.; Goldberger, J; Yang, P. Semiconductor nanowires and nanotubes. *Ann. Rev. Mater. Res.* **2004**, *34*, 83–122.

11. Lim, J. A.; Liu, F.; Ferdous, S.; Muthukumar, M.; Briseno, A. L. Polymer semiconductor crystals. *Mater. Today* **2010**, *13*, 14–24.

12. Kim, F. S.; Ren, G. Q.; Jenekhe, S. A. One-dimensional nanostructures of π-conjugated molecular systems: Assembly, properties, and applications from photovoltaics, sensors, and nanophotonics to nanoelectronics. *Chem. Mater.* **2011**, *23*, 682–732.

13. Yu, H.; Kim, D. Y.; Lee, K. J.; Oh, J. H. Fabrication of one-dimensional organic nanomaterials and their optoelectronic applications. *J. Nanosci. Nanotechnol.* **2014**, *14*, 1282–1302.

14. Dai, Z. R.; Pan, Z. W.; Wang, Z. L. Novel nanostructures of functional oxides synthesized by thermal evaporation. *Adv. Funct. Mater.* **2003**, *13*, 9–24.

15. Lee, J.-K.; Koh, W.-K.; Chae, W.-S.; Kim, Y.-R. Novel synthesis of organic nanowires and their optical properties. *Chem. Commun.* **2002**, 138–139.

16. Chiu, J. J.; Kei, C. C.; Perng, T. P.; Wang, W. S. Organic semiconductor nanowires for field emission. *Adv. Mater.* **2003**, *15*, 1361–1364.

17. Liu, H.; Li, Y.; Xiao, S.; Gan, H.; Jiu, T.; Li, H.; Jiang, L.; Zhu, D.; Yu, D.; Xiang, B.; Chen, Y. Synthesis of organic one-dimensional nanomaterials by solid-phase reaction. *J. Am. Chem. Soc.* **2003**, *125* (36), 10794–10795.

18. Zhao, Y. S.; Di, C.; Yang, W.; Yu, G.; Liu, Y.; Yao, J. Photoluminescence and electroluminescence from tris(8-hydroxyquinoline) aluminum nanowires prepared by adsorbent-assisted physical vapor deposition. *Adv. Funct. Mater.* **2006**, *16* (15), 1985–1991.

19. Zhang, F.; Yang, P.; Matras-Postołek, K. Au datalyst decorated silica spheres: Synthesis and high-performance in 4-nitrophenol reduction. *J. Nanosci. Nanotechnol.* **2016**, *16*, 5966–5974.

20. Yu, H.; Kim, D.Y.; Lee, K. J.; Oh, J. H. Fabrication of one-dimensional organic nanomaterials and their optoelectronic. *J. Nanosci. Nanotechnol.* **2014**, *14*, 1282–1302.

21. Liu, B.; Zeng, H. C. Hydrothermal synthesis of ZnO nanorods in the diameter regime of 50 nm. *J. Am. Chem. Soc.* **2003**, *125* (15), 4430–4431.

22. Byrappa, K.; Yoshimura, M. *Handbook of Hydrothermal Technology*, Elsevier, Oxford, UK, **2013**.

23. O'Hare, D. Hydrothermal synthesis, *Encyclopedia of Materials: Science and Technology* (2nd ed.), Elsevier, Amsterdam, New York, **2001**, pp. 3989–3992.

24. Kanti Kole, A.; Sekhar Tiwary, C.; Kumbhakar, P. Morphology controlled synthesis of wurtzite ZnS nanostructures through simple hydrothermal method and observation of white light emission from ZnO obtained by annealing the synthesized ZnS nanostructures. *J. Mater. Chem. C* **2014**, *2* (21), 4338–4346.

25. Ji, S.; Liu, H.; Sang, Y.; Liu, W.; Yu, G.; Leng, Y. Synthesis, structure, and piezoelectric properties of ferroelectric and antiferroelectric NaNbO$_3$ nanostructures. *CrystEngComm* **2014**, *16* (32), 7598–7604.

26. Pal, U.; Santiago, P. Controlling the morphology of ZnO nanostructures in a low-temperature hydrothermal process. *J. Phys. Chem. B* **2005**, *109* (32), 15317–15321.

27. Tam, K. H.; Cheung, C. K.; Leung, Y. H.; Djurišić, A. B.; Ling, C. C.; Beling, C. D.; Fung, S.; Kwok, W. M.; Chan, W. K.; Phillips, D. L.; Ding, L.; Ge, W. K. Defects in ZnO nanorods prepared by a hydrothermal method. *J. Phys. Chem. B* **2006**, *110* (42), 20865–20871.

28. Chen, S.-W.; Wu, J.-M. Nucleation mechanisms and their influences on characteristics of ZnO nanorod arrays prepared by a hydrothermal method. *Acta Mater.* **2011**, *59* (2), 841–847.

29. Long, Y. Z.; Yan, X.; Wang, X. X.; Zhang, J.; Yu, M. *Micro and Nanotechnologies, Electrospining: Nanofabrication and Application*, Elsevier, Amsterdam, **2019**, pp. 21–52.

30. Teo, W. E.; Ramakrishna, S. Electrospun fibre bundle made of aligned nanofibres over two fixed points. *Nanotechnology* **2005**, *16*, 1878–1884.

31. Bhardwaj, N.; Kundu, S. C. Electrospinning: A fascinating fiber fabrication technique. *Biotechnol. Adv.* **2010**, *28* (3), 325–347.

32. Yuh, J.; Nino, J. C.; Sigmund, W. M. Synthesis of barium titanate (BaTiO$_3$) nanofibers via electrospinning. *Mater. Lett.* **2005**, *59* (28), 3645–3647.

33. McCann, J. T.; Chen, J. I. L.; Li, D.; Ye, Z.-G.; Xia, Y. Electrospinning of polycrystalline barium titanate nanofibers with controllable morphology and alignment. *Chem. Phys. Lett.* **2006**, *424* (1), 162–166.

34. Possin, G. E. A method for forming very small diameter wires. *Rev. Sci. Instrum.* **1970**, *41* (5), 772–774.

35. Martin, C. R. Nanomaterials: A membrane-based synthetic approach. *Science* **1994**, *266* (5193), 1961–1966.

36. Cao, B.; Ji, M.; Sun, S. Synthesis and optical properties of organic fluorescent nanowires and nanotubes. *J. Nanosci. Nanotechnol.* **2013**, *13* (8), 5827–5831.

37. Li, L.; Pan, S.; Dou, X.; Zhu, Y.; Huang, X.; Yang, Y.; Li, G.; Zhang, L. Direct electrodeposition of ZnO nanotube arrays in anodic alumina membranes. *J. Phys. Chem. C* **2007**, *111* (20), 7288–7291.

38. Zhao, Y.; Hong, H.; Gong, Q.; Ji L. 1D nanomaterials: Synthesis, properties, application (Editorial), *J. Nanomater.* **2013**, *2013*, 1.

39. Weng, B.; Liu, S.; Tang, Z.-R.; Xu, Y.-J. One-dimensional nanostructure based materials for versatile photocatalytic applications. *RSC Adv.* **2014**, *4* (25), 12685–12700.

40. Luo, J.; Xie, Z.; Lam, J. W. Y.; Cheng, L.; Chen, H.; Qiu, C.; Kwok, H. S.; Zhan, X.; Liu, Y.; Zhu, D.; Tang, B. Z. Aggregation-induced emission of 1-methyl-1,2,3,4,5-pentaphenylsilole. *Chem. Commun.* **2001**, *18*, 1740–1741.

41. Lim, S. J.; An, B. K.; Jung, S. D.; Chung, M. A.; Park, S. Y. Photoswitchable organic nanoparticles and a polymer film employing multifunctional molecules with enhanced fluorescence emission and bistable photochromism. *Angew. Chem. Int. Ed. Engl.* **2004**, *43* (46), 6346–6350.

42. Djurisic, A. B.; Leung, Y. H. Optical properties of ZnO nanostructures. *Small* **2006**, *2* (8–9), 944–961.

43. Gu, Y.; Kuskovsky, I. L.; Yin, M.; O'Brien, S.; Neumark, G. F. Quantum confinement in ZnO nanorods. *Appl. Phys. Lett.* **2004**, *85* (17), 3833–3835.

44. Zhao, Y. S.; Fu, H.; Hu, F.; Peng, A.; Yang, W.; Yao, J. Tunable emission from binary organic one-dimensional nanomaterials: An alternative approach to white-light emission. *Adv. Mater.* **2008**, *20*, 79–83.

45. Kido, J.; Kimura, M.; Nagai, K. Multilayer white light-emitting organic electroluminescent device. *Science* **1995**, *267* (5202), 1332–1334.

46. Jalili, S.; Rafii-Tabar, H. Electronic conductance through organic nanowires. *Phys. Rev. B* **2005**, *71* (16), 165410.

47. Xiao, K.; Tao, J.; Pan, Z.; Puretzky, A. A.; Ivanov, I. N.; Pennycook, S. J.; Geohegan, D. B. Single-crystal organic nanowires of copper-tetracyanoquinodimethane: synthesis, patterning, characterization, and device applications. *Angew Chem. Int. Ed. Engl.* **2007**, *46* (15), 2650–2654.

48. Kasai, H.; Kamatani, H.; Okada, S.; Oikawa, H.; Matsuda, H.; Nakanishi, H. Size-dependent colors and luminescences of organic microcrystals. *Jpn. J. Appl. Phys.* **1996**, *35* (Part 2, No. 2B), L221–L223.

49. Auweter, H.; Haberkorn, H.; Heckmann, W.; Horn, D.; Luddecke, E.; Rieger, J.; Weiss, H. Supramolecular structure of precipitated nanosize beta-carotene particles. *Angew Chem. Int. Ed. Engl.* **1999**, *38* (15), 2188–2191.

50. Fu, H.; Loo, B. H.; Xiao, D.; Xie, R.; Ji, X.; Yao, J.; Zhang, B.; Zhang, L. Multiple emissions from 1,3-diphenyl-5-pyrenyl-2-pyrazoline nanoparticles: evolution from molecular to nanoscale to bulk materials. *Angew Chem. Int. Ed. Engl.* **2002**, *41* (6), 962–965.

51. Xiao, D.; Yang, W.; Yao, J.; Xi, L.; Yang, X.; Shuai, Z. Size-dependent exciton chirality in (R)-(+)-1,1'-bi-2-naphthol dimethyl ether nanoparticles. *J. Am. Chem. Soc.* **2004**, *126* (47), 15439–15444.

52. http://www.eai.in/ref/ct/ee/led.html, Energy Alternatives India (EAI), (access on 08 August 2019).

53. Ju, S.; Li, J.; Liu, J.; Chen, P. C.; Ha, Y. G.; Ishikawa, F.; Chang, H.; Zhou, C.; Facchetti, A.; Janes, D. B.; Marks, T. J. Transparent active matrix organic light-emitting diode displays driven by nanowire transistor circuitry. *Nano Lett.* **2008**, *8*, 997–1004.

54. Kim, K.-K.; Lee, S.-D.; Kim, H.; Park, J.-C.; Lee, S.-N.; Park, Y.; Park, S.-J.; Kim, S.-W. Enhanced light extraction efficiency of GaN-based light-emitting diodes with ZnO nanorod arrays grown using aqueous solution. *Appl. Phys. Lett.* **2009**, *94* (7), 071118.

55. Zhang, Z.; Sun, X.; Dresselhaus, M. S.; Ying, J. Y.; Heremans, J. Electronic transport properties of single-crystal bismuth nanowire arrays. *Phys. Rev. B* **2000**, *61* (7), 4850–4861.

56. Duan, X.; Huang, Y.; Cui, Y.; Wang, J.; Lieber, C. M. Indium phosphide nanowires as building blocks for nanoscale electronic and optoelectronic devices. *Nature* **2001**, *409* (6816), 66–69.

57. Wu, Y.; Fan, R.; Yang, P. Block-by-block growth of single-crystalline Si/SiGe superlattice nanowires. *Nano Lett.* **2002**, *2* (2), 83–86.

58. Zhou, Y.; Freitag, M.; Hone, J.; Staii, C.; Johnson Jr., A. T. Fabrication and electrical characterization of polyaniline-based nanofibers with diameter below 30 nm. *Appl. Phys. Lett.* **2003**, *83* (18), 3800–3802.

59. Pinto, N. J.; Johnson Jr., A. T.; MacDiarmid, A. G.; Mueller, C. H.; Theofylaktos, N.; Robinson, D. C.; Miranda, F. A. Electrospun polyaniline/polyethylene oxide nanofiber field-effect transistor. *Appl. Phys. Lett.* **2003**, *83* (20), 4244–4246.

60. Berson, S.; De Bettignies, R.; Bailly, S.; Guillerez, S. Poly(3-hexylthiophene) fibers for photovoltaic applications. *Adv. Funct. Mater.* **2007**, *17* (8), 1377–1384.

61. Kim, J. S.; Lee, J. H.; Park, J. H.; Shim, C.; Sim, M.; Cho, K. High-efficiency organic solar cells based on preformed poly(3-hexylthiophene) nanowires. *Adv. Funct. Mater.* **2011**, *21* (3), 480–486.

62. Xin, H.; Guo, X.; Ren, G.; Watson, M. D.; Jenekhe, S. A. Efficient phthalimide copolymer-based bulk heterojunction solar cells: How the processing additive influences nanoscale morphology and photovoltaic properties. *Adv. Energy Mater.* **2012**, *2* (5), 575–582.

63. Xia, Y.; Yang, P.; Sun, Y.; Wu, Y.; Mayers, B.; Gates, B.; Yin, Y.; Kim, F.; Yan, H. One-dimensional nanostructures: Synthesis, characterization, and applications. *Adv. Mater.* **2003**, *15* (5), 353–389.

64. Yang, J.; Heo, M.; Lee, H. J.; Park, S.-M.; Kim, J. Y.; Shin, H. S. Reduced graphene oxide (rGO)-wrapped fullerene (C60) wires. *ACS Nano* **2011**, *5* (10), 8365–8371.

65. https://en.wikipedia.org/wiki/Laser_diode, accessed on 27 January 2020.

66. Konstantatos, G.; Badioli, M.; Gaudreau, L.; Osmond, J.; Bernechea, M.; Garcia de Arquer, F. P.; Gatti, F.; Koppens, F. H. Hybrid graphene-quantum dot phototransistors with ultrahigh gain. *Nat. Nanotechnol.* **2012**, *7* (6), 363–368.

67. Cao, H.; Xu, J. Y.; Zhang, D. Z.; Chang, S.; Ho, S. T.; Seelig, E. W.; Liu, X.; Chang, R. P. Spatial confinement of laser light in active random media. *Phys. Rev. Lett.* **2000**, *84* (24), 5584–5587.

68. Huang, M. H.; Mao, S.; Feick, H.; Yan, H.; Wu, Y.; Kind, H.; Weber, E.; Russo, R.; Yang, P. Room-temperature ultraviolet nanowire nanolasers. *Science* **2001**, *292* (5523), 1897–1899.

69. Lan, C.; Li, C.; Yin, Y.; Guo, H.; Wang, S. Synthesis of single-crystalline GeS nanoribbons for high sensitivity visible-light photodetectors. *J. Mater. Chem. C* **2015**, *3* (31), 8074–8079.

70. Saran, R.; Curry, R. J. Lead sulphide nanocrystal photodetector technologies. *Nat. Photonics* **2016**, *10*, 81–92.

71. Jin, Y.; Wang, J.; Sun, B.; Blakesley, J. C.; Greenham, N. C. Solution-processed ultraviolet photodetectors based on colloidal ZnO nanoparticles. *Nano Lett.* **2008**, *8* (6), 1649–1653.

CHAPTER 4

2D Nanomaterials and Their Optoelectronic Applications

H. ELHOSINY ALI[1,2], M. ASLAM MANTHREMMEL[1],
KAMLESH V. CHANDEKAR[3], S. ALFAIFY[1], and MOHD. SHKIR[1*]

[1]*Advanced Functional Materials & Optoelectronics Laboratory (AFMOL), Department of Physics, Faculty of Science, King Khalid University, Abha 62529, Saudi Arabia*

[2]*Physics Department, Faculty of Science, Zagazig University, 44519 Zagazig, Egypt*

[3]*Department of Physics, Rayat Shikshan Sanstha's, Karmaveer Bhaurao Patil College, Vashi, Navi Mumbai 400703, Maharashtra, India*

Corresponding author. E-mails: shkirphysics@gmail.com; shkirphysics@kku.edu.sa.

ABSTRACT

In this chapter, we present shortly a general classification, synthesis, and fabrication of nanoparticles (NPs). Subsequently, we focus more attention on the classification of nanomaterials in two-dimensional (2D) for optoelectronic applications. 2D materials, such as nanocomposite materials, have become the most powerful semiconductor materials in the field of optoelectronic devices due to their extraordinary properties. The anisotropic-chemical functions of 2D nanomaterials have engrossed a huge interest from all around the world that involves electrooptic, catalysis, supercapacitor, and energy-related materials. Due to the layer-dependent and properly shaped bandgaps, photodetectors depending on different 2D components are intended and produced in a rational manner. Using the distinctive characteristics of 2D components, many unexpected physical events of collisions relying on 2D components can be achieved when separate 2D components are placed together. This allows hetero junctions more famous than 2D components themselves, and the development of 2D components for humans is easier than ever before. In this work, we focused on the polymeric nanocomposite as it is one of the important famous type of nanocomposites.

4.1 INTRODUCTION

Two-dimension (2D) nanomaterials are very attractive with respect to their integration and potential uses in optoelectronic devices. Currently, nanocomposites have concerned a lot of concentration due to their widespread candidates to enhance the structural, optical, mechanical, spectral, and nonlinear optical features of the various inorganic–organic materials functioned in the range of UV–VIS–IR spectral. These nanocomposites based on organic–inorganic materials have fashioned widespread attention in developing new devices/materials [1].

Based on whether the light is formed from the charges or charge is generated using the light falling on the material/device, the optoelectronics is split into two kinds of electrical devices. While the light-emitting diodes (LEDs) and lasers are the examples for the former, and photovoltaic devices and photodetectors are the examples for the later. Depending on the physical processes, accountable for the electron or charging phase used for the service of machines, the domain can, also, be split into several fields. Photoemission, photoconductivity, stimulated emission, photoelectrical, and radiative recombination are the common mechanisms that most optoelectronic devices are looking to integrate with it.

The word nanocomposite is commonly used for describing a very variety of products where one of the parts has a submicron aspect. A closer and far more rigid concept would involve a nanocomposite to be an essentially fresh fabric (hybrid) in which the element in the scale of nanometer or composition provides rise to inherently different characteristics that are not available in the corresponding macroscopic composites or natural materials. This concept requires a lower size of the nanostructure than a characteristic dimension underpinning a physical property of the substance. This scale would, for example, be related to the wavelength of de Broglie (electrons) and is responsible for the characteristics of a conductor or semiconductor in this scale and would be related to the mechanical properties of a polymer with the size of a polymer or crystal, as both extend from a few nanometers to hundreds of nanometers. Special emphasis is provided to the models that are applied to quasi-2D patterned inorganic additives (e.g., 2:1 aluminosilicates

[1–10], from which most of the research studies are extracted, along with layered dual hydroxides [11]), and to a lesser degree to pseudo-dimensional fillers [e.g., carbon nanotubes (CNTs) [12]]. Concurrent gains across various characteristics—with simultaneous improvement of mechanical, heat, and thermomechanical reaction—are typically accomplished, in relation to different characteristics, such as enhanced resistance, flammability, and biodegradability, opposed to unfilled polymers. Therefore, the subsequent nanocomposite structure is best defined by the word "hybrid" (the word denoting the major modifications in various product characteristics) rather than composite (the word traditionally linked to incremental enhancement of more than one property [13–15]). The fundamental new characteristics for such NPs typically arise due to the variations happening to the polymer environment around the fillers, as these polymers are typically adsorbed on the filler surfaces or enclosed between fillers and thus govern the effective surface area of the filler—for example, the available surface area is different when the fillers are totally dispersed or the surfaces are attached with clusters. Therefore, a real "nanocomposite" at small filler volumes could lead to the excellent spread of the filler, near to that elevated image ratio's limit [16]. The resulting composition is said to be a normal composite if the nanometer dimension of the inorganic fillers is not diffused or utilized. For concrete nanometer-thin coated fillers, plastics are recognized for the effectiveness of diffuse minerals of wax when properly modified [1, 2]. The sector has lately achieved momentum, primarily owing to two significant results that led to the resurgence of these products: first, the study by Unitika and Toyota scientists on nylon-6/montmorillonite fabric [17, 18] where very mild inorganic loadings have led in concurrent and notable improvements in heat and mechanical characteristics. Vaia et al. [19] have secondly discovered that the use of organic compounds is feasible for melt-mixing polymers with clays. Since then, the elevated demand for agricultural applications has given rise to energetic studies, which has disclosed at the same time drastic increases in researches on nanosized thin inorganic fillers dispersed polymers [11, 20–30]. Here, changes in properties arise from the nanocomposite frameworks that are usually relevant across a broad spectrum of polymers [7, 11], and the subject is very broad.

We have been discussing about the nanocomposite assembly of the polymer filled with high aspect-ratio nanoscale inorganic filler to improve the polymer properties or vice-versa. We will continue our discussion in this chapter by restricting on polymer/inorganic-filler nanocomposites. This chapter will also deal with few methods for synthesizing polymeric nanocomposite materials and some of their functional properties for optoelectronic applications.

4.2   CLASSIFICATION OF NANOMATERIALS

Nanomaterials are classified by the sizes in the nanometer area ($\leq$100 nm), as illustrated in Figure 4.1a.

1. zero dimensional (0D)—confined in all three dimensions.
2. one dimensional (1D) nanomaterial—one dimension not in nanoscale, confined in the other two dimensions [NW, NT, nanorods (NRs), etc.].
3. 2D nanomaterial—two dimensions not in nanoscale, confined in one dimension (nanosized films/layers, graphene, etc.) (see Figure 4.1b).
4. three-dimensional (3D) nanomaterial—growth in three dimensions within nanoscale (NP dispersions, nanogranules, etc.).

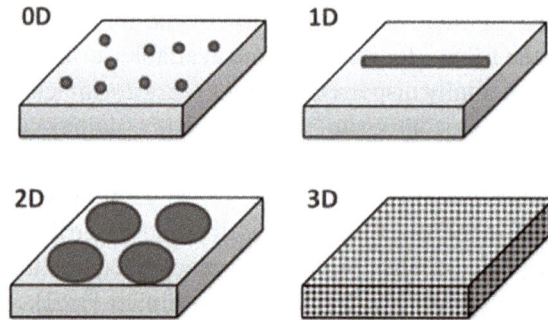

(a)

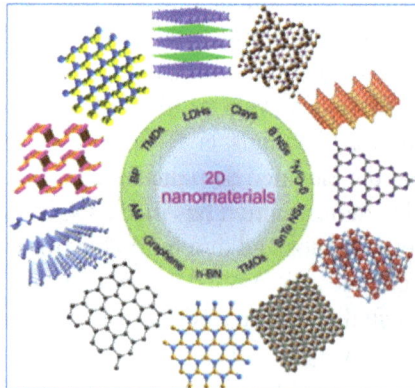

(b)

FIGURE 4.1(a)   The schematic illustrates the types of nanomaterials [31]; (b) the schematic illustrates 2D nanomaterials [32].

Sources: (a) Reprinted with permission from Ref. [31]. © 2015 Elsevier. (b) Reprinted with permission from Ref [32]. © 2019 Elsevier.

Crystalline or amorphous or multicrystalline can be present in nanomaterials. They can compose a chemical element of one or multiple phases. Also, they can be metallic, ceramic, or polymer in various forms and aspects.

4.2.1 NANOMATERIALS

Nanomaterials are materials designed in such a way that they have magnitudes of fewer than 100 nm in at least 1D (dimensions < 100 nm) like a nanosized particle, wire, rod, tubes, and other forms.

4.2.1.1 NANOPARTICLES

In general, NPs are nanostructured particles having their dimensions within a 100 nm scale and can be synthesized through a series of physical, chemical, or biological procedures [33]. The physical and chemical characteristics of NPs are distinct from those of their bulk counterparts (for example, mechanical strength, melting point, surface area, etc.). The ability to tailor their specific characteristics makes them attractive for different manufacturing requirements. The advancement in the field of new NPs increased rapidly in the last few decades. The categories are very broad in general and include metal oxide-based NPs, various quantum dots, silicon oxides, dendrimers, and certain layered assemblies. They can be classified based on their physicochemical characteristics, size, morphology, etc. A simple classification of NPs based on their organic origin is depicted in Figure 4.2.

4.2.1.2 FULLERENES

Fullerenes constitute the basic form of a hollow sphere, tube, ellipsoids, and different other forms of the carbon molecules, and they are naturally present in such carbon forms. They are comparable in design to graphite that is formed of a layer of hexagonal sheets, but comprising pentagonal (or seldom heptagonal) chains which stop the graphene layer becoming flat [35]. Spheres are known as buckyballs. CNT is one of the most studied forms of fullerene, which is nothing but cylindrical fullerene. Again, as it is a fullerene, the compositions of CNTs are comparable to that of graphite. Moreover, they were first produced by graphite destination via laser ablation in He gas. Combustion, thermal decomposition and heat plasma pyrolysis of hydrocarbons [36], and laser spraying are the other methods used for the fullerene manufacturing.

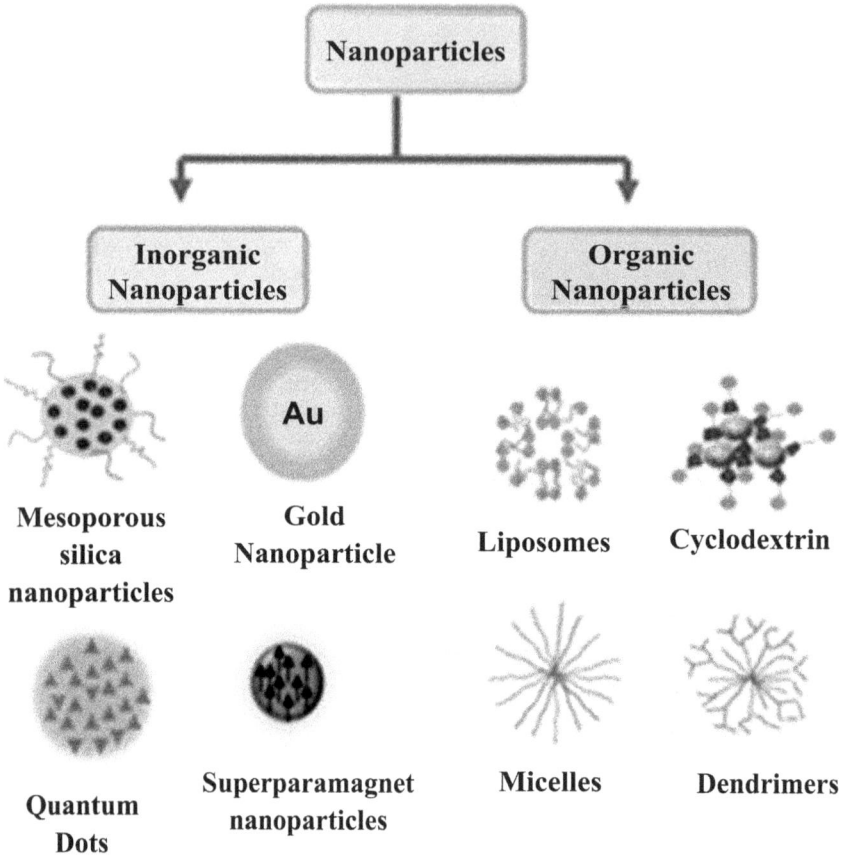

FIGURE 4.2 Classification of NPs [34].

Source: Reprinted with permission from Ref. [34]. © 2016 Springer Nature.

4.2.1.3 CARBON NANOTUBES

The lengthy slender cylinders of CNTs have been found by Ljima for the first time in Japan in 1991 with a technique of arc discharge. They are the unique type of fullerenes; composed of concentric multiwalled (MW) graphite sheets (MWCNTs) parted through van der Wall forces or of single-walled itself (SWCNTs). They are macromolecules of distinctive magnitude, form and display notable physical and mechanical characteristics, and are comparable in composition to C60 (buckyballs) having a configuration of the extended curved framework [34, 37]. The diameter of a CNT with single walls is 0.6–5 nm, whereas the internal diameter of a CNT with several walls is 1.5–15 nm while the exterior is of 2.5–50 nm diameter size. CNTs are generated in different sizes and proportions based on the production procedure [37]. CNTs have highly tunable heat, digital, and mechanical characteristics depending on the radius, width, chirality, or bend of the distinct types of the nanotube. They are an excellent illustration of real nanotechnology because they are formed in the 0.6 < diameter < 100 nm ranges.

4.2.1.4   NANOWIRES

It is the form of 1D material, in which quantum confinement is restricted at the nanoscale in another 2D. The motion of carriers is unrestricted along the length of the wire and is quantized at nanoscale along with the other two orthogonal directions of motion of carriers. It is possible to estimate the quantized energies along with the directions independently, which are perpendicular to the direction of motion of carriers (i.e., along the length of wire). The energy states are discrete along with the directions where electron confinement occurred. These quantized energies are added to energy contribution along the length of the wire. The NW is a special class of NR in which its diameter lies at the nanometer scale and the ratio of diameter to length is called the aspect ratio. The NWs have special attraction because it exhibits the unique physical and chemical characteristics that are different from those of the NPs. NWs can be made of metals or semiconductors or different kinds of semiconductor material within a single wire. The metallic or semiconducting NWs with a high aspect ratio can be fabricated for various applications. It can be used as nano-interconnections in electronic devices because it has a direction for efficient electron transport. Various deposition and suspension techniques are being used to prepare the NWs for numerous applications.

4.2.1.5   QUANTUM DOTS

Small semiconductor nanostructures in which the motion of carriers is quantized along all three directions with the energy states are discrete. Quantum dots have a diameter of 2–10 nm (10–50 atoms) with new optical, kinetic, and catalytic characteristics. The quantized energy range of quantities is discreet and their spatially confined analogous wave functions located in the quantum states spread over several periods.

4.2.1.6   METALLIC NPS

The metallic NPs are the nanosized metals whose length, breadth, and thickness within the nanometer scale. The quantum confinement occurs in all directions on metallic NPs. The metallic NPs exhibit a higher surface-to-volume ratio as compared to bulk counterpart with large surface energy make diffusion faster even at low temperature. It shows a large number of kinks [38], a large number of low coordination sites such as corner and edges, short-range ordering, and having a large number of dandling bonds available on the surface of metallic NPs. The metallic NPs exhibit an outstanding surface plasmon resonance and optical property. The metallic NPs can be stabilized in an inert atmosphere. The steric (polymeric) or electrostatic stabilization is employed to stabilize the metallic NP by coating of a capping agent. The metallic NPs are easily synthesized and reproducible using the minimum number of precursors, easily tuned in a particular shape and economical. Bottom-up and top-to-down-methods are used for the growth of metallic NPs. The Au and Ag metallic NPs exhibit the surface plasmon absorption that can change the optical property of metallic NPs. The metallic NPs of Pt or Rh with an average size of 1–20 nm dispersed in various substrates SiO_2 or Al_2O_3 have been used in heterogeneous catalysis from the last five decades [39–41].

4.2.1.7   CARBON BLACK

It is a colloidal particle form of solid carbon containing over 96% of pure carbon with a limited amount of O_2, N_2, H_2, and S to import the solubility or better dispersion, synthesized through an extremely controlled process. They are produced by two processes namely (1) the oil furnace processes and (2) the thermal processes. They are the wide range of particles having their sizes within 10–300 nm range, obtained from fossil (hydrocarbon) fuels by the anthropogenic combustion or thermal decomposition under a small amount of combustion air at a temperature of 1300–1500 °C [34, 39]. Carbon black is basically found useful to improve the optical, electrical, and physical characteristics of various classes of materials. They have also found applications in rubber compounds, printing, surface coating, etc., as reinforcing agents and black pigments.

4.2.2 NANOCLAYS

They are a category of mixed organic–inorganic nanomaterials composed of layered mineral silicates with a prospective use of rheological modifiers, gaseous absorbents, and medication suppliers in plastic nanocomposites. Based on their morphology and chemical composition, they can be classified into different categories like kaolinite, halloysite, montmorillonite, hectorite, bentonite, etc. Layered silicate (mixed silicate) is a common word for both real and synthetic silicates (laponite, montmorillonite, and hectorite). In a number of material implementations, montmorillonite is the most prevalent nanoclay. The plaque-like montmorillonite contains an aluminum-silicon coating of 1 nm thickness, with cations estimated in hundreds of nanometers in size and breadth [42, 43].

4.2.3 NANOCRYSTALS

They are crystalline-rich solids with a minimum of 1D in nanosize. The characteristics of nanocrystals can be altered based on the methods. They are available everywhere and can be found in solar cells and can be implemented on flexible substrates to fabricate many useful devices. They can be integrated into digital systems for energy-efficient applications, like LEDs, and used in filtering for raw petroleum reinforcement in gasoline engines. They are applicable in various other applications like catalysts, solar cells, detectors, etc.

4.2.4 NANOCOMPOSITES

They are composite material in which at least one phase or constituent is in the nanoscale or nanosized grout ingredients are supplied to the nanomaterial for improving the novel characteristics of the final products [44], and an overview of 2D nanocomposite and their usages can be seen in Figure 4.3. Nanocomposites are the collection of two/more discrete immiscible phases separated by one another by interface region and possess varying physical and chemical characteristics. They are also the best suitable emerging material with outstanding properties to overcome the limitation of microcomposites. However, maintaining the stoichiometry and elemental compositions of the material in the nanoscale is remains as a challenge in preparation techniques and requires many modifications in the preparation techniques to retain the elemental composition and stoichiometry of nanocomposite material.

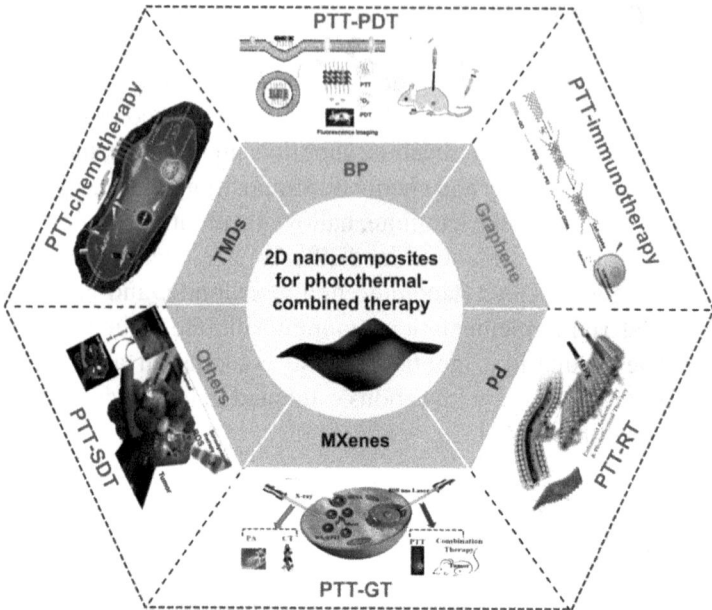

FIGURE 4.3   2D nanocomposites and their applications in one scheme [45].
Source: Reprinted with permission from Ref. [45]. © 2019 Elsevier.

The higher amount of constituent or phase that generally present is called the matrix. The mechanical property of nanocomposite material can be improved by adding another constituent into the matrix material through a process called reinforcement, and it can be identified in the form of nanosized filler materials. The surface-to-volume ratio of the reinforcement material utilized during the growth of nanocomposite is crucial to understand its structural characteristics. The nanocomposite exhibits the anisotropy in all crystallographic directions because the constituent or phases have distinct properties and inhomogeneous distribution of reinforcement.

Nanocomposite materials have more advantages over conventional composites [6, 46–48].

First, an additional small amount of nanofiller materials is required to improve the properties of the matrix in the nanocomposite. But, in the case of conventional composite material the higher concentration of NPs is necessary to improve the properties (see Figure 4.4 and Table 4.1).

Second, the nanocomposite materials are smaller in weight as compared to conventional composite materials due to the addition of little quantity of nanofiller material in the matrix of nanocomposite.

Third, compared to conventional composites, the size dependency of nanocomposite materials is more useful due to their enhanced mechanical, thermal, electrical, optical, chemical, and magnetic properties [49].

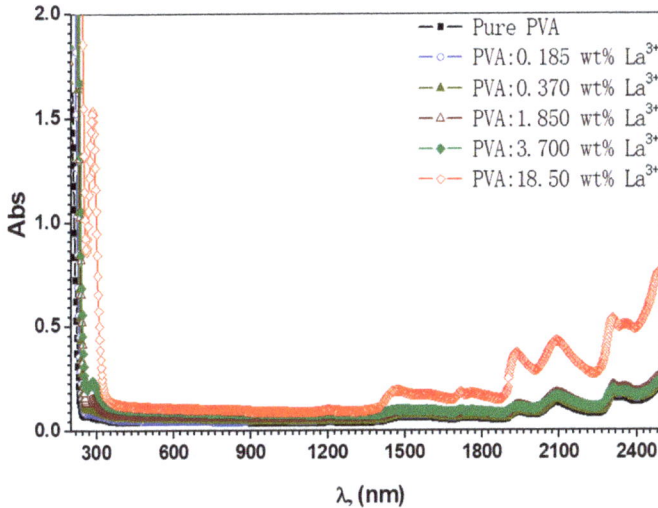

FIGURE 4.4 Absorbance spectra of La^{3+}/polyvinyl alcohol (PVA) polymeric composite films.

TABLE 4.1 Refractive Index and Energy Gap Parameters

Composition	E_{d1}, (eV)	E_{d2}, (eV)	E_{id1}, (eV)	E_{id2}, (eV)	n	E_d, (eV)	E_s, (eV)	S_o, (10^{-6} nm^{-2})	λ_o, (nm)	n_∞
Pue PVA	5.60	–	5.16	–	1.57	7.02	5.15	30.6	221	1.59
PVA+0.185 wt.% La^{3+}	5.48	–	5.14	–	1.65	7.58	5.14	29.3	231	1.60
PVA+0.370 wt.% La^{3+}	5.31	–	5.09	–	1.81	8.80	5.06	34.1	233	1.69
PVA+1.850 wt.% La^{3+}	5.19	–	4.96	–	1.86	10.32	4.96	37.4	241	1.79
PVA+3.700 wt.% La^{3+}	5.13	–	4.88	–	2.09	12.44	4.89	47.4	244	1.95
PVA+18.50 wt.% La^{3+}	4.89	4.07	4.56	3.65	2.46	15.93	4.69	64.0	245	2.20

Nanocomposite materials have several advantages over unadventurous composite material owed to their fascinating characteristics and they possess extensive applications in numerous fields. The higher surface-to-volume ratio of reinforcement NPs and superior aspect ratio of nanocomposite materials make them distinct from the conventional composite.

The reinforcement materials could be NPs (metallic/mineral/CNTs, etc.), nanofibers (electrospun nanofiber), or nanosheets (graphene, exfoliated clay stack), etc. The interface area connecting the nanoreinforcement and matrix is usually higher in order of magnitude as compared to that in conventional composite materials.

Clay minerals are nanofillers having wide applications in polymer nanocomposites due to their enhancement in properties in comparison to the conventional composite. In this new class of polymer–clay composites, the matrix is polymer and the dispersed phases constitute silicate NPs.

The polymer/clay nanocomposite is prepared by different synthesis methods. The first method is in-situ polymerizations wherein the polymerization between the clay layers is carried out under favorable conditions using a monomer clay dispersion medium. Clays hold higher surface energy and hence they attract monomer until reaching equilibrium and cause polymerization between the lower polarity layers, which eventually dislocate the equilibrium and diffuse newer polar species through the layer stacks [39].

In the second method, a polymer (or pre-polymer) soluble solvent is used, and single-layered silicate is exfoliated inside by means of the solvent. A solvent has higher entropy in which layered silicates can be effortlessly dispersed due to the disorganization of a layer that exceeds the organization's entropy of the lamellae. These delaminated layers are then served with polymer, and the layers filled with the polymer are unified by the precipitation of the mixture upon solvent evaporation [39].

4.2.4.1 TYPES OF NANOCOMPOSITES

Based on the matrix, the nanocomposites can be commonly categorized into three classes [47]:

1. polymer matrix nanocomposite (PMC),
2. ceramic matrix nanocomposite, and
3. metal matrix nanocomposite.

In polymer nanocomposite (PMC) material, the matrix is an organic polymer with nanoadditives used as reinforcing material. An example related to the conventional procedure and bioinspired procedures to obtain PMCs and related applications is displayed in Figure 4.5a and b, respectively. The elastic modulus and strength of stiffness for the nanoadditive material with small concentration are normally higher than the matrix materials. The nanoadditive materials can be 1D (NTs, NRs, fibers, etc.), 2D (layered structure like clay), or 3D (spherical NPs) structures. The polymer matrix composite offers high specific elastic modulus and strength that defines the ratio of modulus (or strength for specific strength) with the density of that material. Such type of characteristics of PMC can be used in aerospace structural materials and military aircraft. The properties of PMC are determined by the matrix, the reinforcement, and their interphase. The many physical parameters are to be considered to design the PMC. It not only relied just on the type of reinforcement and matrix but also the proportion, the morphology of reinforcement, and the kind of interphases. The PMC is strongest when stressed longitudinally parallel to the direction of the axis and becomes weakest on facing stress in the transverse direction perpendicular to the axis.

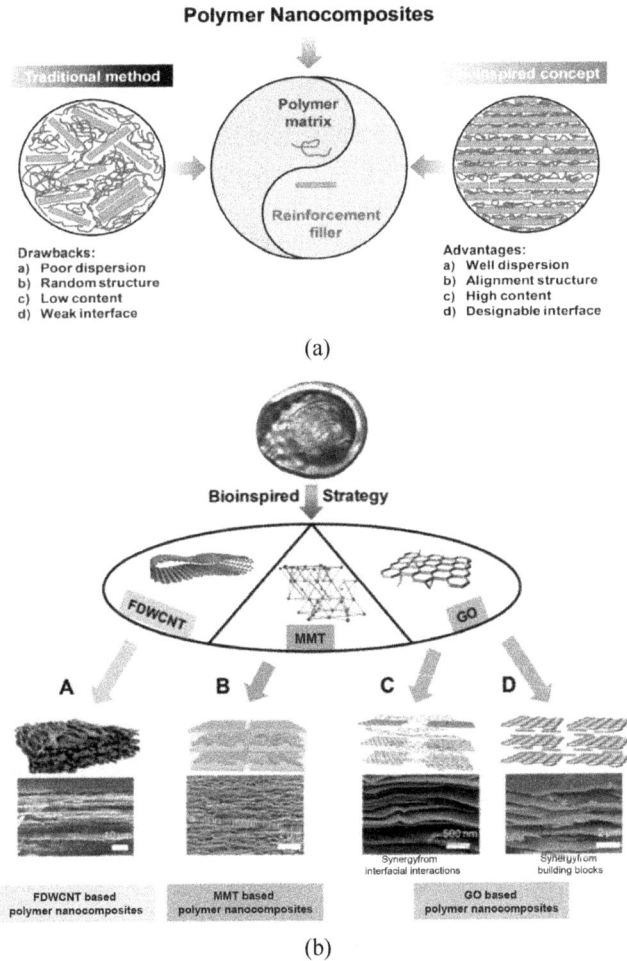

(a)

(b)

FIGURE 4.5 (a) Schemes for comparison between conventional and bioinspired synthesis of PMCs, and (b) bioinspired schemes for synthesizing PMCs [50].

Source: Reprinted with permission from Ref. [50]. © 2017 Elsevier.

The resistance of PMC is determined by the properties of the matrix to most deteriorative processes causing eventual structural failure. They can be impact failure, chemical variations, H_2O absorption, delamination, and temperature creeps. So, in general, the matrix could be considered as the weaker link in the polymer nanocomposite structure. However, compared to conventional composite, the polymer matrix in nanocomposite material exhibits enhanced barrier resistance, wear resistance and flame retardancy, and improvement in optical, electrical, and magnetic characteristics. PMCs include thermosetting and thermoplastic resin-based composites, one component PMCs, and polymer blended matrix composites.

A thermoplastic polymer constitutes a polyamide matrix with glass and carbon fibers as reinforcing components. Carbon fibers have found extensive use as a reinforcement element in the aerospace industry. A weak intermolecular force is developed between the polymer matrix and fillers. A remarkable improvement in the mechanical properties of a composite is developed if the filler dispersion happened at the atomic or molecular level. As we have mentioned earlier, the clay material (minerals) can be one of the best choices as a filler material in composite. Compared to typical silicate clay layers with 6.5 μm radius and 0.3 mm length, composed of single layers of thicknesses ~1 nm and high aspect ratio platelets of ~100 nm around them [51], glass fibers are ~4×10^9 times bigger. Though the polymers are inferior in their electrical, thermal, and mechanical properties in comparison with that of ceramic and metal, they are far superior in many other aspects like better durability, lightweight, low cost and easier processing, corrosion and heat resistance, ductility, lesser density, etc., as their backbone is hydrogen supported by lightweight C atoms with low-coordination number. These properties make them suitable for many applications in automobile, aerospace, defense, electronic and structural components, and construction materials [52].

The PMC doped with a little quantity (0.03%–0.04%) of mica-type silicate in epoxies has reported about 450% enhancement in the modulus values in the rubbery regions. It has been observed experimentally that the nanoscale additive material with higher surface energy alters the material property drastically. The additive nanoscale particle of high surface energy with anisotropy along the crystallographic directions in the polymer matrix lowers the interparticle separation resulting in the increased interaction strength in the polymer matrix and expands the performance space to replace conventionally filled polymers [6].

Now, when talking about reinforcement material, the properties of composite was found size-independent of reinforcement in the case of microcomposites but found size dependent in the case of nanocomposite. The nanocomposites are simply the final product of mixing fillers and polymer matrix. For example, Vollenberg and Heikens used 35- and 400-nm-sized alumina beads as fillers with styrene–acrylonitrile-copolymer, polystyrene, polypropylene, and polycarbonate [52].

4.2.4.2 PMC PROCESSING TECHNIQUES

Many processes have been employed to prepare the polymer nanocomposites including layered materials such as (1) direct polymer-filler mixing, (2) in-situ polymerization, (3) polymer intercalation using solution, (4) intercalation by melting, (5) template growth, and (6) sol–gel technique.

The following physical parameters are responsible to change the properties of nanocomposites drastically [53–57]. (1) Procedure adopted for the preparation, (2) size, shape, nature, and orientation of nanofillers, (3) characteristics and volume fraction of NPs, (4) degree of mixing, (5) nature of the interphase, (6) kind of adhesion at interphase, and (7) overall system morphology.

The properties of nanocomposite will be deteriorated drastically if the NPs are not evenly dispersed in the matrix as they tend to form NPs agglomerates which in turn act like defect centers deforming the properties of NPs and thereby nanocomposites.

The nature of filler-matrix interphase also plays a vital role in the enhancement of nanocomposite properties which is needless to say and is different from the individual properties of the filler and matrix in composition and microstructure. A good bonding between the filler and matrix at the interface yields nanocomposites with enhanced properties. The interfacial properties are dependent on the characteristics of the bound surfaces and hence the resulting nanocomposites' characteristics can be suitably controlled by the optimization of the filler-matrix interphase bonding. At interphase, the between the two is strongly depends on the individual surface energies and their ratio. Since the NPs possess high surface area and hence surface energy, their contribution to the interphase phenomenon is higher to enhance the nanocomposite characteristic properties [58, 59].

The chemical or mechanical process has also been employed to fabricate the PMC. The uniformity and homogeneity of NP are required to fabricate the PMC. In-situ polymerization has occurred via chemical processes and NPs or nanoplatelets are dispersed in monomers (matrix) where a molecule that is bound to another identical molecule form a polymer. The polymerization happens mainly between the intercalated sheets.

In a typical mechanical process, the solution mixing approach is followed for the direct polymer–nanoplatelet intercalation. In this process, the polymer and nanoplatelet sheets are dissolved one by one to mix in the solvent where they become swollen by displacing the solvent as a result of the interaction and intercalation of polymer chains in the solution into the nanoplatelet layers.

As already mentioned, nanofillers possess great nature to aggregate due to their high surface energy unless evenly dispersed in the polymer matrix or vice-versa, which deteriorate nanocomposite properties. Such types of difficulties can be minimized by making the polymerization reaction or surface modification of nanofiller materials [60]. Since the nanoplatelets are plate-like fillers, surface modifications become a requirement for the homogeneous dispersal in the matrix of polymers during the top-down intercalation approach. The monodisperse of nanoplatelets is achievable by (1) intercalation of melt and (2) in-situ polymerization methods [7, 35, 38].

First, the melt intercalation: this involves the direct mixing of the nanosized fillers into the polymer matrix at extremely high temperatures above the melting point. A static or under shear thermal annealing is employed on the polymer-nanofiber mixture.

Melt bending is a similar technique in which the melting of polymers (powder/pellet) to a viscous solution form happens first. The fillers are then diffused into this molten solution at a high shear rate. The compression and injection moldings and fiber production techniques are followed to give the final shape of the component.

Melt compounding is another approach in which nanofibers are added to polymer melt above T_c of glass. The viscous drag induces a hydrodynamic shear stress in the molten polymer which in turn results in the disruption of the nanofillers aiding their homogeneous dispersion inside the matrix.

Second, in an in-situ polymerization, a solution of low molecular weight monomers that can effortlessly permeate through the layers of nanofillers causes their (nanofillers') swelling [38]. The polymerization of this swollen mixture can be achieved by heat treatment, radiation exposure, organic initiator, or initiator diffusion. The polymerization of the monomers happens between interlayers resulting in either intercalated or exfoliated nanocomposite polymers.

Third, in a typical in-situ template synthesis process, the polymer and clays are dissolved in an aqueous medium causing the trapping of polymer chains between nanoclay layers. The nucleation and evolution of layered nanoclays on the polymer chain happens upon the application of elevated temperatures [7, 38].

Fourth, sol–gel technique follows a bottom-up approach. Here, a homogeneous colloidal suspension of NPs dispersed in a solution of monomers (matrix) forms the sol and the 2D integrated networks formed between phases by polymerizing reactions, and then the hydrolysis process forms the gel [7]. In this process, the polymer has the role of a nucleation agent as well as the growth limiter of NP crystal layers. During the growth process, the polymers seep inside the NP layers and control the further growth of NPs, and eventually form nanocomposite material.

Fifth, straightforward fraternization of polymers and nanofillers is key method of nanocomposite fabrication [7, 58]. Here, the mixing of the polymer and nanofillers can be done in two ways—without and with the involvement of a solvent/solution. The former approach follows the melt compounding technique explained in the previous page, wherein the polymer is mixed with nanofillers greater than T_c of glass [7, 58]. In the latter case, they are directly mixed in a suitable solvent to form a solution and the approach is known as solution mixing or solvent method [7].

Sixth, the solvent process comprises the NPs dispersal in a solvent and dissolving polymer in a cosolvent. Then, the solvent evaporation or coagulation technique is carried out to form the nanocomposite. Compared to melt compounding, the developed shear stresses in this approach are very small [7].

4.3 CONCLUSION AND FUTURE PROSPECTS

In this chapter, the latest developments in synthesis/fabrication of 2D NP and nanocomposites have been discussed comprehensively. It is well-known that the chemical and physical properties of NPs are significantly different from the bulk material. The optimizations of the physical parameters have been discussed in order to broaden the understanding of nanocomposite materials in the present study. In the previous section, the NPs are classified into four major sections according to their dimensions: 0D, 1D, 2D, and 3D nanostructures.

The synthesis and fabrication of 2D nanomaterials have been discussed in the second section. Besides this, the section also covers the physical and chemical routes for the synthesis of 2D polymeric nanocomposite materials. The involved elements and synthesis procedures of the nanocomposites require optimization to overcome the various factors limiting the optical and electrical properties of nanocomposites. The ongoing research works are, thus, on the exhaustive level to grow active nanoscale particles and composites. A novel but efficient growth process allows tuning of the particle size and morphology of resulting nanocomposites and any improvement in properties of nanocomposite in turn is helpful for its performance in a solar cell or other optoelectronic applications. The improvements in magnetic, electrical, and optical properties of nanocomposite materials are certain compared to conventional composite materials. The developments of 2D-based nano-materials are helpful for the progression from the old technologies, and the inspiring results in this field are beneficial to industries and society. For that matter, from a research point of view, we should also acquire grater experience regarding the physical and chemical characteristics of individual 2D-based nanomaterial.

ACKNOWLEDGMENT

The authors would like to express their gratitude to King Khalid University, Saudi Arabia, for providing administrative and technical support.

CONFLICT INTEREST

It is declared that there is no conflict interest in current work.

KEYWORDS

- 2D nanostructures/nanocomposites
- polymer nanocomposites 2D

REFERENCES

1. M. Mohsen-Nia, F.S. Mohammad Doulabi, PVAc Microspheres via Semicontinuous Emulsion Polymerization: Synthesis, Characterization, Kinetic, and Surface Morphology Studies. The Journal of Adhesion, 87(2011) 1020–1037.
2. B.K.G. Theng, Formation and Properties of Clay–Polymer Complexes. Amsterdam: Elsevier, 2012.
3. B.K.G. Theng, The Chemistry of Clay-Organic Reactions. London: Adam Hilger Ltd., Rank Precision Industries, 1974.
4. T.J. Pinnavaia, G.W. Beall, Polymer–Clay Nanocomposites. Hoboken, NJ: John Wiley, 2000.

5. L.A. Utracki, Clay-Containing Polymeric Nanocomposites. Shropshire: iSmithers Rapra Publishing, 2004.

6. Y.-W. Mai, Z.-Z. Yu, Polymer Nanocomposites. Cambridge: Woodhead Publishing, 2006.

7. M. Alexandre, P. Dubois, Polymer-Layered Silicate Nanocomposites: Preparation, Properties and Uses of a New Class of Materials. Materials Science and Engineering: R: Reports, 28 (2000) 1–63.

8. E.P. Giannelis, R. Krishnamoorti, E. Manias, Polymer-silicate nanocomposites: model systems for confined polymers and polymer brushes, in: Polymers in Confined Environments. Berlin: Springer, 1999, 107–147.

9. S.S. Ray, M. Okamoto, Polymer/Layered Silicate Nanocomposites: A Review From Preparation to Processing. Progress in Polymer Science, 28 (2003) 1539–1641.

10. P.C. LeBaron, Z. Wang, T.J. Pinnavaia, Polymer-Layered Silicate Nanocomposites: An Overview. Applied Clay Science, 15 (1999) 11–29.

11. F. Leroux, J.-P. Besse, Polymer Interleaved Layered Double Hydroxide: A New Emerging Class of Nanocomposites. Chemistry of Materials, 13 (2001) 3507–3515.

12. E.T. Thostenson, Z. Ren, T.-W. Chou, Advances in the Science and Technology of Carbon Nanotubes and Their Composites: A Review. Composites Science and Technology, 61 (2001) 1899–1912.

13. D.H. Solomon, D.G. Hawthorne, Chemistry of Pigments and Fillers. New York, NY: Wiley, 1983.

14. S. Al-Malaika, A. Golovoy, C.A. Wilkie, Chemistry and Technology of Polymer Additives. Oxford: Blackwell Science, 1999.

15. H. Karian, Handbook of Polypropylene and Polypropylene Composites, Revised and Expanded. Boca Raton: CRC Press, 2003.

16. C.A. Mitchell, J.L. Bahr, S. Arepalli, J.M. Tour, R. Krishnamoorti, Dispersion of Functionalized Carbon Nanotubes in Polystyrene. Macromolecules, 35 (2002) 8825–8830.

17. Y. Kojima, A. Usuki, M. Kawasumi, A. Okada, Y. Fukushima, T. Kurauchi, O. Kamigaito, Mechanical Properties of Nylon 6–Clay Hybrid. Journal of Materials Research, 8 (1993) 1185–1189.

18. Y. Kojima, A. Usuki, M. Kawasumi, A. Okada, T. Kurauchi, O. Kamigaito, Synthesis of Nylon 6–Clay Hybrid by Montmorillonite Intercalated with ε-caprolactam. Journal of Polymer Science Part A: Polymer Chemistry, 31 (1993) 983–986.

19. R.A. Vaia, H. Ishii, E.P. Giannelis, Synthesis and Properties of Two-dimensional Nanostructures by Direct Intercalation of Polymer Melts in Layered Silicates. Chemistry of Materials, 5 (1993) 1694–1696.

20. T. Lan, P.D. Kaviratna, T.J. Pinnavaia, Mechanism of Clay Tactoid Exfoliation in Epoxy-Clay Nanocomposites. Chemistry of Materials, 7 (1995) 2144–2150.

21. M.S. Wang, T.J. Pinnavaia, Clay–Polymer Nanocomposites Formed From Acidic Derivatives of Montmorillonite and an Epoxy Resin. Chemistry of Materials, 6 (1994) 468–474.

22. T.J. Pinnavaia, Intercalated Clay Catalysts. Science, 220 (1983) 365–371.

23. R. Krishnamoorti, R.A. Vaia, E.P. Giannelis, Structure and Dynamics of Polymer-Layered Silicate Nanocomposites, Chemistry of Materials, 8 (1996) 1728–1734.

24. E.P. Giannelis, Polymer Layered Silicate Nanocomposites. Advanced Materials, 8 (1996) 29–35.

25. S.D. Burnside, E.P. Giannelis, Nanostructure and Properties of Polysiloxane-Layered Silicate Nanocomposites. Journal of Polymer Science Part B: Polymer Physics, 38 (2000) 1595–1604.

26. R.A. Vaia, K.D. Jandt, E.J. Kramer, E.P. Giannelis, Microstructural Evolution of Melt Intercalated Polymer–Organically Modified Layered Silicates Nanocomposites. Chemistry of Materials, 8 (1996) 2628–2635.
27. R.A. Vaia, G. Price, P.N. Ruth, H.T. Nguyen, J. Lichtenhan, Polymer/Layered Silicate Nanocomposites as High Performance Ablative Materials. Applied Clay Science, 15 (1999) 67–92.
28. M.G. Kanatzidis, C.G. Wu, H.O. Marcy, D.C. DeGroot, C.R. Kannewurf, Conductive Polymer/Oxide Bronze Nanocomposites. Intercalated Polythiophene in Vanadium Pentoxide (V2O5) Xerogels. Chemistry of Materials, 2 (1990) 222–224.
29. Y.J. Liu, D.C. DeGroot, J.L. Schindler, C.R. Kannewurf, M.G. Kanatzidis, Intercalation of Poly(ethylene oxide) in Vanadium Pentoxide (V2O5) Xerogel. Chemistry of Materials, 3 (1991) 992–994.
30. Y.J. Liu, J.L. Schindler, D.C. DeGroot, C.R. Kannewurf, W. Hirpo, M.G. Kanatzidis, Synthesis, Structure, and Reactions of Poly(ethylene oxide)/V_2O_5 Intercalative Nanocomposites. Chemistry of Materials, 8 (1996) 525–534.
31. P.I. Dolez, Nanomaterials Definitions, Classifications, and Applications, in: Nanoengineering. Amsterdam: Elsevier, 2015, 3–40.
32. T. Hu, X. Mei, Y. Wang, X. Weng, R. Liang, M. Wei, Two-Dimensional Nanomaterials: Fascinating Materials in Biomedical Field. Science Bulletin, 64 (2019) 1707–1727.
33. R. Aitken, K. Creely, C. Tran, Nanoparticles: An Occupational Hygiene Review: Health Safety Executive, in, Research Report 274. London: HSE Books, 2004.
34. A. Mageswari, R. Srinivasan, P. Subramanian, N. Ramesh, K.M. Gothandam, Nanomaterials: classification, biological synthesis and characterization, in: Nanoscience in Food and Agriculture 3, Berlin: Springer, 2016, 31–71.
35. R.J. Reddy, Preparation, Characterization and Properties of Injection Molded Graphene Nanocomposites, Department of Mechanical, College of Engineering, Wichita State University, Wichita, KS, 2010.
36. H.-M. Yang, W.-K. Nam, D.-W. Park, Production of Nanosized Carbon Black from Hydrocarbon by a Thermal Plasma. Journal of Nanoscience and Nanotechnology, 7 (2007) 3744–3749.
37. S. Ranjan, N. Dasgupta, E. Lichtfouse, Nanoscience in Food and Agriculture. Berlin: Springer, 2016.
38. H. Kumar, N. Venkatesh, H. Bhowmik, A. Kuila, Metallic Nanoparticle: A Review. Biomedical Journal of Scientific & Technical Research, 4 (2018) 3765–3775.
39. S. Palsule, Polymers and Polymeric Composites: A Reference Series. Berlin: Springer, 2016.
40. N. Mizuno, M. Misono, Heterogeneous Catalysis. Chemical Reviews, 98 (1998) 199–218.
41. H.-B. Pan, C.M. Wai, Facile Sonochemical Synthesis of Carbon Nanotube-Supported Bimetallic Pt–Rh Nanoparticles for Room Temperature Hydrogenation of Arenes. New Journal of Chemistry, 35 (2011) 1649–1660.
42. J. Zhang, R.K. Gupta, C.A. Wilkie, Controlled Silylation of Montmorillonite and its Polyethylene Nanocomposites. Polymer, 47 (2006) 4537–4543.
43. H.A. Patel, R.S. Somani, H.C. Bajaj, R.V. Jasra, Nanoclays for Polymer Nanocomposites, Paints, Inks, Greases and Cosmetics Formulations, Drug Delivery Vehicle and Waste Water Treatment. Bulletin of Materials Science, 29 (2006) 133–145.
44. R. Asmatulu, W.S. Khan, R.J. Reddy, M. Ceylan, Synthesis and Analysis of Injection-Molded Nanocomposites of Recycled High-Density Polyethylene Incorporated with Graphene Nanoflakes. Polymer Composites, 36 (2015) 1565–1573.

45. X. Wang, L. Cheng, Multifunctional Two-Dimensional Nanocomposites for Photothermal-based Combined Cancer Therapy. Nanoscale, 11 (2019) 15685–15708.

46. R.A. Vaia, H.D. Wagner, Framework for Nanocomposites. Materials Today, 7 (2004) 32–37.

47. H.E. Ali, Y. Khairy, H. Algarni, H.I. Elsaeedy, A.M. Alshehri, H. Alkharis, I.S. Yahia, The Visible Laser Absorption Property of Chromium-Doped Polyvinyl Alcohol Films: Synthesis, Optical and Dielectric Properties. Optical and Quantum Electronics, 51 (2019) 47.

48. P. Liu, A.L. Cottrill, D. Kozawa, V.B. Koman, D. Parviz, A.T. Liu, J. Yang, T.Q. Tran, M.H. Wong, S. Wang, M.S. Strano, Emerging Trends in 2D Nanotechnology that are Redefining our Understanding of "Nanocomposites". Nano Today, 21 (2018) 18–40.

49. W.S. Khan, N.N. Hamadneh, W.A. Khan, Polymer nanocomposites–synthesis techniques, classification and properties, in: Science and Applications of Tailored Nanostructures. Cheshire: One Central Press (OCP), 2016.

50. C. Huang, Q. Cheng, Learning From Nacre: Constructing Polymer Nanocomposites. Composites Science and Technology, 150 (2017) 141–166.

51. D.D. Chung, Composite Materials: Functional Materials for Modern Technologies. Berlin: Springer Science & Business Media, 2013.

52. E.T. Thostenson, C. Li, T.-W. Chou, Nanocomposites in Context. Composites Science and Technology, 65 (2005) 491–516.

53. D.R. Paul, L.M. Robeson, Polymer Nanotechnology: Nanocomposites. Polymer, 49 (2008) 3187–3204.

54. K.I. Winey, R.A. Vaia, Polymer Nanocomposites. Mrs Bulletin, 32 (2007) 314–322.

55. J. Jordan, K.I. Jacob, R. Tannenbaum, M.A. Sharaf, I. Jasiuk, Experimental Trends in Polymer Nanocomposites—A Review. Materials Science and Engineering: A, 393 (2005) 1–11.

56. I.-Y. Jeon, J.-B. Baek, Nanocomposites Derived from Polymers and Inorganic Nanoparticles. Materials, 3 (2010) 3654–3674.

57. D. Ciprari, K.I. Jacob, R. Tannenbaum, Characterization of Polymer Nanocomposite Interphase and its Impact on Mechanical Properties. Macromolecules, 39 (2006) 6565–6573.

58. S.G. Miller, Effects of Nanoparticle and Matrix Interface on Nanocomposite Properties, University of Akron, Akron, OH, 2008.

59. M. Tanahashi, Development of Fabrication Methods of Filler/Polymer Nanocomposites: With Focus on Simple Melt-Compounding-based Approach Without Surface Modification of Nanofillers. Materials, 3 (2010) 1593–1619.

60. F. Yang, Y. Ou, Z. Yu, Polyamide 6/Silica Nanocomposites Prepared by In Situ Polymerization. Journal of Applied Polymer Science, 69 (1998) 355–361.

CHAPTER 5

3D Nanomaterials and Their Optoelectronic Applications

SAJID ALI ANSARI[1*] and NAZISH PARVEEN[2]

[1]*Department of Physics, College of Science, King Faisal University, P.O. Box 400, Hofuf, Al-Ahsa 31982, Saudi Arabia*

[2]*Department of Chemistry, College of Science, King Faisal University, P.O. Box 380, Hofuf, Al-Ahsa 31982, Saudi Arabia*

Corresponding author. E-mail: sansari@kfu.edu.sa.

ABSTRACT

Over the last few decades, various materials especially zero-, one-, and two-dimensional have attracted much attention due to its various exceptional properties compared to the bulk materials. However, these materials had various issues including structural and assembly inhomogeneity. Though, constructing three-dimensional (3D) using simple techniques has become the foundations of nanotechnology and nanoscience, which is due to the unique 3D structure. In addition, the development of 3D nanomaterials has one of the unique assemblies that have been used for various applications from energy to environmental applications in the past few years. In this chapter, we summarize the synthesis of various 3D material and its optoelectronic applications.

5.1 INTRODUCTION OF 3D NANOMATERIALS

Three-dimensional (3D) nanomaterials have become cornerstones of nanotechnology and nanoscience. Nanostructure science and technology has emerged as one of the broad and interdisciplinary areas of research and development (R&D) activities that have been increasing explosively across the world in the past few years [1]. Three-dimensional nanomaterialss have the potential to revolutionize the ways of developing and creating materials and products and the nature and range of accessing functionalities. Three-dimensional nanomaterials have a remarkable commercial impact, which is expected to increase significantly in the future [1].

Nanoscale materials, also known as nanomaterials, refer to substances that have their dimensions or at least one dimension, less than about 100 nanometers. A nanometer is equivalent to 1 millionth of a millimeter or about 100,000 times lesser than the radius of a human hair. Figure 5.1 shows a comparison of a single-walled carbon nanotube (CNT) with other objects. The use of nanomaterials is increasing fast for their unique optical, electrical, magnetic, and other properties. Gradually new properties of nanomaterials are emerging, which have the potential of revolutionizing and greatly impacting medicine, electronics, and other fields.

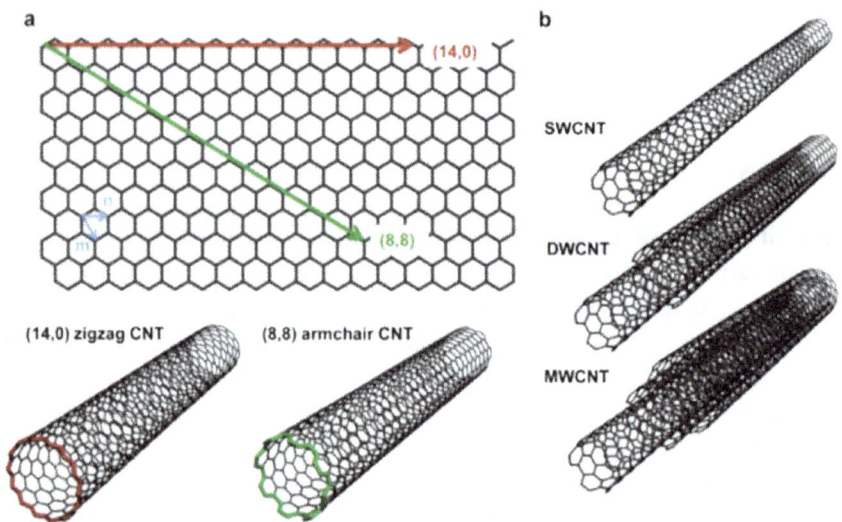

FIGURE 5.1   A comparison of a 3D single-walled CNT, made of nanomaterials with other objects [42].

Source: Reprinted with permission from Ref. [42]. © 2011 American Chemical Society.

The dimension of the structural elements of nanomaterials defines the main type of nanostructured material, such as zero dimensional (0D), one dimensional (1D), two dimensional (2D), and three dimensional (3D) nanomaterials. Zero-dimensional nanomaterials include nanodispersions and nanocluster materials, that is, materials that have isolated nanoparticles from each other [13]. Two-dimensional nanomaterials are used in films, multilayers, plates, or networks, while 3D nanomaterials refer to nanophase materials comprising equiaxed nanometer-sized grains.

Grieger et al. [13] underline that some of the nanomaterials occur naturally but most of them are engineered nanomaterials (ENs) that are developed and designed for particular purposes and are often used in commercial processes and products, such as sunscreens, sporting goods, cosmetics, tires, electronics, stain-resistant clothing, as well as most of the other products used in everyday life. Products and processes made of nanomaterials are also used for medicinal purposes, such as imaging, diagnosis, and drug delivery. ENs are resources developed at the molecular (nanometer) levels for taking advantage of their novel properties and small size, which their bulk and conventional counterparts do not have.

Lu [21] highlights the two main reasons responsible for the different specific properties of nanomaterials at the nanoscale. These two reasons include new quantum effects and increased relative surface area of the nanomaterials. The surface area to volume ratio of nanomaterials is greater than the conventional forms of nanomaterials. The higher surface area to volume ratio of nanomaterials leads to not only greater chemical reactivity but also affects their strength. Similarly at the nanoscale, the quantum effects of nanomaterials play a crucial role in determining the characteristics and properties of the materials to estimate their novel electrical, optical, and magnetic behaviors.

Nanomaterials are used commercially for many years or decades and there is a broad range of commercial nanomaterials that are used for producing wrinkle-free and stain-resistant textiles, sunscreens, cosmetics, electronics, varnishes, and paints. Besides, nanomaterials are also used in nanocomposites and nanocoatings that are further used for other consumer products and services, such as sports equipment, bicycles, windows, and automobiles. There are glass bottles that have novel ultraviolet (UV)-blocking coatings for protecting beverages from damages by sunlight. Long-lasting tennis balls are made by using nanoclay/butyl-rubber composites. Nanoscale titanium dioxide is used in cosmetics, self-cleaning windows, and sun-block creams, and nanoscale silica is also used in a wide range of products, such as dental fillings and cosmetics [21].

5.2 CLASSIFICATION OF NANOMATERIALS

There are many ways of classifying 3D nanomaterials. For example, in general terms, 3D nanomaterials, which are intentionally produced, are classified as follows:

1. carbon-based 3D nanomaterials,
2. metal-based 3D nanomaterials,
3. dendrimers, and
4. composites.

5.2.1 CARBON-BASED 3D NANOMATERIALS

These 3D nanomaterials mostly comprise carbon and take the form of hollow tubes, ellipsoids, or spheres. Ellipsoidal and spherical carbon 3D nanomaterials are known as fullerenes, while the cylindrical tubes are known as nanotubes. These 3D nanoparticles have multiple potential applications, such as stronger and lighter materials, improved films and coatings, and applications in electronics [12].

5.2.2 METAL-BASED 3D NANOMATERIALS

Metal-based 3D nanomaterials comprise quantum dots, nanosilver, nano-gold, and metal oxides, like titanium dioxide. Quantum dots are closely packed semiconductor crystals comprising hundreds or thousands of atoms but their size is only a few nanometers to a few hundred nanometers. The optical properties of quantum dots change with change in the size of quantum dots [12].

5.2.3 DENDRIMERS

Dendrimers refer to nano-sized polymers made of branched units. The surface of dendrimers contains a number of chain ends that are tailored for performing specific chemical functions. This property is also used for catalysis. The 3D dendrimers comprise interior cavities that enable us to place other molecules useful for drug delivery [12].

5.2.4 COMPOSITES

Composites refer to a combination of nanoparticles with other nanoparticles or with bulk-type large materials. Nanomaterials, such as nano-sized clays, are added to products, such as auto parts, packaging materials, and so on, for enhancing their mechanical, flame-retardant and thermal properties, and removing barriers.

The above mentioned intentionally produced 3D nanomaterials have many unique properties, which enable them to be used for different applications in a number of medical, commercial, environmental, and military sectors triggered by the novel electrical, magnetic, mechanical, thermal, catalytic, or imaging features of the nanomaterials. These 3D nanomaterials also have the potential to be used in multiple complex systems and nanostructures.

3D nanomaterials are of extremely small sizes that have at least one of the dimensions 100 nm or less. Grieger et al. [13] classify 3D nanomaterials on the basis of their dimension. For example, one dimension nanomaterials at the nanoscale, such as surface films, 2D nanomaterials, such as strands or fibers, and 3D nanomaterials, such as particles. These nanomaterials exist in a single, aggregated, fused, or agglomerated forms with tubular, spherical, and irregular shapes. Common types of 3D nanomaterials include dendrimers, quantum dots, nanotubes, and fullerenes. Nanomaterials are also applied in the field of nanotechnology, and they display different chemical and physical characteristics, such as CNT, silver nano, fullerene, carbon nano, photocatalyst, silica, and so on.

Lu [21] adds one more type of nanomaterials on the basis of their dimension—0D nanomaterials. Hence, according to Lu [21], nanomaterials are classified as follows:

- 0D nanomaterials,
- 1D nanomaterials,
- 2D nanomaterials, and
- 3D nanomaterials or nanostructures.

Figure 5.2 displays the image of different nanomaterials.

Zero-dimensional nanomaterials also refer to atomic filaments, clusters, and cluster assemblies, and their most common representations are the nanoparticles that are crystalline or amorphous and they exhibit various forms and shapes. One-dimensional nanomaterials refer to multilayers and their most common representation includes nanowires, nanotubes, and nanorods, which are crystalline or amorphous. Two-dimensional nanomaterials refer to ultrafine-grained over buried layers or layers and their most common representation includes nanolayers, nanofilms, and nanocoatings, which are crystalline or amorphous. Three-dimensional nanomaterials refer to nano-phase materials comprising equiaxed nanometer-sized grains. Their dimensions are not confined to the nanoscale. 3D nanomaterials contain bundles of nanowires, dispersions of nanoparticles, nanotubes, and multinanolayers.

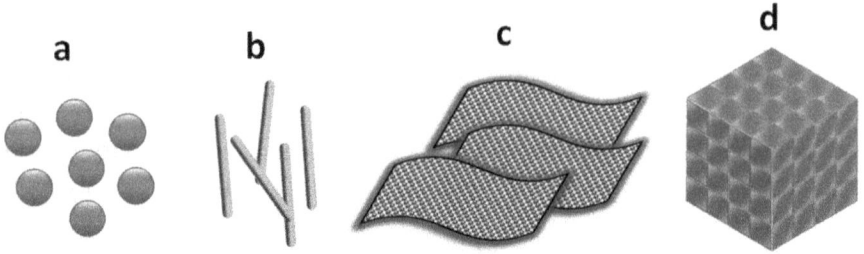

FIGURE 5.2   Classification of nanomaterials (a) 0D spheres and clusters; (b) 1D nanofibers, nanowires, and nanorods; (c) 2D nanofilms, nanoplates, and networks; (d) 3D nanomaterials.

Another explanation of different types of nanomaterials is: every dimension of 0D nanomaterials is at the nanoscale, one dimension of 1D nanomaterials is at the nanoscale, one dimension of the 2D nanomaterials is at the nanoscale, while two dimensions are at the macroscale. In the case of 3D nanomaterials, neither any dimension is at the nanoscale nor every dimension is at the macroscale.

Abraham [9] classifies nanomaterials as:

1. carbon black, CNTs, graphene, fullerene nanofibers, fullerene, and nanofibers,
2. silica fumes,
3. clay,
4. metal/alloys, and
5. ceramics.

Abraham [9] adds that 3D nanomaterials are used in rubber products/tyres, fillers, pigments, electronic components, semiconductors, pharma additives, polishing slurries, synthetic bone, polymer composites, and sunscreen lotion.

However, 3D nanomaterials are also classified as organic and inorganic 3D nanomaterials. Organic nanomaterials include micelles, liposomes, hybrid, dendrimers, and compact polymeric, while inorganic nanomaterials include fullerenes, silica, quantum dots, and gold nanomaterials [7].

5.3   HISTORY OF NANOMATERIALS AND THEIR IMPORTANCE

Nanoparticles are neither considered as a new scientific discovery of science nor they are considered as an innovation, and they are defined on the basis of their technical achievement or their particle size. They have always been part of the smoke. They have been used in the prehistoric cave paintings as carbon black particles and natural iron oxide, for example, in Altamira, and so on. Nano-sized particles are different from coarser particles, as nanoparticles have the increasing tendency of forming agglomerates. Macroscopically, agglomerates are perceived as one particle but they can be broken down into the primary particles in biological materials; however, this dissociation is often of toxicological relevance [12].

The history of nanomaterials is believed to be started just after the big bang with the formation of nanostructures in the early meteorites. Later on, many other nanostructures evolved by nature were found, such as seashells, skeletons, and so on. The formation of nanoscaled smoke particles took place when the early humans used fire [12]. However, the scientific story of nanomaterials started much later. One of the first known scientific studies of nanomaterials was done by Michael Faraday in 1857, when he synthesized the colloidal gold particles. The investigation of nanostructured catalysts has been done for more than 70 years. Manufacturing and sale of fumed and precipitated silica nanoparticles began in the USA and Germany by the early 1940s. These nanoparticles were manufactured as a substitute for ultrafine carbon black to be used for rubber reinforcements. This marks the beginning of the 3D nano-materials. The advent of 3D nanomaterials, for example, amorphous silica particles that were not only nano-sized to be used in a number of large-scale applications of day to day consumer products, which ranged from automobile tires to nondiary coffee creamer, catalyst supports, and optical fibers.

The development of remarkable 3D nanomaterials, metallic nanopow-ders, took place in the 1960s and 1970s, and these metallic nanopowders were useful for the magnetic recording tapes. The production of nanocrystals by using the popular technique of evaporating inert gas took place in the year 1976. Another remarkable finding of 3D nanomaterials is the fact that the Maya blue paint is made up of nanostructured hybrid materials. However, the origin of its ability to resist biocorrosion and acids and its color is yet to be understood, but a research study done on an authentic sample from Jaina Island shows that the hybrid materials are made of palygorskite (clay) crystals (needle-shaped) forming a superlattice having a period of 1.4 nm, and contains amorphous silicate substrate and metal (Mg) nanoparticles. The presence of superlattice and nanoparticles is responsible for the beautiful tone of the blue color, as shown by fabricating synthetic samples [12].

Presently, the nanophase engineering has expanded in an increasingly growing number of functional and structural materials, both organic and inorganic, which for manipulating mechanical, electric, catalytic, optical, magnetic, and electronic functions. The production of cluster-assembled or nanophase materials is done by creating separated small clusters that are fused into bulk-like materials or embedded into compact solid or liquid matrix materials, such as nanophase silicon, which is different from normal silicon in electronic and physical properties, and which can be used in macroscopic semiconductor process for creating new devices. For example, doping ordinary glass with quantized semiconductor "colloids" converts it into a high performing optical medium, which has the potential to be applied in optical computing [21].

Steels can be made more resistant to corrosion, stronger, or tougher if their composition is changed (by adding other elements or more carbon) or if their microstructures are modified. One of the extreme microstructural routes to strengthen materials is reducing their crystallite sizes from the micrometer scale ("coarse-grained") to the nanoscale. For example, nanograined copper (Cu) or aluminum becomes harder and tougher than high strength steels, but these materials are very brittle and easily crack due to deformed tension when pulled apparently because of localization of strain which resists deformation. However, nanograined metals may be deformed plastically by rolling at ambient temperature or under compression, it implies that moderate deformation may occur by suppressing the cracking process. Tremendous efforts are made for exploring the ways of suppressing strain localization in tensioned nanomaterials and making them ductile. Gradient microstructures, which have increased grain size at the surface and coarse-grain size in the core, are found to be an efficient effective approach used to improve ductility [21].

Nanomaterials played an important role in the successful production of Hydroxyapatite (HAp) from a recycled seashell, eggshell, and phosphoric acid. The phases obtained were dependent on the ratio of calcined seashell/eggshell to phosphoric acid, the mechano-chemical activation method (attrition milling or ball milling), and the calcination temperature. The factors which characterized the HAp structures included scanning electron microscopy, X-ray diffraction, and infrared spectroscopy. Compared to ball milling, attrition milling was found to more effective in yielding pure, homogeneous, and nanosize Hap [11].

The importance of nanomaterials has been increasing rapidly in the world leading to a substantial increase in its production. Presently, nanomaterials are widely used applications in optical, electronic, and biomedical fields. This has substantially increased the exposure toward nanoparticles, but there is a lack of respective occupational exposure limit [12]. Hence, it is imperative to ensure the occupational exposure limit so that workplace safety can be ensured. Additionally, there is a lack of appropriate risk assessment framework for the evaluation of risks associated with the use of 3D nanomaterials. Besides, the use of 3D nanomaterials also involves several challenges that need to be addressed, and then only, their use can be more beneficial for the humanity in the long run.

5.3.1 RISKS AND CHALLENGES ASSOCIATED WITH THE USE OF NANOMATERIALS

The use of 3D nanomaterials has become mandatory today and its global sale has crossed the mark of €500 billion in the year 2010 despite the fact that regulators and scientists are still struggling with many questions related to 3D nanomaterials. One of the key questions is the risk posed by nanomaterials to the environment and/or the human health. The only decision supporting tool to evaluate the risk posed by the 3D nanomaterials is the risk assessment framework that has been developed to evaluate the risks posed by chemicals. This framework underpins regulations that cover nanomaterials or provides information for making the case for regulatory amendment. This framework is widely used for many purposes, such as identifying the hazard, assessing hazard, and characterizing risk (including consideration of exposure). But this risk assessment framework has several associated uncertainties in the process, which need to be managed efficiently by applying safety factors. However, several scientists have acknowledged that the methodologies which underpin the conceptual risk assessment framework (e.g., Organisation for Economic Cooperation and Development Test Guidelines, EU Technical Guidance Document) may not be entirely suitable to assess the risk associated with the use of 3D nanomaterials, but the general consensus is that the chemical risk assessment framework is safe and sound, in principle, to be used for assessing the risk associated with nanomaterials, provided some modifications are done [13]. However, in principle, the risk assessment framework seems a sound approach for assessing the risk associated with the use of nanomaterials, but practically it is very difficult as it involves several very significant challenges that need to be overcome first [13].

5.4 DIFFERENT STRUCTURE AND COMPOSITION OF 3D NANOMATERIALS

The structure of the 3D nanomaterials is highly complex and it is often the result of gradual growth from a parent structure. Hierarchical nanostructures of 3D nanomaterials are quite complex and its construction is considered as an important topic in nanotechnology and nanoscience. The 3D nanostructures or nanomaterials are synthesized with the help of building blocks, such as organic molecules or biomolecules. In fact, there are many variants of 3D nanostructures or 3D nanomaterials, such as 3D oxide nanostructures, 3D hydroxide nanostructures, 3D sulfide nanostructures, 3D selenide nanostructures, 3D carbonate nanostructures, and so on.

Van Gough et al. [36] state that programming 3D nanostructures into materials have become very important, due to the increasing role of highly functional solids. Nanomaterials are specifically programmed and structured to be used in solar energy harvesting, molecular separation, energy storage, sensors, nanoreactors, pharmaceutical agent delivery, and advanced optical devices. In recent years, the development of science and technology enabled for programming properties and structures on the nanoscale with the precision needed for a wide range of important applications [17]. This development has also enabled for defining the complete 3D structures of functional solid, with nanometer accuracy, using different templating approaches. This rapid progress also enabled to design of colloidal and molecular templates for an inorganic solid at length scale varying from a few to hundreds of nanometers [16]. Despite the fact that technological developments enabled to form colloidal and molecular building blocks with increased complexity, the resultant self-assembled templates and structures often fail in generating the desired complex nanostructure from these materials. In such cases, there is a need to use other structure programming approach like multibeam optical interference for creating 3D nanostructured solids having unlimited structural complexity [14]. Optical interference has many abilities and one of them is creating a periodic structure by controlling the phase, polarization, direction, and intensity of the interfering laser beams. It has been observed that the minimum characteristic dimensions of the structure formed through optical interference are more than the structure formed through the molecular-based templating strategy [10].

5.5   3D NANOMATERIALS SYNTHESIS AND PROCESSING

5.5.1   *SYNTHESIS OF 3D NANOMATERIALS*

Various techniques are used to synthesize 3D nanomaterials and some of them are briefly described below.

In order to synthesize the 3D nanostructures, many techniques are used such as the gas diffusion technique, the hydrothermal technique, and so on. Secondly, the 3D nanomaterials are synthesized for various objectives that require various specific properties, such as the catalytic chemiluminescence properties, optical properties, superhydrophobicity, magnetic properties, photocatalytic activity, catalytic activity, flammability, and so on. Additionally, the synthesis of 3D nanomaterials is achieved to be applied in Ni-MH batteries, lithium-ion batteries, and so on. The above applications require different techniques of synthesizing 3D nanomaterials [7].

The synthesis of nanomaterials requires both "top-down" and "bottom-up" approaches as it is essential to deal with very fine structures and both the approaches are necessary either for assembling atoms together or disassembling (breaking or dissociating) bulk solids into fine pieces until they constitute only a few atoms. Hence, the synthesis of 3D nanomaterials is a pure and great example of one of the interdisciplinary works, which encompass chemistry, physics, and engineering [1].

There are many more techniques for synthesizing 3D nanomaterials. Some of the commonly used methods include the following.

5.5.2 MECHANICAL GRINDING OR MECHANICAL SYNTHESIS OF NANOMATERIALS

Mechanical attrition or mechanical grinding is a "top-down" method used to synthesize nanomaterials. In this method, the nanomaterial is prepared by structural decomposition of coarse-grained structure and not by cluster assembly [19].

5.5.3 WET CHEMICAL SYNTHESIS OF NANOMATERIALS

Wet chemical synthesis of nanomaterials is broadly classified as follows:

The "top-down" method—in this method, single crystals are etched to produce nanomaterials in an aqueous solution, for example, synthesizing porous silicon by using electrochemical etching [20].

The bottom-up method—this method includes sol–gel method, precipitation, and so on, where materials that contain the desired precursors are mixed in a controlled manner for forming a colloidal solution [1].

5.5.4 GAS-PHASE SYNTHESIS OF NANOMATERIALS

The gas-phase synthesis method is of great interest as it allows an elegant way of controlling process parameters for producing a chemical composition, size, and shape-controlled nanostructures [1].

5.5.5 FURNACE

This is one of the simplest ways of producing nanomaterials by heating the desired materials in a crucible, heat resistant, which contains the desired material. This method is adequate for materials having a high vapor pressure when the heated temperature is as high as 2000 °C [1].

5.5.6 FLAME-ASSISTED ULTRASONIC SPRAY PYROLYSIS

In this process of synthesizing nanomaterials, precursors are nebulized and a flame is used to burn unwanted components for getting the required material, for example, ZrO_2, which is obtained by using the precursor of Zr $(CH_3CH_2CH_2O)_4$. Another variant of flame assisted ultrasonic spray pyrolysis is flame hydrolysis that is used for manufacturing fused silica [1].

5.5.7 PROCESSING OF NANOMATERIALS

Various methods are used to process 3D nanomaterials and some of the key methods are briefly described below.

5.5.7.1 GAS CONDENSATION PROCESSING

In this processing technique, an inorganic or metallic material, such as a suboxide, is vaporized with the help of thermal evaporation sources like crucibles, sputtering sources, or electron beam evaporation devices or in an atmosphere containing 1–50 mbar He (or any other inert gas, such as Ar, Ne, Kr) [23].

5.5.7.2 CHEM'ICAL VAPOR CONDENSATION

The chemical vapor condensation (CVC) process or the CVC is intended for adjusting the parameter fields during the synthesis process for suppressing film formation and enhancing the homogeneous nucleation of materials in the gas flow [24].

5.5.8 SPUTTERED PLASMA PROCESSING

Sputtered plasma processing is also known as a variant of the gas condensation process except for the fact that the source materials are sputtering targets and these targets are sputtered utilizing rare gases and the constituents are enabled to agglomerate for producing nanomaterials. Both radio-frequency and direct current sputtering are used for synthesizing nanoparticles. Again multitarget sputtering or reactive sputtering is used for making alloys and/or carbides, nitrides, and oxides of nanomaterials. This method is particularly ideal for preparing nonagglomerated and ultrapure nanoparticles of metal [28].

5.5.9 MICROWAVE PLASMA PROCESSING

This technique is not very different from the CVC method, which has been discussed previously, except in this process plasma is employed in place of high temperature for decomposing the metal-organic precursors. In this method, microwave plasma is used in a quartz reaction vessel of 50 mm diameter and placed in a cavity with a connection with a microwave generator. A precursor-like a chloride compound is used in the front end of the reactor [30].

5.5.10 LASER ABLATION

Laser ablation is extensively used to prepare nanomaterials and particulate films. A laser beam is used in this process as a primary excitation source of ablation to generate clusters directly from solid samples in a number of applications [1].

5.6 DIFFERENT PROPERTIES OF 3D NANOMATERIALS

Three-dimensional nanomaterials have unique properties, such as large surface area, particle small size distribution, and so on, which enable nanomaterials to be used in different types of matrices like ceramic matrices, cementitious matrices, and polymeric matrices. The prime reason for dispersing nanomaterials into a matrix is improving the properties of the composites, such as thermomechanical properties, mechanical properties, and so on [7]. Besides the above-mentioned basic properties of nanomaterials, for example, they have the structural features in between the bulk materials and those of atoms. Although most of the microstructured materials have the same properties corresponding to the bulk materials, but the properties of materials having nanometer dimensions are remarkably different from the bulk materials and those of the atoms. This is primarily due to the fact that the nanometer size of the materials enables them:

1) large fraction of surface atoms,
2) high surface energy,
3) spatial confinement, and
4) reduced imperfections [32].

The above-mentioned properties are not found in the corresponding bulk materials. The small dimensions of nanomaterials enable them to have a very high surface area to volume ratio, which enables more "surface" dependent material properties. Specifically, when the size of nanomaterials is comparable to length, the surface properties of nanomaterials affect the entire material. This in turn enhances or modifies the properties of the bulk material. For instance, metallic nanoparticles are used as a very active catalyst. Chemical sensors from nanowires and nanoparticles enhance the sensor selectivity and sensitivity. The nanometer feature sizes of nanomaterials have spatial confinement effects on the material, which lead to the quantum effects [13].

The charge carrier density and energy band structure in the materials may be modified differently from the bulk and as a result, they modify the optical and electronic properties of the materials. For instance, light-emitting diodes (LEDs) and lasers from both of the quantum wires and quantum dots are very useful for the future optoelections. One of the fast-developing areas is the high-density information storage that uses the quantum dot device [32]. Reduction in the imperfections is also one of the important factors in determining the properties of the nanomaterials. These materials and nanostructures favor a self-purification process to remove the intrinsic material defects and impurities, upon thermal annealing, to near the surface. This increase in materials perfection greatly impacts the properties of nanomaterials. For instance, it enhances the chemical stability of several nanomaterials, it makes the mechanical properties of nanomaterials better and so on. The CNTs are known for their superior mechanical properties. The nanometer size of nanomaterials enables them to have multiple novel properties. These novel properties of nanomaterials enable them for multiple novel applications [1].

5.7 VARIOUS APPLICATIONS OF 3D NANOMATERIALS

In the case of cement matrices, various types of nanomaterials may be utilized as supplementary cementitious materials (SCMs) in cement mortars, pastes, and concretes. It has been found that the addition of nanomaterials greatly influences the performance of the materials containing cement as well as their microstructure [35]. This is possible due to many reasons such as the filler effect of nanomaterials of filling the voids available between cement grains. In addition, nanomaterials have the ability of participating in the pozzolanic reactions or accelerating such reactions. This leads to consuming calcium hydroxide and forming an "additional" hydrated calcium silicate (C–S–H) gel [32, 33]. Different nanomaterials are useful as SCM for cement paste, mortars, or concrete. The best improvement in the properties is the uniform dispersion of nanomaterials into the matrix. Well-dispersed nanomaterials work as the center of crystallization of cement hydrate leading to accelerate the hydrations of cement [2]. Depending on the quantity and type of nanomaterials, different mixing techniques are used for dispersing such nanoscale materials into cement composites. A number of studies have been done for exploring the dispersion method in the case of CNTs. One of the most popular methods is sonication [5], which has been employed by a group of researchers for sonicating CNTs and ordinary portland cement together in isopropanol and then the liquid is dried. However, this type of mixing involves several drawbacks such as it damages the surface grain of OPC, lowers the rate of hydration of cement [22]. In order to mitigate drawbacks, researchers suggested using dispersing agents, such as superplasticizers [26]. For increasing the repulsive force between adjacent colloidal particles, scientists suggest to add nanoparticles to water. In these cases, materials tend for agglomerating when they come in contact with water [7].

Sharma [31] underlines potential applications of nanomaterials in some important areas where adding nanomaterials enhances the existing systems. The identified important areas include electrical systems, biomedical systems, electronics, and energy efficiency.

5.7.1   ELECTRICAL SYSTEMS

Electrical systems refer to electrical components connected for carrying out some operations and they include batteries, solar cells, passive components, and so on. The common applications of nanomaterials to improve the performance of the conventional electrical systems are as follows:

1) nanobatteries,
2) surface coatings of electrical goods, and
3) fuel cells [37].

5.7.2   BIOMEDICAL SYSTEMS

Adding nanomaterials to a biomedical system improves the performance of the system by enabling early diagnostics, better medical treatment, and preventing deadly diseases. In a nutshell, they improve the overall efficiency of the biomedical systems [38, 39].

5.7.3   ELECTRONICS

Nanomaterials play a promising role in the area of electronic applications. The most common nanomaterial used for electronic applications is graphene that is used in communication devices, high-performance sensors, resonators, high-frequency amplifiers, and field-effect transistors. Addition of the nanomaterials significantly improves the performance of the electronic systems [41].

5.7.4   ENERGY EFFICIENCY

Nanomaterials also have the potential of enhancing energy efficiency and they are effectively used in various areas, such as power transmission, heat transfer systems, fuel cells, fossil fuels, gas turbines, and so on [31].

The emerging need for renewable energy and high-speed electronics has encouraged researchers for discovering, developing, and assembling new groups of nanomaterials in unconventional device architects. Carbon base nanomaterials, among these materials have attracted great attention from the researchers for their unique physical and structural properties. Carbon nano-materials are entirely composed of sp^2 bonded graphitic carbon, and they are often found in all reduced dimensions including 0D fullerenes, 1D CNT, and 2D grapheme [25]. These low-dimensional nanomaterials are grouped together to form 3D carbon nanomaterials. The first carbon nanomaterial was C_{60} (i.e., buckminsterfullerene), which was successfully isolated with the help of laser ablation of graphite in a very high flow of helium [3].

The sp^2 hybridized carbon nanomaterials have greatly delocalized electronic structure that is responsible for their use as high mobility electronic elements. Furthermore, the ability of tuning the band gap of semiconductor CNTs through controlling diameter brings unique opportunities to customize optical and optoelectronic properties of carbon nanomaterials [8]. Therefore, for this reason, carbon nanomaterials are often referred as a potential successor of conventional semiconductors, such as silicon in optoelectronic and electronic applications [4]. The following section presents an overview of employing 3D nanomaterials for optoelectronic applications.

5.8   3D NANOMATERIALS AND THEIR OPTOELECTRONIC APPLICATIONS

Tiwari et al. [34] state that 3D nanomaterials have become very important. They are used in a number of applications in catalysis, electrode, and magnetic fields. The superior properties and large surface area of 3D nanomaterials have enabled them to be used for optoelectronic applications. It is already known that the characteristics of 3D nanomaterials depend on their shape, size, morphology, and dimensionality, which are crucial for their ultimate performance. Synthesizing 3D nanomaterials with controlled morphology and structure is crucial for their optoelectronic applications. Kamanina et al. [18] add that optimization of the nanomaterials has paved the way for optoelectronic application of these materials that have optimized the inorganic system's mechanical hardness and increased the organic system's photorefractive parameters. In fact, the 3D nanomaterials have drastically improved the inorganic material's surface mechanical properties. There are different types of 3D nanomaterials, which can be used for optoelectronic applications, such as CNTs,

5.9   CNTS FOR OPTOELECTRONICS

CNTs have been found to be one of the ideal 3D nanomaterials for the optoelectronic applications. As discussed above, CNTs are 1D nanomaterials; therefore, they are bundled together to make 3D nanomaterials so that their performance in optoelectronic applications is improved. Semiconducting CNTs refer to direct band gap materials incorporating a wide range of optoelectronic devices, like light detectors, transparent conductors, and light emitters. As discussed above, van Hove singularities in the 3D density and strongly bound excitons are responsible for making CNTs an ideal alternative for optoelectronic applications. It must be noted that the exciton binding energy is dependent on the diameter of the 3D CNTs as well as the surrounding environment. The earlier studies on CNT optical excitations were done in dispersed aqueous solution which contained individual 3D CNTs coated with surfactants. In most of the cases, excitons were generated by photoexcitation in the second sub-band followed by radioactive decay to the first sub-band. 2D plots of photoluminescence as a function of emission energies and excitation presents peaks that uniquely identify the chirality of the CNT.

Besides photoluminescent applications, 3D semiconducting CNTs are ideal materials for electroluminescent and photocurrent devices. For instance, 3D CNTs have been effectively used as the basis of photodetectors and light-emitting transistors. The early studies of electroluminescence in ambipolar 3D CNT presented a wide range of interesting observations:

1. Unlike traditional p–n junctions, ambipolar 3D CNT's electroluminescence does not need extrinsic doping.
2. The off state demonstrated maximum electroluminescence efficiency.
3. Polarization of the emitted light and the 3D CNT axis were parallel to each other.

Additionally, the positions of light emitted from the 3D CNT's channel may be tuned by recombining site through biasing conditions, as shown later in different studies. The applications of semiconducting 3D CNT to photodetector have been extensively studied. Photodetection is an inverse of electroluminescence, in which optically generated exciton is separated into holes and free electrons for producing photocurrents with applied fields or open circuit photovoltages in an asymmetric configuration of field. 3D CNT has been considered as a component of nonlinear optics and it is observed that dispersed 3D CNTs serve as saturated absorbers having up to 40% passive mode-lockers and attenuation efficiency for femtosecond lasers. In a nutshell, for instance, the advantages of 3D CNTs in the optoelectronic application are that they are capable of offering facile fabrication having wavelengths which can be widely tuned on the basis of the 3D CNT diameter.

5.10 GRAPHENE FOR OPTOELECTRONICS

The specific attributes of graphene make it an ideal nanomaterial for optoelectronic applications. They are bundled together to make 3D nanomaterials having linear dispersion attribute with zero band gap that ensures the possibility of optical excitations which can be widely tuned. A single layer of graphene is capable of absorbing 2.3% of the incident light having the least reflection (<0.1%) over this range of wavelength. However, graphene is not luminescent in itself, but graphene oxide (GO), one of the chemical derivatives grapheme, exhibits photoluminescence over a broader range. As one of the light-emitting materials, graphene has limited potential, but as a transparent conductor, it is very effective in organic TFTs, organic photovoltaics and flexible organic LEDs. Graphene-based transparent conductors are capable of showing performance metrics closer to the performance metrics of indium tin oxide plus superior mechanical flexibility. Graphene is also used in photodetectors in which optically generated pairs of electron–hole are separated by an external applied bias. Unlike conventional bulk semiconductors and 3D CNTs, the linear dispersions of graphene provide uniform photoresponses from the THz to the ultraviolet range. Additionally, the high mobility of graphene is capable of yielding ultrafast photodetection as much as 40 GHz. Since the parasitic capacitance limited the operational speed in the early report, it is expected that the intrinsic photoresponse speed can be as much as 500 GHz. The internal built-in electric fields at metal–graphene contacts are capable of an efficient way of achieving photodetection. Metal–graphene contacts show 15%–30% internal photocurrent efficiency and 6 mA/W external photocurrent yield. In the recent years, photo responses were also found in a p–n junction, which were created through top-gate architecture on graphene. This device is designed to exploit hot carriers in graphene, the nonlocal transports of hot carriers contribute to the photo response other than the anticipated photovoltaic effect. Bioimaging is one of the crucial related optical applications of graphene-derived materials. Photoluminescence in GO is used for live-cell imaging. In the case of vivo applications, issues related to the potential toxicity of graphene are needed to be addressed, as suggested in a recent report, the encapsulation of graphene in the biocompatible block copolymer Pluronic is capable of reducing inflammation and toxicity in the lungs of mice. This biocompatible dispersion method provides an additional opportunity for graphene-based biomedical applications, which include drug delivery and imaging contrast agents.

Study and optimization of new nanoscale materials useful for the optoelectronic application have been considered.

The 3D nanomaterials are ideal for optoelectronic applications due to the optimized mechanical hardness of their inorganic systems and increased photorefractive parameters, which drastically improve the treatment process of the 3D nanomaterials. For instance, the surface mechanical hardness of the infrared and UV range soft materials is enhanced up to 3–10 times for optoelectronic applications [6]. The 3D nanomaterials are effectively used in the modern semiconductor optoelectronic devices triggered by engineering electrons, photons, and thermal properties of the 3D nanomaterials at the nanolevel [29]. Reich et al. [27] add that the new hybrid 3D nanomaterials have optoelectronic properties which make it ideal for not only optoelectronic applications but also photovoltaic applications. The unusual quantum effects of 3D nanomaterials at the nanoscale have paved the way for an excellent performance of optoelectronic devices [15]. The use of nanodome structure and grapheme/ZnO nanocomposite provided an antireflection layer for solar cells, which decreased the Fresnel reflection and enhanced the transmitted light through the solar cell. These developments paved the way for the efficient use of 3D nanomaterials for optoelectronic devices [40].

5.11   CONCLUSION

Three-dimensional nanomaterials play a crucial role in nanotechnology and nanoscience. There are different types of nanomaterials, such as 0D nanomaterials, 1D nanomaterials, 2D nanomaterials, and 3D nanomaterials, and all of them contribute significantly to the development of nanostructure science and nanotechnology, which have emerged as one of the broad and interdisciplinary areas of R&D activities. Nanomaterials have the potential of revolutionizing the ways of developing and creating materials and products and the nature and range of accessing functionalities. Nanomaterials are materials having at least one of their dimensions, less than about 100 nanometers. The classification of nanomaterials is done on the basis of their dimensions and nanostructured materials, such as 0D, 1D, 2D, and 3D nanomaterials.

Nanomaterials not only occur naturally but they are also engineered and ENs are developed and designed for specific purposes and commercial objectives. Nanomaterials are used in various processes and products, such as sunscreens, sporting goods, cosmetics, tires, electronics, stain-resistant clothing, and so on. The specific properties of nanomaterials depend on their new quantum effects and increased relative surface area. The specific properties of nanomaterials enable them for various applications, such as cement matrices for improving performance of the cementitious materials, electrical systems like batteries, solar cells, passive components, and so on, biomedical system to improve the performance of the system like diagnostics, medical treatment, and preventing deadly diseases, electronic applications for improving the performance of communication devices, high-performance sensors, resonators, high-frequency amplifiers, and field-effect transistors and also enhancing the energy efficiency of power transmission, heat transfer systems, fuel cells, fossil fuels, gas turbines, and so on.

The emerging need for renewable energy and high-speed electronics led to the development of new groups of nanomaterials based on carbon. Carbon-based nanomaterials are widely used for optoelectronic applications for their greatly delocalized electronic structure that makes them suitable for high mobility electronic elements. The superior properties and large surface area of 3D nanomaterials have enabled them to be used for optoelectronic applications.

DECLARATION OF INTEREST

The authors report no conflicts of interest.

ACKNOWLEDGEMENTS

This study was supported by King Faisal University, Saudi Arabia.

KEYWORDS

- **nanomaterials**
- **optical properties**
- **optoelectronic devices**

REFERENCES

1. Alagaraci, A. (2016). *Introduction to Nano Materials*. Madras: Indian Institute of Technology.
2. Al-Safy, R. A. (2015). Effect of incorporation techniques of nanomaterials on strength of cement-based materials. *The 2nd International Conference of Buildings, Construction and Environmental Engineering, BCEE2*, Beirut.
3. Al-Safy, R. A., Al-Mahaidi, R., Simon, G. P., & Habsud, J. (2012). Experimental investigation on the thermal and mechanical properties of nanoclay—modified adhesives used for bonding CFRP to concrete substrates. *Construction and Building Materials*, *28*(10), 769–778.
4. Ashby, M. F., Ferreira, M. F., & Schodek, D. L. (2009). *Natomaterials, Nanotechnologies and Design*. Oxford, UK: Elsevier Ltd.
5. Bharj, J., Singh, S., Chander, S., & Singh, R. (2014). Role of dispersion of multiwalled carbon nanotubes on compressive strength of cement paste. *International Journal of Mathematical, Computational, Natural and Physical Engineering*, *8*(2), 340–343.
6. Chen, C. (2012). *Nanomaterials in Optics, Electronics and Energy Applications*. Taiwan: National Taiwan University.
7. Darweesh, H. H. (2018). Nanomaterials: classification and properties—part I. *Nanoscience*, *1*, 1–11.
8. Darweesh, H. M. (1992). *Utilization of Cement Kiln Dust in Ceramics to Minimize Environmental Pollution*. Institute of Environmental Studies and Researches, Ain Shams University, Cairo.
9. Abraham, T. (2012). *Nanotechnology and Nano Materials: Types, Current/Emerging Applications and Global Markets*. Innovative Research and Products, Inc., Stamford.
10. Drexler, K. E., Peterson, C., & Pergamit, G. (1991). *Unbounding the Future: The Nanotechnology Revolution*. New York, NY: William Morrow.
11. Gergely, G., Wéber, F., Lukács, I., Illés, L., Tóth, A. L., Horváth, Z. E., et al. (2010). Nano-hydroxyapatite preparation from biogenic raw materials. *Central European Journal of Chemistry*, *8*(2), 375–381.
12. German Research Foundation. (2013). *Nanomaterials*. Weinheim: WILEY-VCH Verlag GmbH & Co. KGaA.
13. Grieger, K. D., Owen, R., & Baun, A. (2010). Redefining risk research priorities for nanomaterials. *Journal of Nanoparticle Research*, *12*, 383–392.
14. Guo, W., Han, L., Xia, R., Cui, F., Chen, S., Ma, J., et al. (2015). Repair of mandibular critical-sized defect of minipig using in situ periosteal ossification combined with mineralized collagen scaffolds. *Journal of Biomaterials and Tissue Engineering*, *5*, 439–444.
15. Harris, P. J. (2011). *Carbon Nanotube Science: Synthesis, Properties and Applications*. Cambridge: Cambridge University Press.
16. He, W., Li, W., Su, J., Wang, H., Ye, Z., Cai, M., et al. (2015). In vitro evaluation of the cytocompatibility of an acellular rat brain matrix scaffold with neural stem cells. *Journal of Biomaterials and Tissue Engineering*, *5*, 628–634.
17. Jiang, X., Wu, H., Zheng, L., & Zhao, J. (2015). Effect of in-situ synthesized nano-hydroxyapatite/collagen composite hydrogel on osteoblasts growth in vitro. *Journal of Biomaterials and Tissue Engineering*, *5*, 523–531.
18. Kamanina, N. V., Vasilyev, P. Y., Serov, S. V., Uskokovic, D. P., Savinov, V. P., & Bogdanov, K. Y. (2010). Nanostructured materials for optoelectronic applications. *Acta Physica Polonica A*, *117*(5), 786–790.

19. Kumar, M. S., Raju, N. M., Sampath, P. S., & Jayakumari, L. S. (2014). Effect of nanomaterials on polymer composites—an expatiate view. *e-Journal of Reviews on Advanced Materials Science, 38*, 40–45.

20. Liu, J., Mao, K., Wang, X., Guo, W., Zhou, L., Xu, Z., et al. (2015). Calcium sulfate hemihydrate/mineralized collagen for bone tissue engineering: in vitro release and in vivo bone regeneration studies. *Journal of Biomaterials and Tissue Engineering, 5*, 267–274.

21. Lu, K. (2014). Making strong nanomaterials ductile with gradients. *Science, 345*, 1455–1456.

22. Makar, J. M., & Chan, G. W. (2009). Growth of cement hydration products on single walled carbon nanotubes. *Journal of the American Ceramic Society, 92*, 1303–1310.

23. Morsy, M. S., Alsayed, S. H., & Aqel, M. (2010). Effect of nano-clay on mechanical properties and microstructure of ordinary portland cement mortar. *International Journal of Civil & Environmental Engineering, 10*(1), 21–25.

24. Nalwa, H. S. (2000). *Handbook of Nanostructured Materials and Nanotechnology.* San Diego, CA: Academic Press.

25. NSTC. (2007). *The National Nanotechnology Initiative—Strategic Plan.* Washington, DC: Executive Office of the President of the United States.

26. Qing, Y., Zenan, Z., Li, S., & Rongshen, C. (2008). A comparative study on the pozzolanic activity between nano-SiO_2 and silica fume. *Journal of Wuhan University of Technology—Materials Science Edition, 21*, 153–157.

27. Reich, S., Thomsen, C., & Maultzsch, J. (2004). *Carbon Nanotubes: Basic Concepts and Physical Properties.* Weinheim: Wiley-VCH.

28. Sadiq, M. S. (2013). *Reinforcement of Cement-based Matrices with Graphite Nanomaterials.* East Lansing, MI, USA: Michigan State University.

29. Saito, R., Dresselhaus, G., & Dresselhaus, M. S. (1998). *Physical Properties of Carbon Nanotubes.* London: Imperial College Press.

30. Samal, S. S., & Bal, S. (2008). Carbon nanotube reinforced ceramic matrix composites—a review. *Journal of Minerals & Materials Characterization & Engineering, 7*(4), 355–370.

31. Sharma, P. (2015). Potential applications of nanomaterials. *International Journal for Research in Applied Science & Engineering Technology, 3*(IX), 302–304.

32. Sobolev, K., & Ferrada-Gutiérrez, M. (2005). How nanotechnology can change the concrete world: part 1. *American Ceramic Society Bulletin, 84*(10), 14–17.

33. Sobolev, K., & Ferrada-Gutiérrez, M. (2005). How nanotechnology can change the concrete world: part 2. *American Ceramic Society Bulletin, 11*, 16–19.

34. Tiwari, J. N., Tiwari, R. N., & Kim, K. S. (2012). Zero-dimensional, one-dimensional, two-dimensional and three-dimensional nanostructured materials for advanced electrochemical energy devices. *Progress in Materials Science, 57*, 724–803.

35. Tseng, T. Y., & Nalwa, H. S. (2015). *Handbook of Nanoceramics and Their Based Nanodevices, Volume 1–5.* Los Angeles, CA: American Scientific Publishers.

36. Van Gough, D., Juhl, A. T., & Braun, P. V. (2009). Programming structure into 3D nanomaterials. *Materials Today, 12*(6), 28–35.

37. Venkatesan, J., Jayakumar, R., Anil, S., Chalisserry, E. P., Pallela, R., & Kim, S. (2015). Development of alginate-chitosan-collagen based hydrogels for tissue engineering. *Journal of Biomaterials and Tissue Engineering, 5*, 458–464.

38. Wang, J., Cheng, N., Yang, Q., Zhang, Z., Zhang, Q., & Biomater, J. (2015). Double layered collagen/silk fibroin composite scaffold that incorporates TGF-1 nanoparticles for cartilage tissue engineering. *Journal of Biomaterials and Tissue Engineering, 5*, 357–363.

39. Wei, X., He, K., Yu, S., Zhao, W., Xing, G., Liu, Y., et al. (2015). RGD peptide modified poly(lactide-*co*-glycolide)/-tricalcium phosphate scaffolds increase bone formation after transplantation in a rabbit model. *Journal of Biomaterials and Tissue Engineering,* *5*, 378–386.
40. Wong, H. S., & Akinwande, D. (2011). *Carbon Nanotube and Graphene Device Physics.* Cambridge: Cambridge University Press.
41. Yuvarani, I., Senthilkumar, S., Venkatesan, J., Kim, S., Al-Kheraif, A. A., Anil, S., et al. (2015). Chitosan modified alginate-polyurethane scaffold for skeletal muscle tissue engineering. *Journal of Biomaterials and Tissue Engineering, 5*, 665–672.
42. Schnorr J. M., & Swager T. M. (2011). Emerging applications of carbon nanotubes. *Chemistry of Materials, 23*, 646–657.

CHAPTER 6

Advances of Nanostructured Thin Films and Their Optoelectronic Applications

M. ASLAM MANTHRAMMEL*, MOHD SHKIR*, and SALEM A. ALFAIFY

Advanced Functional Materials and Optoelectronics Laboratory (AFMOL), Department of Physics, King Khalid University, Abha 61413, Kingdom of Saudi Arabia

Corresponding author. E-mail: muhd.aslam@gmail.com.

ABSTRACT

This chapter emphasizes the basics and synthesis of nanostructured thin films using different techniques. The properties that were specific to the material turn out to be size dependent in nanoscale. Thus, the structural, electrical, optical, and morphological characteristics of the sample show different behavior that can be highly tuned for the device-oriented properties depending upon the growth mechanism involved. In this chapter, we will deal more with such changes regarding the samples grown in thin film form. Fundamentals behind the various vacuum and wet chemical techniques for the nanothin film synthesis are discussed. This chapter also covers the substrate selection as well as the importance of tuning the synthesis procedures for device-oriented applications. The essential optical characterization details have been briefed out at the end to determine the band structure of semiconductor thin film samples.

6.1 INTRODUCTION

Though the thin film technology is not at all a new phenomenon for the research community, the invention of nanotechnology has brought it to new heights making it old wine packed in a new bottle. The technology can be broadly referred as surface engineering where the material layers define the surface with thicknesses ranging from nanometers to few micrometers. The technology is useful to alter the chemical as well as physical characteristics and morphologies of the substrates. The thin films have found huge applications in our day-to-day life, and they are not limited to sensors, photovoltaics, photocatalysis, self-cleaning planes, chip systems, and many other optoelectronic and energy generations and storage devices. The choice of the application is depending upon their growth conditions and film properties, such as optical, structural, electrical, magnetic, and electron transportation behaviors. The common synthesis techniques involve vapor deposition techniques, such as physical vapor deposition, chemical vapor deposition, and wet chemical techniques such as chemical bath deposition (CBD) and sol–gel procedure.

In the current era, the technology is progressively evolving with newer thin film materials and advanced characterization methods offering engineered micro- and nano-structures having highly tunable properties. It means the coating procedures also have to be rapidly progressing in order to meet up with the innovative requirements demanding improved performance on application levels. A mere comparison of the film characteristics of the same material with that of just one decade old will tell you how fast the technology is advancing. The idea is to produce more innovative nanostructures like nanocomposites, nanolaminates, and superlattices with cost-effectiveness. These can be in the forms of nanoparticles, nanowires (NWs), nanorods (NRs), or nanotubes grown in a controlled manner in thin film structure offering different properties. The NWs and NRs are one-dimensional (1D) nanostructured growth having an arbitrary length, but diameter restricted in few nanometre scales. Nanotubes are also falling in this 1D nanostructured growth, but rather in a tube shape. They can be grown vertically aligned, or in a randomized manner. The NRs sometimes take the shapes of spherical agglomerates or even in a beautiful flower kind of growth. The quantum dots (QDs) are 0D nanomaterials having their sizes <10 nm. However, they are not usually directly part of a thin film system but are sometimes intentionally incorporated into the nanostructured films in order to tune the film properties for the specific needs. For example, successive ionic layer adsorption and reaction (SILAR) technique is used to incorporate CdS, PbS, CdSe, etc., QDs on the TiO_2 matrix for QD-sensitized solar cell application [1]. The thin film can be a monolayer or a multilayer. The monolayers are usually composed of a homogeneous distribution of crystallites or nanostructures whereas the multilayers can be composites and inhomogeneous depending upon the applications. The tandem solar cells are the best example of the multilayered thin film. In a multilayer thin film, the layers are chosen such that each layer can contribute for a single application in a different manner and the layer properties are tailored beforehand for the same. They can be periodic or randomly defined patterns.

6.2 THIN FILMS VERSUS BULK MATERIALS

Compared to their bulk counterparts, thin film forms have many advantages. The thin films properties can be highly tailored by varying their thicknesses, substrate material, and other deposition parameters to alter their physical, electrical, and optical characteristics. In particular, nanostructured thin films offer highly exploiting research opportunity. For example, it is now well-known that the band gap of material displays a blue shift in the band gap after growing in nanoscale. In a similar manner, the right choice of the dopants and deposition conditions can produce a red shift in the band gap of the sample. Though this is true in the case of nanostructured powder sample as well, the phenomenon is more distinct in thin film forms even at the microscale. In thin film forms, not just the grains but the deposition parameters and thicknesses also offer unique properties allowing far advanced tunability for the device-oriented properties. The same film prepared by different technologies offers different properties. In many procedures, the substrate temperature also greatly influences the quality and adhesion properties of the film. Thus, the same sample can be used in different applications based on the particular optimum properties of the sample that have been acquired by fine-tuning of the deposition parameter as well as the procedure.

6.3 DEVICE ENGINEERING IN THIN FILMS

The invention of nanotechnology has opened more rooms for the device-oriented applications that would not have been feasible before a decade. Engineering the nanomaterials with various nanostructures (such as NWs, nanotubes, nanocomposites, superlattices) and fine-tuned properties for smart materials (band gap tailoring for photovoltaic applications, structured materials), etc., have proved to enhance the functionality of thin films for better device performance. In the recent past, dye sensitized and quantum sensitized nanostructured thin films attracted more research interest for solar cell applications that eventually lead the research community into another exciting research field of perovskite-based solar cells. Thus, the tailored materials are the future of the thin film industry. The device-oriented thin film engineering starts from the choice of the type of substrate and the procedures followed to attain the film.

6.4   SUBSTRATE SELECTION FOR NANOSTRUCTURED THIN FILMS

The selection of proper substrate for the nanostructured thin film is very important for suitable applications. For example, a window layer of a thin film solar cell has to be coated on a highly conductive transparent glass (transparent conducting oxide) substrates. On the elementary level, the glass substrates can be used to study the properties of the nanocoated films. But on the application level, the nanocoating must be made on the right substrate. The substrates can be semiconductors, metallic, alloys, and ceramics depending upon the applications. The thin film properties such as structure, conductivity, and adhesion are strongly dependent on the substrate property as well. A conducting substrate is required when the thin film has to be highly conducting and especially when a back contact is required. The features of end product are highly reliant on the properties of the substrate as well.

6.5   SYNTHESIS TECHNIQUES FOR NANOSTRUCTURED THIN FILMS

Based on the varieties of techniques that overlap in procedures and mechanisms forming hybrid deposition processes, it is not easy to classify the deposition parameters into a few generalized categories. However, most of them fall under vacuum coating and wet chemical coating categories. The vacuum coating units are the coating techniques that require a high vacuum and yield very quality thin films. They are further categorized into physical vapor deposition and chemical vapor deposition (PVD or CVD) techniques based on the physical or chemical processes involved. On the other hand, wet chemical techniques are solution-based techniques that do not involve a high vacuum for the deposition of thin films. Dip coating, sol–gel spin coating, and hydrothermal techniques are examples of the wet chemical techniques.

6.5.1   PVD TECHNIQUES

These are the most common vacuum-coating techniques used by the research community. The technique is based on the deposition of the material placed on a source in vacuum onto a substrate placed above it. The material/procedure thus undergoes three phases during this process: (1) conversion of the source material into the vapor phase or knocking the atoms or molecules out of the source, (2) source to substrate transportation of these vapor species—these vapors might be of ions or plasma and are sometimes chemically reacted on their path with other gas species intentionally introduced into the chamber to alter the deposition on the substrate into a new category termed as reactive deposition, and (3) impingement of these vapor species and growth of the thin film on the substrate surface that is placed above the source at a suitable distance, by condensation and recondensation at the surface by the bombardment of further incoming vapor species. This growth mechanism typically proceeds via nucleation and growth phases, such as adsorption, diffusion at the surface, making chemical binding with nearby nucleation sites, and further surface level atomic processes. Initially, they grow through different island growth formed from the small clusters nucleated on the substrate and join together via diffusion mechanisms. Few examples of this technique include thermal evaporation, flash evaporation, sputtering, electron beam evaporation, pulsed laser deposition (PLD), and cathodic arc deposition. The differences in the processes are mainly the way the atoms and molecules are ejected from the source material. For example, thermal evaporation uses the resistive heating method to eject the species out of the source while electron beam evaporation makes use of high beam electron energy for the same purpose. However, the latter technology can effectively eject the atoms and molecules out of a material that has a high boiling temperature as well. We will discuss few PVD systems to elaborate on the technique.

6.5.1.1 THERMAL EVAPORATION TECHNIQUE

Among all other vacuum techniques, the thermal evaporation unit is so popular that it is generally referred as a vacuum coating unit.

A schematic diagram of a thermal evaporation unit is shown in Figure 6.1. It works on the principle of resistive heating of the source material from a heated boat at very low pressure ($\sim 10^{-6}$ Torr) and extremely clean conditions onto the substrate surface where it is allowed to condense to form a film. Here, the overall process can be divided into three steps. First, sublimation of the source material by quick vaporization of the solid source material above its evaporation temperature at very low vacuum or gaseous plasma followed by transportation of the vapors in vacuum or partial vacuum to the substrate surface and finally condensation of the vapors onto the substrate to generate thin films [2]. The high vacuum condition inside the chamber helps to prevent the particles from scattering and minimize the residual gas impurities.

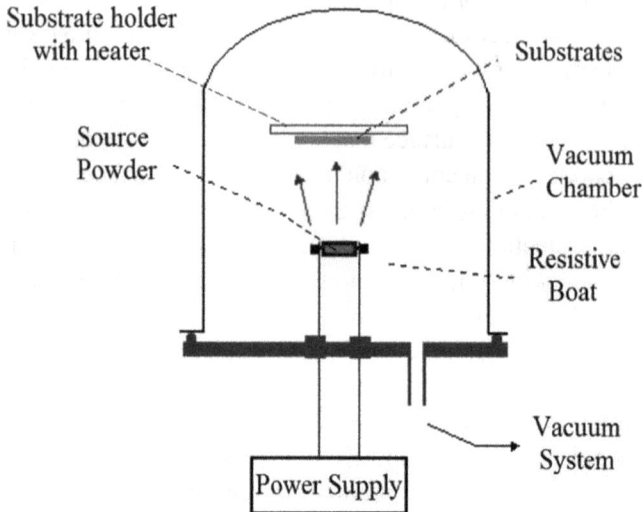

FIGURE 6.1 Schematic diagram of vacuum coating unit.

The thermal evaporation units are backed up by rotary and diffusion pumps to attain the required vacuum inside the chamber. Initially, the low vacuum up to 10^{-3} Torr would be achieved using a rotary backing pump and then the higher vacuum ($\sim 10^{-6}$ Torr) would be achieved using the diffusion pump attached with the unit. Most of the units also provide a glow discharge facility inside the chamber that facilitates extra cleaning of the substrates. The units also have provisions for the substrate heating and thickness monitoring. Usually, the films prepared at lower substrate temperatures (e.g., V_2O_5 thin films prepared below 300 °C) are amorphous in nature [3]. Thus, most of the time, the films are deposited above some particular substrate temperature to attain crystallinity. Substrate temperature also provides good adhesion property to the film.

The main disadvantage of the resistive-heating-based technique arises when we need to prepare composite films from the materials that have a different boiling point. While the temperature of the boat rises because there is a chance that the material has a lower vaporizing point evaporates first and thus drops the stoichiometry of the composite film. This difficulty can be overcome by bringing the stoichiometric material directly into the sublimation and forms the basis for the techniques, such as electron beam evaporation, pulsed laser beam deposition, etc. So, PVD techniques are basically the same but differ in terms of the vaporizing technique and thereby the resulting applications.

6.5.1.2   FLASH EVAPORATION TECHNIQUE

As we said, the thermal evaporation technique fails to achieve good quality composite films; however, it can be modified to a small extent to achieve directly the sublimation point of the material by allowing it to fall into a preheated boat and flashes out into the vapor phase to form the film. This technique is called the flash evaporation technique.

FIGURE 6.2   Flash evaporation feeder of HindHIVAC vacuum coating unit for representation.

This is a modified version of thermal evaporation and is used to prepare composite films out of two materials having dissimilar evaporation point. The source powder is prepared as a stoichiometric mixture of finely ground powders in the required ratio from both the materials. Unlike thermal evaporation where the source powder is prefilled inside the boat before heating the boat, an electrical pulse relay attached with a feeder (Figure 6.2) mechanism is used to ensure the uniform falling of the material into the preheated tungsten boat continuously. The powders with very fine uniform grains will be falling repeatedly on the preheated boat. Sometimes, the falling can be controlled in such a way that the powders will fall on the hot boat in steps and are used when the boiling points of the two materials are far apart (referred to as stepwise flash evaporation). As soon as the powder falls on the heated boat, it flashes out and condenses onto the substrate kept at the required temperature to form the film. The tungsten spiral boats are beneficial when there is a requirement of high evaporation temperature for one of the material used.

Thermal (and flash) evaporation technique is considered as the simplest physical deposition procedure for thin film deposition. Other important physical deposition techniques include electron beam evaporation, PLD, molecular-beam epitaxy (MBE), and sputtering techniques.

6.5.1.3   ELECTRON BEAM EVAPORATION

Here, the resistive heating method to vaporize the source material (usually a rod or ingot) is replaced by a highly intense and energetic electron beam created from a charged filament (referred as electron gun) under high vacuum that is directed with the aid of magnetic and electric fields to the substrate (anode). While the magnetic field directs the electron beam from the filament toward the source material (kept at a positive potential), the electric field steers the beam over the substrate to ensure uniform heating and eventually results in vapor formation. These vapors are then used for the film deposition on the substrate.

6.5.1.4   PULSED LASER DEPOSITION

Here, the vaporizing agent is a high energetic pulsed laser beam that can ablate the material for the thin film deposition inside the vacuum chamber. The technique is also highly useful for the reactive evaporation of the material for the metal oxide thin film deposition [4].

6.5.1.5   SPUTTERING TECHNIQUE

This is PVD technique used for the thin film deposition by sputtering (a phenomenon by which atomic particles of a solid sample are ejected from its surface by the bombardment of another energetic particles on it) material from a target to deposit it onto a substrate. Here, instead of the energetic electron beam or laser, energetic particles (or ions) from inert gas/plasma is used to eject the source material from the target (material to be coated) to the substrate. The ejected particles enter into the vapor phase and condense on the substrate to form a very thin layer. The technique is useful for the deposition of nitrides, oxides, metals, and alloys.

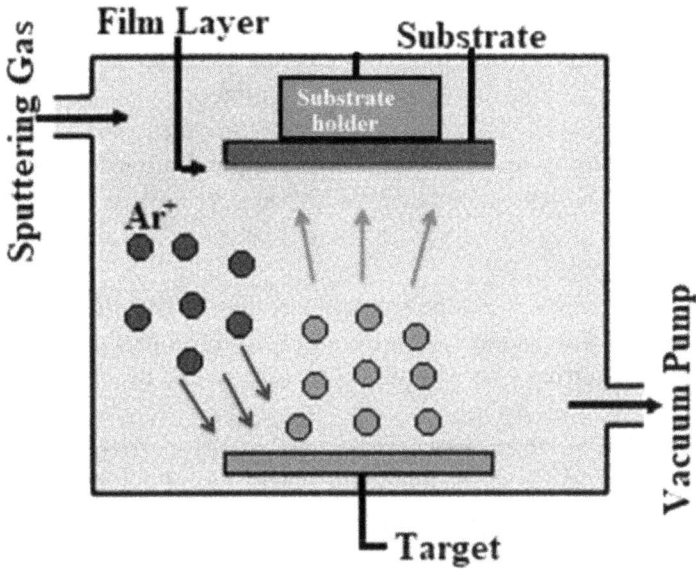

FIGURE 6.3 Schematic of a typical sputtering unit.

A typical sputtering unit (Figure 6.3) consists of a vacuum chamber containing metallic anode and cathode in order to obtain a glow discharge in the residual gas in the chamber [5, 6]. The front surface of the cathode is the slab of a target or the source material to be deposited.

The technique involves the following steps.

1. Introduction of controlled gas (like inert argon) into the evacuated chamber.
2. Establishing a self-sustaining plasma by electrically energizing the cathode.
3. Ionization of the gas species inside the plasma to become positively charged ions.
4. Accelerated movement toward the target (source) slab and striking with sufficient kinetic energy (greater than the target's surface binding energy), and dislodging the atoms/molecules out of the target slab— sputtering. (In order to achieve efficient momentum transfer, the atomic

weights of sputtering (bombarding) gas and the target material should be closely matching. Thus, neon is preferred for the lighter elements while krypton or xenon is chosen for heavy elements).

5. Forming vapor stream of the sputtered material, striking, adsorbing, and finally coating on the substrate to form the thin film.

Based on the type of power used on the cathode, the sputtering can be DC sputtering (DC power—mostly for conducting samples) and RF sputtering (RF power for nonconducting samples). Sputtering can be made using noninert gases as well. This type of sputtering is called reactive sputtering, where the coating is not the original pure target material but its reactive species like oxide. An example of reactive sputtering is the growth of the silicon oxide layer using oxygen gas as the sputtering gas and Si (elemental target) as the cathode [7].

Two main advantages of the sputtering technique over thermal evaporation are its ability to sputter even materials with high vapor pressure (boiling point) and transferring a very close composition as that of the target material to the deposited layer. A simple but common practice of sputtering deposition is used in conventional scanning electron microscopy, where the samples are usually sputter-coated with a conducting layer of Au/ Pd that avoid charging of the sample with an electron beam.

6.5.1.6   *MOLECULAR-BEAM EPITAXY*

Sometimes, we may require high quality single-crystalline thin films for some specific and improved device-oriented application, and the precise solution for the growth is MBE. Highly ordered crystalline overlayer (one atomic layer over another) deposition on a crystalline substrate is referred as epitaxy and the layer is called epitaxial film [8–10]. The word "molecular-beam" designates a unidirectional kinematic stream of molecules (or atoms) without causing any collisions amongst each other.

MBE is the epitaxial growth of thin films comprised of single crystals prepared in high or ultra-high vacuum (10^{-8}–10^{-12} Torr). The epitaxial growth takes place at an extremely slow rate (approximately a few Å/s) [11], and this is the main advantage of the MBE technique.

Figure 6.4 shows the schematic of a typical MBE system.

The processes involved can be summarized as follows:

1. Sublimation of the source material into the gaseous phase by increasing its vapor pressure through a rise in temperature.

2. Transfer of the gaseous particles from the source to the substrate surface, where the condensation takes place.
3. Reaction of the condensate with the surface to align in the epitaxial arrangement and the growth of crystalline structure at an appropriate growth rate. Various steps happening during this process are adsorption of the impinging molecules on the surface followed by surface diffusion, thermal desorption, dissociation and alignment into the crystal lattice, etc.

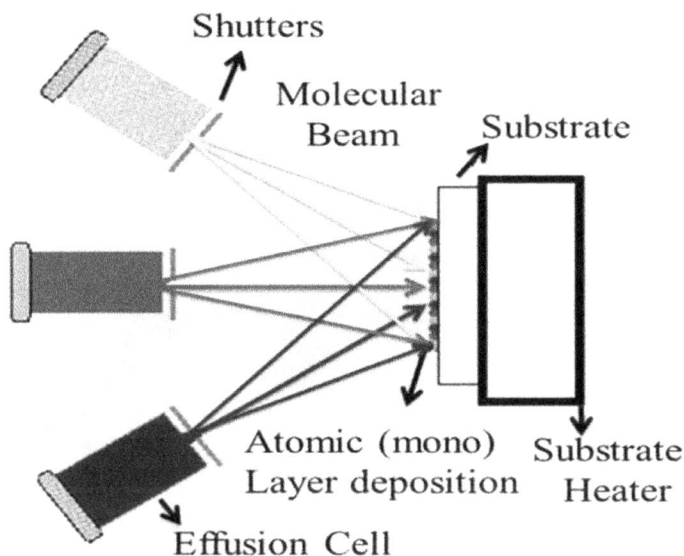

FIGURE 6.4   Schematic of a typical MBE system.

The MBE technique has a lot of distinguished features over other techniques. The ability to prepare high quality single crystalline film by employing molecular or atomic beam is the main attraction of the technique. Also, the impurity level in the epitaxial layers is extremely small owing to the ultra-high vacuum environment, as well as the absence of any type of carrier gases for the growth mechanism [11]. The unit provides the option to synthesize complicated arrays of monolayers comprised of different materials by the controlled use of different effusion cells containing the evaporation sources and their shutters. Such monolayers can be confinement structures like two-dimensional quantum wells [12] or 1D QDs [13]. The MBE technique can also provide a high growth rate for the materials having lower vapor pressure when its high purity is required. The technique also has inbuilt features for the in-situ, real-time monitoring, and controlling of the film growth to meet the optimum growth condition for the high purity stoichiometric and epitaxial layer.

6.5.2 WET CHEMICAL TECHNIQUES

Unlike the vacuum coating, the wet chemical techniques are simple, cost-effective, and do not require costly equipment and very professional hands for the synthesis of nanostructured thin films. They are highly suitable for the large scale production of thin films. The technique is based on the thin film preparation from solution-state precursors by various techniques, such as dip coating, drop-casting, CBD, spin coating, spray pyrolysis, and hydro/solvothermal. The technique is highly useful for the synthesis of metal oxide thin films. Not only the technique but also the precursor solutions, solvents, temperature, etc., also play a significant role in the structure, morphology, and growth mechanism of the films. In view of this as well as recent applications in dye sensitized and QD-sensitized solar cells, we will highlight a few of the wet chemical techniques here.

6.5.2.1 SOL–GEL-ASSISTED SPIN COATING

The spin coating technique is a wet chemical technique and is usually combined with sol–gel method for the preparation of thin films. Here, the deposition technique can be summarized as the drop-wise addition of sol–gel on the substrate, spinning-up, and spinning-off of the spin coater and simultaneous evaporation of the solvent during high-speed rotation followed by annealing the formed film at a suitable temperature.

The thin film is deposited on the substrate by the high angular spin speed via centrifugal force. The solvent, being volatile, simultaneously evaporates. The film's thickness is dependent on the solution's concentration, solvent, and spin speeds [14, 15]. The desired thickness is obtained by multilayer deposition followed by a short-time (2–10 min) annealing at ~120 °C for each deposition. Finally, the formed film is post-annealed at a suitable temperature to vaporize the unwanted components on the layer and to get the proper film. The technique is suitable for high-quality oxide layers. However, the technique is not suited for large scale production as it limits the substrate's size and requires very flat substrates. Programmable spin coaters allow the user to control the spinning-up time [time required to reach the required rotation per minute (rpm)], the dwell time, that is, the total time spent at constant rpm and spinning downtime, that is, the time to reach back into the steady-state.

6.5.2.2   SUCCESSIVE IONIC LAYER ADSORPTION AND REACTION

In the SILAR technique, the anionic and cationic precursors are placed in two beakers distinctly and the substrate is immersed one by one in the two reactors followed by cleaning, heat drying, and cooling the layer during each dip [16–19]. For example, the cationic solution for CuS thin film could be 0.5M $CuCl_2$ solution while for CdS, it could be 0.5M $CdCl_2$ or $(CdNO_3)_2$ solution. In both cases, the anionic solution could be 0.5M Na_2S solution. The procedure can be summarized as follows: (1) immersion of the substrate in the cationic solution for a fixed duration (say 5 min), (2) rinsing inside the proper medium (preferably the solvent used to prepare cation solution, kept at 60–80 °C) and heat drying followed by cooling, (3) immersion inside the anionic medium for the fixed duration (same as step 1), and (4) rinsing inside the proper medium and heat drying followed by cooling. These four steps dipping and drying procedure is called one SILAR cycle. The process has to be repeated several cycles until getting the required thickness of the film.

6.5.2.3   CHEMICAL BATH DEPOSITION

In the CBD method, the anionic and cationic precursors react in a single reactor in the presence of a complexing reagent [20]. Here, the separately heated cation and anion solutions are mixed together in the presence of the complexing agent to initiate the reaction while the substrate is continuously stirring in the solution. During the reaction stage, a thin film is developed on the substrate surface. The role of the complexing agent here is to reduce spontaneous precipitation by controlling the release of the cations and thus causing in slow precipitation of the compound in the solution. In the CBD technique, temperature, as well as pH of the solution plays a crucial role on the film deposition, and often it is required to keep the same pH throughout the deposition process by continuously monitoring and supplying suitable base (ammonia, NaOH, etc.) in droplets [20].

6.5.2.4   DOCTOR BLADE TECHNIQUE

The doctor blade technique has attracted the research interest in the near past when micron level thick mesoporous films were required for the synthesis of dye sensitized and QD-sensitized solar cells.

The technique involves the preparation of a thick paste and spreading the same on the substrate by a suitable glass rod—the doctor blade. Fluorine-doped tin oxide (FTO) or indium-doped tin oxide substrates are usually preferred on the application point of view. Initially, four sides of the FTO substrate are uniformly masked by a suitable tape of appropriate size and geometry, as shown in Figure 6.5. A thick film is formed inside the exposed surface by the doctor blade when a constant relative movement is established between the blade and the substrate. However, the technique requires skill for both the preparation of a well-mixed paste, as well as for the deposition of thin films. Further, the film has to be annealed to remove the presence of any additives present (such as solvents and binders). The resulting film is a highly nanostructured thick mesoporous layer. As mentioned, the films are best suited in *dye-sensitized solar cell* (DSSC) or quantum dot sensitized solar cell (QDSSC) applications, and the 10–25 µm thick films can be prepared by the simple doctor blade technique [21].

6.5.2.5   SCREEN PRINTING TECHNIQUE

A typical screen printing plate and schematic of the prepared films are shown in Figure 6.6. Here, the four-sided tapes used in the doctor blade technique are replaced by a porous printing plate fabricated out of a woven mesh of synthetic fabric threads or metal wires. The printing plate is not just acting as a stencil but also serving as a good mask for selected area coating on the substrate. The printing plate is attached to a firm framework (usually rectangular) and sealed at the ends. After placing the substrate under the porous mesh region of the printing plate, a vacuum has been created between the printing plate and its underlying bottom surface by means of a vacuum pump attached to the system. This enhances the masking provided on the substrate while keeping it steady at its exact place and the tension created in the mesh helps effective fluid spreading on the substrate for the coating. After coating, individual films are cut and separated for various applications.

The individual steps can be summarized as follows.

1. Placing the FTO substrate under the screen below porous mesh holes and seals it with the help of a vacuum pump.

2. Filling the mesh openings with the screen printing paste and spreading the paste using a squeegee over the mesh openings.
3. Drying the film/layer at ambient conditions for 5 min to decrease the tension within the layer.
4. Drying the layer at ~110 °C on a hot plate for 5–10 min to evaporate volatile components in the paste and dry the layer before coating the next layer.

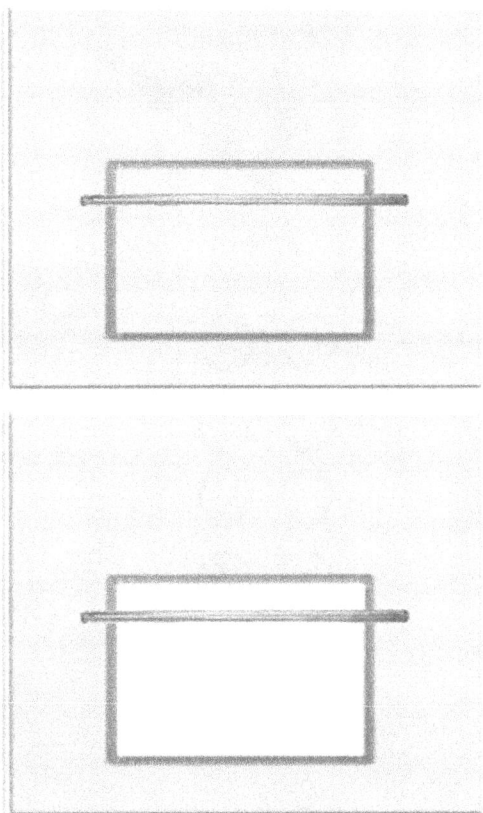

FIGURE 6.5   Schematic of doctor blade technique.

Each typical printing cycle adds the layer thickness by 3–4 μm. Depending on the desired thickness of the layer, the printing cycle has to be repeated and finally annealed at a suitable temperature. The printed glass slides will be then cut into small individual film prints as working electrodes suitable for DSSC and QDSSC [22].

(a)

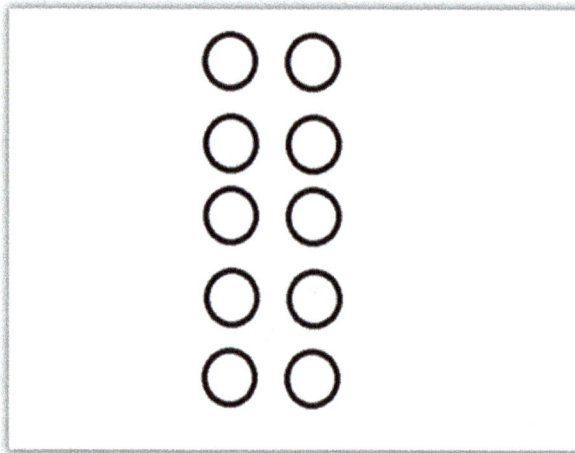

(b)

FIGURE 6.6   Modified version of doctor blade technique. (a) A typical screenprinter and (b) schematic of the film.

6.5.2.6   *HYDROTHERMAL (SOLVOTHERMAL) TECHNIQUE*

The hydrothermal technique is best known for the synthesis of high-quality nanoparticles. However, this technique can be made use for the growth of nanofilms comprising beautiful NRs/NWs assembly [1].

Figure 6.7 shows the autoclave setup used for the hydrothermal synthesis of NR/NW thin films comprising a Teflon liner and an outside protective vessel made of stainless steel as the main components. The optional safety spring attachment above the liner cap is intended to elude building up of pressure inside the Teflon liner beyond a limit. When the pressure exceeds the maximum that the spring can hold, it permits the liner cap to open and avoid an explosion.

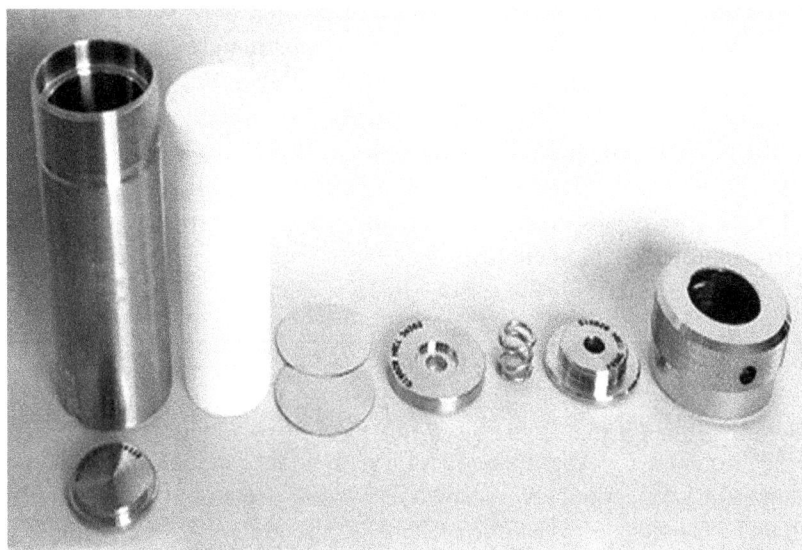

FIGURE 6.7 Individual parts of typical autoclave system used for the hydrothermal technique (acid digestion vessels from Parr instrument).

The film is highly suited for the synthesis of metal oxide (such as TiO_2 and V_2O_5) NR/NW structured thin films. In a typical film synthesis, the substrate (usually transparent conducting oxides such as FTO) is placed at an angle inside the Teflon liner containing the precursor solution and sealed in the high-pressure autoclave. For an aqueous precursor solution, the technique is called hydrothermal and for an organic precursor solution, the technique is referred as solvothermal synthesis. The autoclave is then placed in an oven at a suitable temperature (usually between 150°C and 250 °C) depending on the nature and the quantity of the precursor solution and the type of films expected, etc. The reaction occurs in the aqueous solution at high vapor pressure resulting in the formation of hierarchical NR/NW films on the substrates [1, 17, 23].

6.6   INSIGHT INTO OPTOELECTRONIC APPLICATIONS

We have been talking about nanostructured thin films and various techniques for the effective synthesis of high quality nanostructured thin films. These thin films are useful in many optoelectronic applications that include photodiodes, photovoltaic, sensor, photocatalysis, self-cleaning planes, chip systems, energy generation, and storage devices. It must be noted that one might need to combine two or three techniques in a device-oriented synthesis of nanostructured film for better device performance. One such example will be demonstrated in this section using TiO_2 nanostructured-mesoporous thin film prepared mainly by doctor blading/screen printing or hydrothermal technique for dye sensitized or QD sensitized solar cell application. These wet chemical techniques are referred to as DSSC or QDSSC based on whether sensitized by dye or QD molecule and the technique requires micron level thick film for its active component to function properly which is the electrode known as photoanode..

6.6.1   STRUCTURE OF A DYE OR QD SENSITIZED SOLAR CELL

Schematic diagram of a typical (Q)DSSC is shown in Figure 6.8. The main composition are (1) a photoanode film (a wide bandgap nanostructured mesoporous semiconductor film on a transparent conducting glass—mostly TiO_2 on FTO—sensitized by dye or QDs), (2) a counter electrode (commonly platinum/FTO), and (3) an electrolyte solution (usually iodine based—made out of appropriate amounts of I and KI in acetonitrile [24, 25] or polysulfide electrolytes comprising Na_2S, S, and KCl in 3:7 water/methanol solution [1]) infused in between them. Here, the insertion of dye molecules or QDs into the semiconductor matrix is termed as sensitization.

The cell assembly is made out of the electrodes as shown in Figure 6.8 enclosing the photoanode film inside a spacer. The gap between the electrodes is filled with a suitable electrolyte solution [1].

However, the device performance of this configuration was very poor. This made the researchers to modify the photoanode assembly. (Note that multiple factors affect the cell performance, such as the photoanode, electrolyte, mode of preparation of the counter electrode, the way the electrodes are sealed, the photoanode thickness as well as the spacer thickness for the electrolyte. However, we will limit our attention only to the photoanode film assuming that all other factors remaining the same, to convey the effects happening due to the variations in the synthesized nanostructured film, and to understand how to modify the structure for the best performance of the device. The reader must be aware that the cell performance can be further improved by optimizing other parameters individually, after attaining the optimized photoanode condition). Before continuing the topic, we will just give a small picture of a (Q)DSSC working principle.

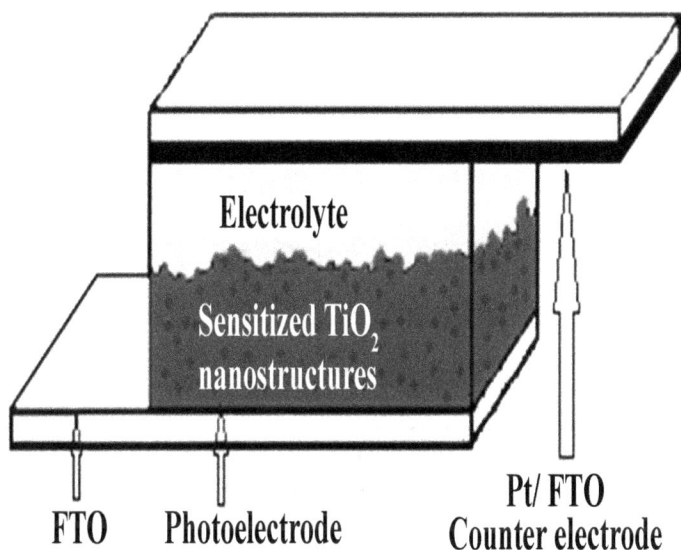

FIGURE 6.8 Schematic of a QDSSC.

The key processes involved in the generation of photocurrent are displayed schematically using the energy diagram in Figure 6.9. The figure can be summarized as:

1. *Absorption:* the dye/QD molecules absorb the photon energy.
2. *Excitation:* dye/QD molecules get excited, go to an unstable excited state, and then inject electrons into the conduction band of TiO_2. During this process, the dye/QD gets oxidized or a hole is left behind.
3. *Transport mechanism:* electrolyte (usually iodine-based electrolyte, that is, containing I^-/I^{3-} ions) acts as the mediator for the electron transport between TiO_2 and (Pt-coated) counter electrode. During the transport mechanism, a mediator (usually iodide ion I^-) from the electrolyte fills the hole with one of its own electron (converting to triiodide I^{3-} ion) and regenerates the oxidized dye molecule back to the normal.

4. Subsequently, the electrons will get injected into the FTO conducting thin film by diffusion of charge-compensating cations in the electrolyte layer close to the TiO$_2$ surface and finally generate a photoelectric current through the external circuit.
5. The I^{3-} reduces back to I$^-$ ion by substituting with the extra electron from the external circuit [26–28]. So the transport mechanism here does not result in any persistent chemical transformation to the individual species involved [28].

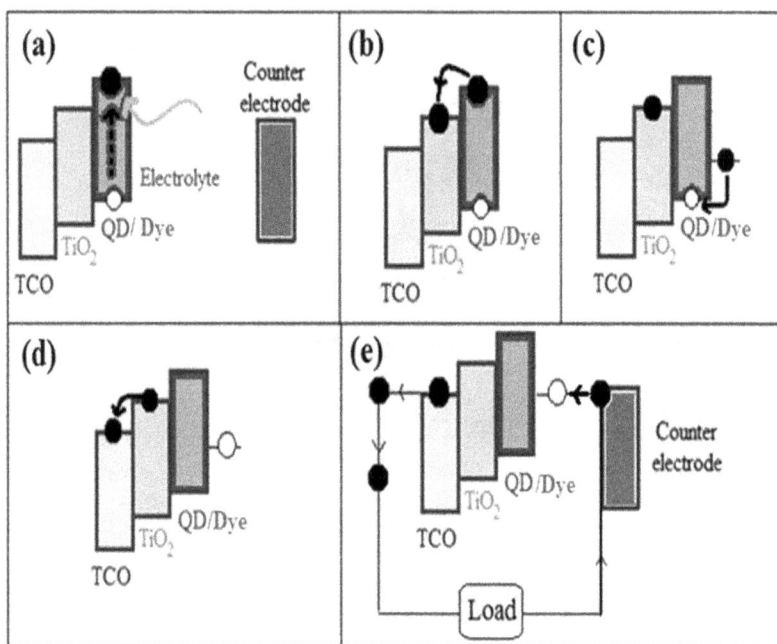

FIGURE 6.9   Schematic of the key working mechanisms involved in the photocurrent generation.

Now, let us come back to our discussion. The studies have shown that the photodiodes made up of TiO$_2$ films having vertically aligned NR structures resulted in better efficiency compared with any flower-like growth or spherically arranged NPs [1]. The films prepared using the doctor blade technique and screen printing resulted in nanoparticle growth, whereas hydrothermal synthesis resulted in NR growth. This example is mentioned to emphasize the importance of growth mechanism for the particular application that we are looking for.

Figure 6.10 shows the growth of TiO$_2$ NR arrays on FTO substrates by hydrothermal technique. The former shows a dandelion-like flower kind of growth [29, 30], whereas the latter shows a somewhat vertically aligned growth of NRs. This gives a clear indication that the same technique can yield different morphologies by suitably varying the deposition parameters. The optimization depends upon the requirement and is based on the particular application.

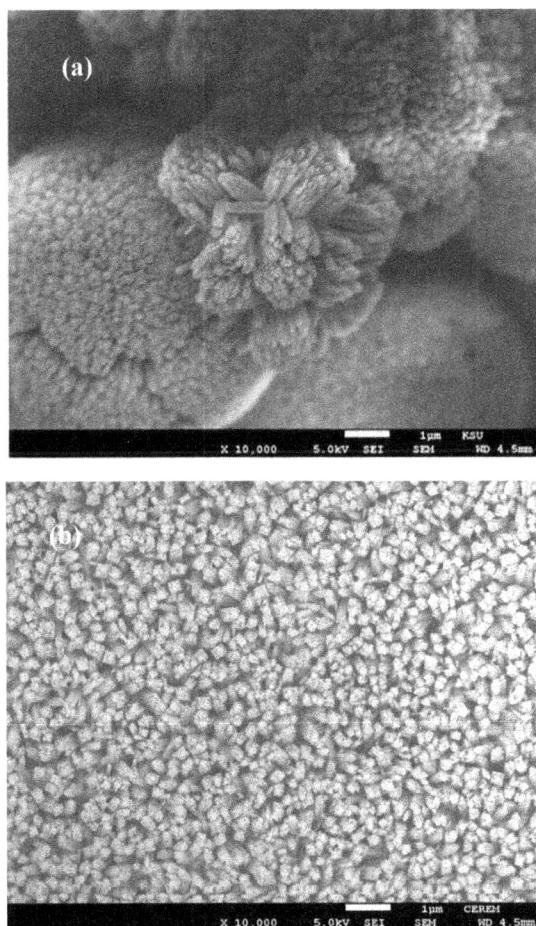

FIGURE 6.10 (a) TiO$_2$ NR arrays over dandelion like flower structures, and (b) vertically aligned growth of NRs grown on FTO substrates by hydrothermal technique.

Source: Reprinted with permission from Ref. [29]. © 2015 by the authors; licensee MDPI, Basel, Switzerland. (http://creativecommons.org/licenses/by/4.0/).

The detail of the hydrothermal technique has been already mentioned in the previous section. Here, we describe a typical synthesis procedure for TiO$_2$ NRs on a cleaned FTO substrate. Initially, a clear transparent precursor solution is set by adding 1 mL of titanium isopropoxide solution dropwise to a 1:1 mixture of DI water and concentrated (35%) HCl solution. The substrate is kept at an angle in the Teflon liner and the precursor solution was slowly added to it (filling about the 30% of the volume). The Teflon liner was loaded in an autoclave and was placed in an oven at 150 °C for 15 h. After synthesis, the autoclave was cooled to room temperature and FTO-coated substrates were taken out, rinsed extensively with deionized water, and allowed to dry in ambient air. TiO$_2$-coated substrates were treated with diluted TiCl$_4$ solution at 60 °C for 15 min and finally annealed at 300 °C for half an hour. The same process was repeated on a pretreated or surface modified FTO substrate to form TiO$_2$ NR films on a blocking layer. The FTO substrate surface was modified by a thin blocking layer of TiO$_2$ prepared by spin coating a solution containing suitable quantities of titanium isopropoxide, DI water and hydrochloric acid, and postannealing at 400 °C for 1 h. The blocking layer was made on FTO substrate leaving 20% of FTO layer masked during the spin coating process in order to use in future applications and/or contact extensions. It was found that the NR films prepared on the bare FTO substrate and on TiO$_2$ blocking layer/FTO substrates differ in geometry, as shown in Figure 6.10. The NR structure on the bare FTO substrate formed as petals of a dandelion kind of flower-like structures, whereas that on the blocking layer took the vertically aligned NR growth. (This has been explained on the basis of SnO$_2$'s role to act as a seed for the flower kind of growth. In the second case, the blocking layer hinders the SnO$_2$ or FTO layer in the vertical NR growth [1]).

Now let us come back to our discussion on (Q)DSSC. The TiO$_2$ layer (mesoporous or NR assembly, thus, prepared by doctor blading/screen printing/hydrothermal) is used as the photoanode of the (Q)DSSC. CdSe or CdS QDs or dye molecules were inserted into the nanostructured TiO$_2$ thin films by proper means (like dip coating at favorable temperature, in-situ growth by SILAR, CBD, etc.). The process of inserting dye molecules QDs into the TiO$_2$ matrix is called QD sensitization. Ruthenium dye, CdS, CdSe, CdTe, PbS QDs, etc., are the widely used sensitizers [31–34]. TiO$_2$ photoanode films were prepared on FTO substrates without and with a blocking layer and sensitized using the same sensitizer to form sensitized solar cells. It was noted that the film with a blocking layer has significant enhancement on the performance of the cell [1]. Also, the vertically aligned

NRs have reported better solar cell efficiency compared to that of flowered and spherical nanoparticle structures. So, controlling the growth conditions, layer structures, and morphologies play an important role in the device performance.

6.7 CHARACTERIZATION TECHNIQUES

The characterization techniques for the nanostructured thin films are more or less the same as that of nanoparticles or powders. So, here we will be describing only the major differences followed in characterizing the film samples compared to that of powder samples and do not elaborate separately on the various techniques. Especially, the optical studies deserve special attention in the case of thin film samples. For the structural characterization is concerned, fixed grazing angle incidence of X-ray diffraction (XRD) technique [1, 35] is preferred for the thin film samples in order to avoid intense signal coming from the substrate. The 2θ position, broadness, and intensity of the XRD data, all bear the vital information regarding the nanostructured growth of the material. The broader the peak, the lower the crystallite sizes of the grains in the film. A well-distinct peak having high scattering intensity originates from the good crystallinity of the film [36].

Here, we describe a few important parameters that can be estimated from the nanocrystalline thin film samples using XRD and optical studies.

$$D = \frac{0.9\lambda}{\beta \cos(\theta)} \tag{6.1}$$

where D is the crystallite size, λ is the X-ray wavelength, β is the full width at half maximum (FWHM), and θ is the Bragg diffraction angle [37, 38].

Dislocation density (δ) that gives the length of dislocation lines per unit volume of the crystal is evaluated using the formula

$$\delta = 1/D^2 \tag{6.2}$$

where D is the crystallite size [39, 40]. Large D or small δ value indicates better crystallinity of the film.

Microstrain (ε) values are calculated by using the relation

$$\varepsilon = \beta \cos(\theta)/4 \tag{6.3}$$

The factors like stress, strain, and film thickness, influence the crystallization and line broadening of an XRD pattern. One of the methods to separate the contributions of particle size and microstrain to the line broadening in the XRD pattern is the Williamson–Hall (WH) plot [41, 42]. Effective grain size and strain values are calculated by using the WH plot from the following relation.

$$\beta \cos(\theta) = \frac{0.94\lambda}{D} + 4\varepsilon \sin(\theta) \tag{6.4}$$

where β is FWHM, θ is the Bragg diffraction angle, λ is the X-ray wavelength used, D is the crystallite size, and ε is the effective microstrain. A plot of $\beta \cos(\theta)$ versus $\sin(\theta)$ will be a straight line and the values for D and ε can then be calculated directly from the Y-intercept and the slope of the straight line, respectively.

6.7.1   OPTICAL ANALYSIS

Optical measurements provide the most direct and perhaps the simplest method for determining the band structure of semiconductor samples. The shape of the transition provides insight into the electron interband transitions (between energy states in the valence band and conduction band). The transition region where the sharp discontinuity occurs in the vicinity of the energy bandgap is called the "absorption edge."

Further, Tauc's relation connects the energy band gap (E_g) with absorption coefficient (α), and wavelength ($\lambda = c/v$, where c is the speed of light) by the formula [43–45]

$$\alpha h v = A\left(h v - E_g\right)^n \tag{6.5}$$

where A is a constant, h is Plank's constant, v the frequency, hv is photon energy of the incident beam, and n is an index whose value determines the type of transition; 1/2 for direct allowed and 2 for indirect allowed transitions, 3/2 for direct forbidden, and 3 for indirect forbidden transitions. Optical band gap can be determined by plotting $(\alpha h v)^{1/n}$ versus hv and extrapolating the linear portion of the curve to the hv axis.

Optical transmission data also provide important information about the thickness and nanocrystalline growth of semiconductor films. Nanocrystalline growth of a film is reflected by a blue shift in the absorption edge or a corresponding increase in band gap [1]. The band gap values of bulk and nanomaterials can be related using an effective mass approximation (EMA) and a hyperbolic band model (HBM) by the following formulae [46–49]:

EAM:

$$E_g(n) = E_b(b) + \frac{\hbar^2 \pi^2}{2m^* R^2} \tag{6.6}$$

HBM:

$$E_g^2(n) = E_g^2(b) + \frac{2\hbar^2 \pi^2 E_b(b)}{m^* R^2} \tag{6.7}$$

where R is the size, m^* is the effective mass of the specimen, $E_b(b)$ is the bulk band gap and $E_g(n)$ is the band gap of the (nano) sample. So an obvious increase in band gap value is expected as we go to nanosize. Also, an increased transmission is expected as we go more and more toward nano.

Thickness information of the film is directly reflected in optical transmission as an interference pattern. More ripples indicate higher thickness for the film. So, transmission data could be used to estimate film thickness and there exist techniques to extract information from the transmission fringe pattern. However, not enough ripples are expected for nanosamples with a thickness <200 nm. Film thickness for such films can be estimated by simulating the transmission data with puma software using the unconstrained optimization technique described by Birgin et al. [50]. The method gives a very accurate thickness estimation for samples with thickness >70 nm.

ACKNOWLEDGMENT

The author (Dr. Aslam) would like to thank Prof. Amanullah Fatehmulla and Prof. Abdullah M. Al-Dhafiri for introducing multiple growth as well as characterization techniques present at King Saud University, Riyadh, and their fruitful contribution as PhD Supervisors. The authors also would like to express their gratitude to King Khalid University, Saudi Arabia, for providing administrative and technical support.

CONFLICT OF INTEREST

The authors confirm that there is no conflict of interest.

KEYWORDS

- thin films
- growth techniques
- optoelectronics
- DSSC
- QDSSC

REFERENCES

1. M.M. Aslam, S.M. Ali, A. Fatehmulla, W.A. Farooq, et al., *Growth and characterization of layer by layer CdS-ZnS QDs on dandelion like TiO₂ microspheres for QDSSC application.* Materials Science in Semiconductor Processing, **2015**. *36:* pp. 57–64.

2. D.M. Mattox, *Foundations of Vacuum Coating Technology*, William Andrew Publishing/Noyes: Norwich, 2003. pp. 19–24.

3. M. Prześniak-Welenc, M. Łapiński, T. Lewandowski, B. Kościelska, et al., *The influence of thermal conditions on V_2O_5 nanostructures prepared by sol–gel method.* Journal of Nanomaterials, 2015. **2015**: pp. 1–8.

4. D.H. Lowndes, D.B. Geohegan, A.A. Puretzky, D.P. Norton, et al., *Synthesis of novel thin-film materials by pulsed laser deposition.* Science, 1996. *273*(5277): pp. 898–903.

5. M. Dawber, *2—Sputtering techniques for epitaxial growth of complex oxides*, in *Epitaxial Growth of Complex Metal Oxides*, G. Koster, M. Huijben and G. Rijnders, Editors, Woodhead Publishing: Cambridge, 2015. p. 31–45.

6. D. Depla, S. Mahieu and J.E. Greene, *Chapter 5—sputter deposition processes*, in *Handbook of Deposition Technologies for Films and Coatings* (Third Edition), P.M. Martin, Editor, William Andrew Publishing: Boston, 2010. pp. 253–296.

7. W.-F. Wu and B.-S. Chiou, *Optical and mechanical properties of reactively sputtered silicon dioxide films.* Semiconductor Science and Technology, 1996. *11*(9): pp. 1317–1321.

8. K. Alavi, *Molecular Beam Epitaxy*, in *Encyclopedia of Materials: Science and Technology*, K.H.J. Buschow, et al., Editors, Elsevier: Oxford, 2001. pp. 5765–5780.

9. M. Guina and S.M. Wang, *Chapter 9—MBE of dilute-nitride optoelectronic devices*, in *Molecular Beam Epitaxy*, M. Henini, Editor, Elsevier: Oxford, 2013. pp. 171–187.

10. S. Franchi, *Chapter 1—Molecular beam epitaxy: fundamentals, historical background and future prospects*, in *Molecular Beam Epitaxy*, M. Henini, Editor, Elsevier: Oxford, 2013. pp. 1–46.

11. L. Morresi, *Basics of molecular beam epitaxy (MBE) technique*, in *Silicon Based Thin Film Solar Cells*, R. Murri, Editor, Bentham Science Publishers: UAE, 2013. pp. 81–107.

12. R. Dingle, W. Wiegmann and C.H. Henry, *Quantum States of Confined Carriers in Very Thin Al_xGa_{1-x} As-GaAs-$Al_xGa_{1-x}As$ Heterostructures.* Electronic Structure of Semiconductor Heterojunctions, 1988. *1:* pp. 173–176.

13. D. Leonard, M. Krishnamurthy, C.M. Reaves, S.P. Denbaars, et al., *Direct formation of quantum-sized dots from uniform coherent islands of InGaAs on GaAs surfaces.* Applied Physics Letters, 1993. *63*(23): pp. 3203–3205.

14. S. Ahmadi, N. Asim, M.A. Alghoul, F.Y. Hammadi, et al., *The role of physical techniques on the preparation of photoanodes for dye sensitized solar cells.* International Journal of Photoenergy, 2014. *2014*: pp. 19.

15. D.B. Hall, P. Underhill and J.M. Torkelson, *Spin coating of thin and ultrathin polymer films.* Polymer Engineering and Science, 1998. *38*(12): pp. 2039–2045.

16. S. Ruhle, M. Shalom and A. Zaban, *Quantum-dot-sensitized solar cells.* Chemphyschem, 2010. *11*(11): pp. 2290–2304.

17. A. Fatehmulla, M. Aslam Manthrammel, W.A. Farooq, S.M. Ali, et al., *Photovoltaic and impedance properties of hierarchical TiO$_2$ nanowire based quantum dot sensitized solar cell.* Journal of Nanomaterials, 2015. *2015*: p. 9.

18. J. Puišo, S. Lindroos, S. Tamulevičius, M. Leskelä, et al., *Growth of ultra thin PbS films by SILAR technique.* Thin Solid Films, 2003. *428*(1): pp. 223–226.

19. H. Güney and D. İskenderoğlu, *Synthesis of MgO thin films grown by SILAR technique.* Ceramics International, 2018. *44*(7): pp. 7788–7793.

20. F.M. Amanullah, A.S. Al-Shammari and A.M. Al-Dhafiri, *Co-activation effect of chlorine on the physical properties of CdS thin films prepared by CBD technique for photovoltaic applications.* Physica Status Solidi (A) Applications and Materials Science, 2005. *202*(13): pp. 2474–2478.

21. A. Fatehmulla, W.A. Farooq, M. Aslam, M. Atif, et al., *Photovoltaic and impedance characteristics of modified SILAR grown CdS quantum dot sensitized solar cell.* Journal of International Scientific Publications: Materials, Methods and Technologies, 2014. *8*: pp. 676–683.

22. M. Atif, W. Farooq, A. Fatehmulla, M. Aslam, et al., *Photovoltaic and impedance spectroscopy study of screen-printed TiO$_2$ based CdS quantum dot sensitized solar cells.* Materials, 2015. *8*(1): pp. 355–367.

23. S.M. Ali, M. Aslam, W.A. Farooq, A. Fatehmulla, et al., *Photovoltaic and impedance spectroscopy of CdS quantum dots Onto Nano Urchin TiO$_2$ structure for quantum dots sensitized solar cell applications.* Journal of Nanoelectronics and Optoelectronics, 2016. *11*(3): pp. 363–367.

24. I.S. Yahia, H.S. Hafez, F. Yakuphanoglu, B.F. Senkal, et al., *Photovoltaic and impedance spectroscopy analysis of p–n like junction for dye sensitized solar cell.* Synthetic Metals, 2011. *161*(13–14): pp. 1299–1305.

25. A. Fatehmulla, M. Atif, W.A. Farooq, M. Aslam, et al., *Photovoltaic properties of ammoniated ruthenium oxychloride dye based solar cell.* Optoelectronics and Advanced Materials-Rapid Communications, 2014. *8*(5-6): pp. 587–592.

26. X. Yang, G. Haoshuang, X. Huating and H. Mingzhe, *TiO$_2$ nanowire dye-sensitized solar cells fabricated by hydrothermal method.* Journal of Physics: Conference Series, 2011. *276*(1): pp. 012196.

27. M. Sokolský and J. Cirák, *Dye-sensitized solar cells: materials and processes.* Acta Electrotechnica et Informatica, 2010. *10*(3): pp. 78–81

28. M. Grätzel, *Solar energy conversion by dye-sensitized photovoltaic cells.* Inorganic Chemistry, 2005. *44*(20): pp. 6841–6851.

29. S. Ali, M. Aslam, W. Farooq, A. Fatehmulla, et al., *Assembly of CdS quantum dots onto hierarchical TiO2 structure for quantum dots sensitized solar cell applications.* Materials, 2015. *8*(5): pp. 2376–2386.

30. M.A. Manthrammel, *Synthesis and characterization of V_2O_5-TiO_2 nano-composite metal oxide thin films and their applications in quantum dot sensitized solar cells & gas sensors*, King Saud University, Doctoral Thesis: Riyadh, 2016.

31. S. Kumar, P. Sharma and V. Sharma, *CdS nanopowder and nanofilm: simultaneous synthesis and structural analysis.* Electronic Materials Letters, 2013. *9*(3): pp. 371–374.

32. P.A. Chate, S.S. Patil, J.S. Patil, D.J. Sathe, et al., *Nanocrystalline CdSe: structural and photoelectrochemical characterization.* Electronic Materials Letters, 2012. *8*(6): pp. 553–558.

33. V. Senthamilselvi, K. Saravanakumar, N.J. Begum, R. Anandhi, et al., *Photovoltaic properties of nanocrystalline CdS films deposited by SILAR and CBD techniques—a comparative study.* Journal of Materials Science–Materials in Electronics, 2012. *23*(1): pp. 302–308.

34. Y. Qin and Q. Peng, *Ruthenium sensitizers and their applications in dye-sensitized solar cells.* International Journal of Photoenergy, 2012. *2012:* pp. 1–21.

35. W.H. Bragg and W.L. Bragg, *The reflection of X-rays by crystals.* Proceedings of the Royal Society of London. Series A, 1913. *88*(605): pp. 428.

36. J. Yi, Y.-l. Liu, Y. Wang, X.-p. Li, et al., *Synthesis of dandelion-like TiO_2 microspheres as anode materials for lithium ion batteries with enhanced rate capacity and cyclic performances.* International Journal of Minerals, Metallurgy, and Materials, 2012. *19*(11): pp. 1058–1062.

37. P. Scherrer, *Bestimmung der Größe und der inneren Struktur von Kolloidteilchen mittels Röntgenstrahlen.* Göttinger Nachrichten Math. Phys., 1918. *2:* pp. 98–100.

38. A.L. Patterson, *The scherrer formula for X-ray particle size determination.* Physical Review, 1939. *56*(10): pp. 978–982.

39. V. Bilgin, S. Kose, F. Atay and I. Akyuz, *The effect of substrate temperature on the structural and some physical properties of ultrasonically sprayed CdS films.* Materials Chemistry and Physics, 2005. *94*(1): pp. 103–108.

40. S.A.-J. Jassim, A.A.R.A. Zumaila and G.A.A. Al Waly, *Influence of substrate temperature on the structural, optical and electrical properties of CdS thin films deposited by thermal evaporation.* Results in Physics, 2013. *3*: pp. 173–178.

41. G.K. Williamson and W.H. Hall, *X-ray line broadening from filed aluminium and wolfram.* Acta Metallurgica, 1953. *1*(1): pp. 22–31.

42. E.J. Mittemeijer and U. Welzel, *The "state of the art" of the diffraction analysis of crystallite size and lattice strain.* Zeitschrift Fur Kristallographie, 2008. *223*(9): pp. 552–560.

43. J. Tauc, *Optical properties and electronic structure of amorphous Ge and Si.* Materials Research Bulletin, 1968. *3*(1): pp. 37–46.

44. J. Tauc and A. Menth, *States in the gap.* Journal of Non-Crystalline Solids, 1972. *8*: pp. 569–585.

45. N. Chopra, A. Mansingh and G.K. Chadha, *Electrical, optical and structural properties of amorphous V_2O_5-TeO_2 blown films.* Journal of Non-Crystalline Solids, 1990. *126*(3): pp. 194–201.

46. C. Vatankhah and A. Ebadi, *Quantum size effects on effective mass and band gap of semiconductor quantum dots.* Research Journal of Recent Sciences 2013. *2*(1): pp. 21–24.

47. R. Das and S. Pandey, *Comparison of optical properties of bulk and nano crystalline thin films of CdS using different precursors.* International Journal of Material Science, 2011. *1*(1): pp. 35–40.

48. S. Tiwari and S. Tiwari, *Electrical and optical properties of CdS nanocrystalline semiconductors.* Crystal Research and Technology, 2006. *41*(1): pp. 78–82.

49. H. Lin, C.P. Huang, W. Li, C. Ni, et al., *Size dependency of nanocrystalline TiO$_2$ on its optical property and photocatalytic reactivity exemplified by 2-chlorophenol.* Applied Catalysis B: Environmental, 2006. *68*(1–2): pp. 1–11.

50. E.G. Birgin, I. Chambouleyron and J.M. Martínez, *Estimation of the optical constants and the thickness of thin films using unconstrained optimization.* Journal of Computational Physics, 1999. *151*(2): pp. 862–880.

CHAPTER 7

Fabrication Methods of Organic Solar Cells

MOHD TAUKEER KHAN[1*], MOHD. SHKIR[2], and
ABDULLAH ALMOHAMMEDI[1]

[1]Department of Physics, Faculty of Science, Islamic University of Madinah, Prince Naif bin Abdulaziz, Al Jamiah, Madinah 42351, Kingdom of Saudi Arabia

[2]Advanced Functional Materials & Optoelectronics Laboratory (AFMOL), Department of Physics, College of Science, King Khalid University, Guraiger, Abha 61413, Kingdom of Saudi Arabia

[*]Corresponding author. E-mail: khanmtk@iu.edu.sa.

ABSTRACT

In the last decade, organic solar cells have emerged as one of the most prominent candidates for the conversion of sunlight into electricity and have already achieved more than 18% efficiency. This chapter provides an insight into the organic photovoltaics (OPVs) technology including device architecture, materials, device physics, and fabrication methods. We believe that this chapter will be worthwhile for the beginner of researcher in the OPV technologies.

7.1 INTRODUCTION

Conjugated-polymers (CPs)-based organic photovoltaics (OPVs) systems hold the promise for an environmentally safe, flexible, lightweight, and cost-effective next-generation solar energy conversion platform [1–3]. A lot of progress has been reported in OPV since last decades and already more than 18% power conversion efficiency (PCE) has been achieved [4]. Among various solar power technologies, the OPV opens up a niche market where silicon solar cells are unable to enter due to its rigidness and weight. Many opportunities are present for the OPV technology due to its lightweight and transparent capabilities. Building-integrated photovoltaics is a market sector that can help the OPVs market grow being able to provide an aesthetic window film in addition to other alternatives and designs to reduce a building's energy usage. The reduction of weight for these OPVs allows them to be easily mounted onto a vertical surface, unlike silicon photovoltaics that require to be mounted securely on rooftops to meet various building codes. A transparent and aesthetic way of harvesting renewable energy on surfaces is already in existence in existing homes, as well as a novel way to design future skylights and windows in future building designs. As the technology is able to progress with the advancements in efficiency, novel applications and uses for OPVs will be created. Future possibilities might be organic charging tablet/laptop screens to films that might charge an electric vehicle. There are many opportunities for OPVs to fall into a niche market category if the panels are able to be developed to be more efficient comparable to silicon-based photovoltaics. In this chapter, we have discussed the various kind of OPV technologies including single layer, bilayer, bulk-heterojunction (BHJ), and organic/inorganic hybrid solar cells. We have also discussed the basic principle of working of devices and their fabrication methods in brief.

7.2   DIFFERENT KINDS OF THIN FILM POLYMER SOLAR CELLS

The polymer solar cells reported in the literature can be categorized by their device architecture as having a single layer, bilayer, blend or BHJs, and tandem structure. The reason behind the development of these structures is to achieve higher efficiencies by maximizing sunlight absorption, enhancing charge separation, and collection processes in the polymer materials.

7.2.1   SINGLE LAYER DEVICES

In the first generation of the OPV devices, a single layer of organic semiconductors was sandwiched between two electrodes with different work functions, such as indium tin oxide (ITO) and Al, as shown in Figure 7.1a. The first report on the photovoltaic behavior of organic semiconductors came in 1959. The photovoltaic cell based on anthracene single-crystal exhibited a photovoltage of 200 mV with an extremely low efficiency [5]. Since then, many years of research have shown that the typical power PCE of

OPV devices based on single-layer organic materials remains below 0.1%, making them unsuitable for any possible application. The low efficiency of these devices is primarily due to the low dielectric constant in the range of 2–4 [6], poor carrier mobility of the order of 10^{-3}–10^{-5} cm²/V s [7–11], and short exciton diffusion length 1–10 nm [12–16]. Due to the low dielectric constant of organic semiconductors, the absorption of sunlight leads to bound electron–hole (e–h) pairs generated (referred to as exciton) instead of free e–h pairs. The Columbic binding energy of an e–h pair separated by 0.6 nm in a system with ε_r = 3 is about 0.6 eV [17–20]. The electric field provided by asymmetrical work functions of the electrodes is not enough to break up these photogenerated excitons into free carriers. Since the exciton diffusion lengths in organic semiconductors are much shorter than the device thicknesses; hence, these excitons diffuse within the organic absorbing layer before reach to the electrode. Though the fabrication of single-layer device is very simple, the device suffers from low quantum efficiency and poor PCE due to the high e–h recombination rate.

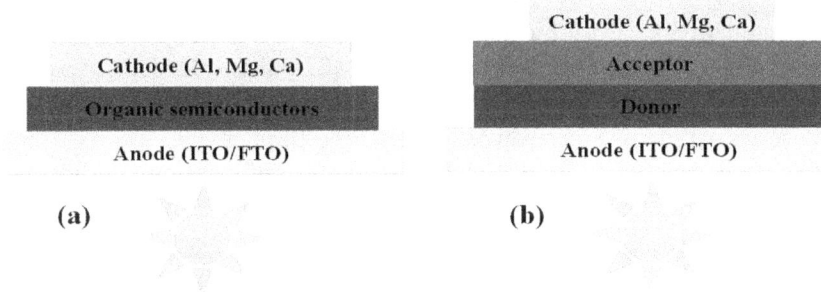

FIGURE 7.1 Schematic of (a) single layer and (b) bilayer OPV devices.

7.2.2 BILAYER DEVICES

The first major breakthrough in OPV performance was achieved in donor/acceptor bilayer device configuration, as shown in Figure 7.1b, demonstrated by Tang [21]. The bilayer device based on copper phthalocyanine (electron donor) and perylene tetra carboxylic (electron acceptor) shows a remarkable PCE ~1%. In bilayer heterojunction device configuration, two materials of different electron affinities and ionization potentials were used. The potential difference between two materials leads to the formation of a potential barrier at the interface and may favor the exciton dissociation. The electron will be accepted by the material with the larger electron affinity and the hole will be accepted by the material with the lower ionization potential. Since the exciton diffusion lengths in the organic materials are much shorter, therefore excitons should be formed within the diffusion length near the interface. Otherwise, the excitons will decay, yielding, and luminescence instead of a contribution to the photocurrent.

7.2.3   *BULK-HETROJUNCTION DEVICES*

7.2.3.1   *POLYMER–FULLERENE BHJ*

The short exciton diffusion length of organic semiconductors limits the performance of the bilayer structure. As discussed above, the exciton should be formed within the diffusion length range near the interface, otherwise they will recombine and give luminescence. Moreover, the short exciton diffusion length also limits the active layer thickness that results in poor sunlight absorption by bilayer OPV devices. To maximize the interfacial surface area for exciton dissociation, bicontinuous network of donor–acceptor (D–A) heterojunction was invented by Heeger et al. [22]. They blended CP poly[2-methoxy-5-(3′,7′-dimethyloctyloxy)-1,4-phenylene vinylene] (MDMO-PPV) with soluble fullerene derivative, namely [6, 6]-phenyl C_{61} butyric acid methyl ester denoted as PCBM in BHJs device configuration, as shown in Figure 7.2a. The device exhibits a PCE of about 2.9% better by more than two orders of magnitude than those that have been achieved with devices made with pure MEH-PPV, decay (2), yielding, for example, luminescence, if they are generated too far from the interface.

In BHJs configuration, donor/acceptor interface is available in bulk of the active layer where exciton dissociation takes place and exciton decay process is dramatically reduced. Hence, charge generation takes place all over in the active layer, provided that there exists a percolation pathway in each material from the interface to the respective electrodes. Later, optimization device structures and morphology of the active layer lead to achieve an efficiency up to 3.3% and an open-circuit voltages (V_{OC}) up to 0.82 V in MDMO-PPV-based devices [23, 24]. However, the relatively larger band gap of MDMO-PPV limits the sunlight absorption, and hence short-circuit current density (J_{SC}) was limited to 5–6 mA/cm². This leads to the development of new polymer poly(3-hexylthiophene) (P3HT), which has relatively low band gap (~1.9 eV), higher crystallinity and charge carrier mobility [25–27]. The optimization of morphology by thermal annealing [27] and solvent annealing [28] delivered an impressive total energy conversion efficiency of 6% [29], with a much higher current density (>10 mA/cm²). Unfortunately, the high highest occupied molecular orbitals (HOMO) (−5.1 eV) energy level of P3HT has restricted the V_{OC} to 0.6 V, which consequently limits the overall efficiency.

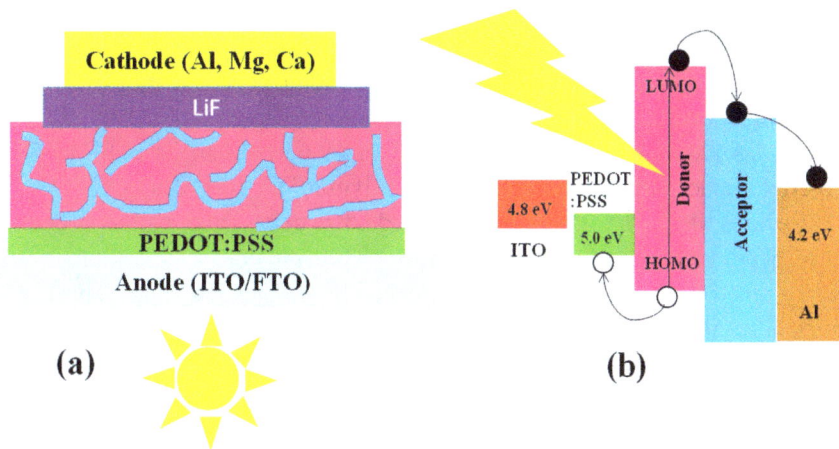

FIGURE 7.2 (a) Device architecture and (b) schematic band diagram for bulk-hetrojunctions OPV device.

To overcome the limited light absorption and low V_{OC}, two different approaches have been adopted to improve the efficiency of low-cost BHJ PV cells. The first approach was to improve V_{OC} by designing polymers with a low HOMO energy level. This approach has resulted in V_{OC} greater than 1 V in a few cases [30–32], though the overall efficiency has been less than 4% because of the mediocre J_{SC}. The second approach was to develop lower band gap of polymers for harvesting more influx photons and enhancing the J_{SC}. The device with poly[(4,4-didodecyldithieno[3,2-*b*:2′,3′-*d*]silole)-2,6-diyl-*alt*-(2,1,3-benzothiadiazole)-4,7-diyl] as donor and PCBM as acceptor exhibits V_{OC} ~ 0.61 V, PCE ~ 5.9% and J_{SC} as high as 17.5 mA/cm² [33]. The device based on a low band gap polymer PBDTTT achieves V_{OC} as high as 0.76 V with a certified PCE of 6.77% [34]. Among these low band gap polymers, PTB7-Th was the most prominent material, displaying an efficiency of over 10% [40, 41]. The BHJ device based on PTB7/PC$_{71}$BM further improved efficiency to 7.4%, with J_{SC} of 14.50 mA cm⁻², V_{OC} = 0.74 V and fill factor (FF) of 0.69 [35]. A PCE of 8.3% was achieved in a device based on PTB7-Th:PC$_{71}$BM when diiodooctane (DIO) was used as a solvent additive in chlorobenzene (CB) solvent [37]. The device based on regioregular PTB7-Th/PC$_{71}$BM exhibits a much higher J_{sc} (19.0 mA cm^{-2}), accompanied excellent V_{oc} (0.81 V) and FF (0.68) afford a champion efficiency of 10.42% [42]. In order to achieve optimum morphology for industrial application of polymer solar cells, several other low band gap polymers with different properties have been reported. The device based on PBnDT-FTAZ/PC$_{61}$BM shows an improved V_{oc} of 0.79 V, J_{sc} of 12.45 mA/cm^2, FF of 72.2%, and overall PCE of 7.1% [36]. Zhou et al. [38] incorporated 5-fluoro-2,1,3-benzothiadiazole into thiophene-based polymer backbone to generate a hole conducting polymer, improve crystal orientation, and the phase-separated morphology. The device achieved a PCE as high as 9.06% with V_{oc} of 0.7 V, J_{sc} of 19.63 mA/cm^2, and FF of 65 in inverted device structure viz. ITO/ZnO/polymer:PC$_{71}$BM/MoO$_3$/Al.

Besides the high PCE of PTB7-based solar cells, these cells suffer from poor performance in thick active layer (>300 nm) devices, which is very important for the industrial application of polymer solar cells. The use of thick film also increases the absorption strength of the solar cell and thus cell efficiency. The main reasons why PTB7 does not perform well in thick-film PSCs are related to the low crystallinity of the PTB7, which results in relatively low hole mobility ($\sim$10^{-4} cm^2 V^{-1} s^{-1}) [35].

The polymer exhibiting the temperature-dependent aggregation (TDA) behavior has achieved a great success in thick-film solar cells. Daize et al. designed chlorinated D–A type polymers PBDTHD-ClBTDD, in which benzo[1,2-b:4,5-b]dithiophene and chlorinated benzothiadiazole units were connected by thiophene π-bridges with an asymmetric alkyl chain [61]. The PSCs based on PBDTHD-ClBTDD:PC$_{71}$BM exhibit a PCEs of 9.11% for a 250-nm-thick active layer. Yan et al. have synthesized PffBT4T-2OD, PBTff4T-2OD, and PNT4T-2OD donor polymers that show a strong TDA effect. By using these polymers, they achieved high PCE up to 10.8% in 300 nm thick active layers based PSCs [51]. In 2017, they achieved a record PCE of 11.7% in inverted configuration, viz., ITO/ZnO/PffBT4T-C$_9$C$_{13}$:PC$_{71}$BM/ V$_2$O$_5$/Al by optimizing the morphology of active layer using the TDA technique and employing nonhalogenated solvents [39].

7.2.3.2 *POLYMER–NONFULLERENE BULK-HETROJUNCTION*

Despite the ultrafast charge separation and efficient charge transport properties, the weak optical absorption, limited tunability over energy levels, and morphological instability of fullerene derivative limits the performance of PSCs. In the past decade, a lot of effort has been put to design the alternate nonfullerene acceptor (NFA) that can retain the advantageous properties of fullerene derivatives and overcome the shortcomings of it. NFA possess the advantages of easy synthesis, strong absorption in the visible region, easily tunable energy levels, and good stability. The most widely investigated NFAs are based on aromatic fused rings and perylene imides. The extended conjugation in fused rings such as pentacene and rubrene enhances the intermolecular charge transport.

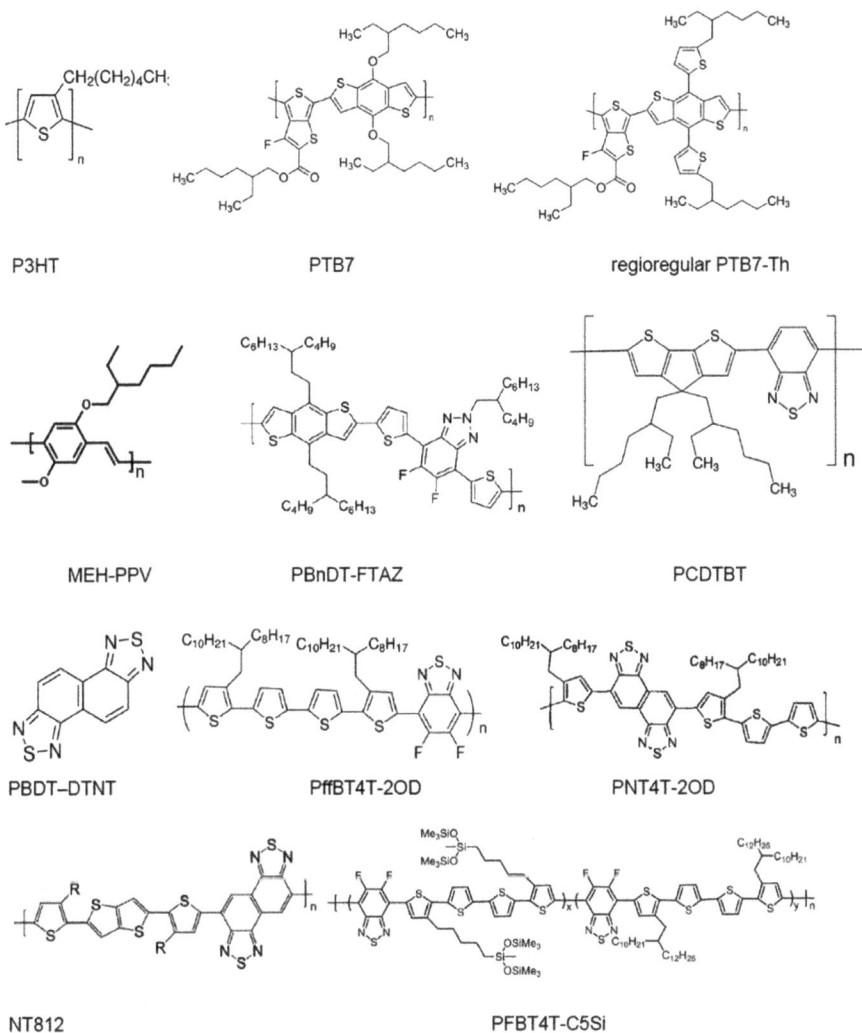

FIGURE 7.3 Chemical structure of donor polymers used in fullerene-based BHJ solar cells.

TABLE 7.1 Year Wise Performance of Polymer BHJ Solar Cells

Active Layer	References	J_{SC} (mA/cm²)	V_{OC} (V)	FF (%)	PCE (%)	Remark	Year
ITO/MEH-PPV:PCBM(1:4)/Ca	[22]	2	0.82	–	2.90	Illumination intensity 20 mW/cm²	1995
ITO/PEDOT:PSS/MEH-PPV:PCBM/LiF/Al	[43]	5.25	0.82	61.00	2.50	–	2001
ITO/PEDOT:PSS/P3HT:PCBM//Al	[44]	1.28	0.48	30.60	0.20	Not annealed	2003
ITO/PEDOT:PSS/P3HT:PCBM (1:1)/LiF/Al	[45]	8.50	0.55	60.00	3.50	Annealed at 75 °C for 4 min	2003
ITO/PEDOT:PSS/P3HT:PCBM (1:0.46)/LiF/Al	[46]	10.1	0.55	81.20	5.20	Annealed at 155 °C for 3 min	2005
ITO/PEDOT:PSS/PCPDTBT:PC71BM/Al	[47]	16.2	0.62	55.00	5.50	Dithiol treatment	2007
ITO/PEDOT:PSS/PBnDT-FTAZ:PC61BM(1:2)/Al	[48]	11.83	0.79	72.90	7.10	–	2011
ITO/PEDOT:PSS/PTB7:PC70BM /Ca/Al	[49]	15.46	0.76	68.00	7.90	Methanol treatment of the active layer	2013
ITO/PFN-OX/PBDT–DTNT:PC71BM/MoO3/Ag	[50]	17.62	0.74	66.10	8.62	Inverted configuration	2014
ITO/ZnO/PffBT4T-2OD:C71BM (1:1.2)/MoO3/Al	[51]	18.80	0.77	75.00	10.80	–	2014
ITO/ZnO/PNT4T-2OD:PC71BM (1:1.2)/MoO3/Al	[51]	19.80	0.76	68.00	10.10	–	2014
ITO/PEDOT:PSS/NT812:PC71BM/PFN-Br-/Ag	[52]	19.09	0.72	72.85	10.33	Solvent additive annealed at 100 °C for 15 min	2016
ITO/ZnO/PFN-OX/NT812:PC71BM/MoO3/Al	[52]	19.91	0.72	71.45	10.23	As above Inverted structure	2016
ITO/ZnO/PffBT4T-C9C13:PC71BM/V2O5/Al	[39]	19.80	0.78	73.00	11.70		2016
ITO/ZnO/PFBT4T-C5Si-25%:PC71BM/MoO3/Al	[53]	19.08	0.76	74.12	11.09		2017
ITO/PEDOT:PSS/PNTT-H:PC71BM/PNDIT-F3N-Br /Ag	[54]	20.20	0.77	71.80	11.30		2017
PffBT4T-C9C11:PC71BM	[55]	20.20	0.79	74.00	11.70		2017
ITO/PEDOT:PSS/P2F-EHp:IT-2F/PNDI-Br /Ag	[56]	19.20	0.92	69.20	12.25		2018

TABLE 7.1 (Continued)

Active Layer	References	J_{SC} (mA/cm^2)	V_{OC} (V)	FF (%)	PCE (%)	Remark	Year
ITO/ZnO/IDT-OB /MoO3/Ag	[62]	16.18	0.88	71.10	10.12		2017
ITO/ZnO/PFN/PTZ1:IDIC/MoO$_3$/Al	[63]	14.90	0.95	68.10	9.60	As cast	2017
		16.24	0.92	76.20	11.50	Thermal annealed at 120 °C for 10 min	
ITO /(PEDOT:PSS/PTPTI-T70:m-ITIC (1:1.25 w/w) /PDINO/Al	[64]	17.20	0.93	69.26	11.02	Thermal annealed at 150 °C for 10 min Solvent: CB:CF:DIO (1:2:0.5 vol%)	2017
ITO/PEDOT:PSS/PM6:IDIC/(PFN-Br/Al	[65]	17.30	0.95	70.00	11.50	Without any treatment	2017
ITO/NP-ZnO/PM7:IT-4F/MoO$_3$/Al	[67]	20.90	0.88	71.10	13.10		2018
ITO/ZnO/PFN/PM6:IT-4F(1:1)/MoO$_3$/Al	[69]	19.4	0.89	66.7	11.50	As cast	2018
		22.2	0.84	72.5	13.50	Thermal annealed at 100 °C for 20 min Solvent additive DIO	
ITO/ZnO/PBDB-T-2F:IT-4F/MoO$_3$/Al	[70]	20.81	0.84	76	13.20		2018
ITO/ZnO/PBDB-T-2F:IT-2Cl/MoO$_3$/Al		21.80	0.86	77	14.40		2018
ITO/PEDOT:PSS/PM6:Y6(1:1.2)/PDINO/Al	[57]	25.30	0.83	74.80	15.70	0.5 wt.% of CN thermal annealing at 110 °C for 10 min	2019
ITO/ZnO/PM6:Y6/MoO$_3$/Ag		25.2	0.82	76.10	15.70		
ITO/PEDOT:PSS/BTPT1-4F/PNDI-Br /Ag	[58]	26.68	0.81	74.11	16.02	Processed with 1.0% dibenzylether	2019

CB: chlorobenzene, CF: chloroform, DIO: 1,8-diiodooctane.

Zhan et al. [60] have synthesized NFA based on a bulky seven-ring fused core. The PSCs based on PTB7-Th:ITIC show better performance (PCE 6.8%) than the devices based on PTB7-Th:PC$_{61}$BM (PCE ~ 6.05%) and are very close to the performance of devices based on PTB7-Th:PC$_{71}$BM (PCE ~ 7.52%%). Zhishan et al. have synthesized fused-ring electron acceptor, IDT-OB, and obtain a PCE of 10.12% in the device configuration ITO/ZnO (30 nm)/active layer (210 nm)/MoO$_3$ (8.5 nm)/Ag (100 nm) [62]. The main advantage of this material was that it can deliver a PCE of 9.17% even the thickness of active layer was up to 210 nm owing to higher carrier mobility, which greatly reduces the charge recombination. Zhang et al. fabricated PSCs based on a wide band-gap polymer PTZ1 as donor and a planar IDT-based narrow band gap small molecule 2,2'-((2Z,2'Z)-((4,4,9,9-tetrahexyl-4,9-dihydro-s-indaceno[1,2-b:5,6-b']dithiophene-2,7-diyl) bis(methanylylidene))bis(3-oxo-2,3-dihydro-1H-indene-2,1-diylidene)) dimalononitrile (IDIC) as acceptor and achieve a PCE of 9.6% for device with active layer thickness of 210 nm. The thermal annealing of the device at 120 °C for 10 min further increase the PCE of up to 11.5% with V_{OC} of 0.92 V, J_{SC} of 16.4 mA cm^{-2}, and a FF of 76.2% [63]. Yang et al. have used ITIC as NFA in combination with random tetra polymerize donor polymers (PTTI-Tx) containing simple thiophene (T) and bithiophene (2T) electron-rich moieties [64]. The different microstructures were formed by variation of T:2T ratio in the polymer backbone. The device based on PTPTI-T$_{70}$:m-ITIC blend exhibits a PCE of 11.02%, with V_{OC} of 0.93 V, J_{SC} of 17.2 mA cm^{-2}, and a FF of 69.26%. Fan et al. have used narrow band gap small molecule acceptor IDIC whose energy levels (lowest unoccupied molecular orbital [LUMO] ~3.91 eV, and HOMO 5.65 eV) were well matched with donor PM6 (LUMO ~3.61 eV and HOMO5.50 eV), as well as possess complementary absorption spectrum, high crystallinity, and strong π–π stacking alignment, which favor charge carrier transport and hence suppress recombination in devices [65]. The device based on PM6:IDIC in conventional device structure, viz., ITO/PEDOT:PSS/PM6:IDIC/(PFN-Br/Al) achieved a PCE as high an 11.90% without any solvent treatment or thermal annealing [65]. The introduction of a fluorine atom into the NFA enhances the carrier mobility, optical absorption, morphology of active layer, and tune the molecular energy levels according to donor materials. By replacing IDIC to tetrafluorinated n-OS acceptor IT-4F, the PCE of the PSCs based on PM6:IT-4F was further enhanced to 13.5% [66]. Fan et al. [69] have fabricated PSCs based on PM6:IT-4F in the device configuration, viz., ITO/ZnO/PFN/PM6:IT-4F(1:1)/MoO$_3$/Al. The device shows a PCE of

11.5%. The device performance was further improved to 13.5% by adding 1,8-DIO as solvent additive and thermal annealing at 100 °C for 20 min. Moreover, when they replace donor polymer PM6 (chlorinated on the main chain) with PM7 (chlorinated on the conjugated thienyl side-chains), device based on PM7:IT-4F shows a PCE of 13.1% with a high V_{OC} of 0.88 V, J_{SC} of 20.9 mA cm^{-2} and FF of 71.1% [67]. The acceptor IT-4F also shows a high PCE of 13.1% with fluorinated donor polymer PBDB-T-SF [68]. The fluorine-contained is rather costly due to their complicated synthesis and low yields in the preparation of components. Therefore, Zhang et al. replaced the fluorine substituents with chlorine. It was found that the PSCs based on PBDB-T-2Cl:IT-4F exhibit higher PCE of 14.4% while the device based on fluorinated polymer PBDB-T-2F-and IT-4F shows relatively low efficiency of 13.2% [70]. Yuan et al. [57] have designed a ladder-type multifused ring, Y6, a new class of NFA, containing an electron-deficient core as a central unit to match with polymer PM6. To tune the electron affinity, a charge-deficient moiety was introduced in the middle of the central core. The fused central unit preserves conjugation along the length of the molecule that allows tuning of the electron affinity. Moreover, long alkyl side chains were introduced on the terminal of the central unit to increase the solubility of the resulting small molecule acceptor.

The Y6-based PSCs delivers a high PCE of 15.7% with both conventional and inverted configuration [57]. Recently, Y6-based PSCs with wide band gap donor polymer P2F-EHp have achieved a record PCE as high as 16.02% in the device configuration viz. ITO/PEDOT:PSS/active layer/PFNDI-Br/Ag, which is the highest PCE based on the single junction PSCs [58]. The champion efficiency of 16.02% was achieved with 1.0% dibenzylether (DBE) added, with a V_{OC} of 0.81 V, J_{SC} of 26.68 mA cm^{-2}, and FF of 74.11%. In contrast, a device with an inverted structure based on the same recipe in the photoactive layer gives a lower PCE of 13.13%, which might be due to the specific vertical components distribution that is unfavorable for downward electron extraction.

7.2.4 *ORGANIC–INORGANIC HYBRID SOLAR CELLS*

Despite the continuous growth in the performance of PSCs, they suffer from limited lifetime due to the use of environment degradable organic materials. Therefore, to make PSCs compatible with the presently available solar cells, the efficiency, as well as lifetime of these devices must be improved. The low performance of PSCs is mainly due to the higher exciton binding energy, poor charge carrier mobility, and short exciton diffusion length in organic semiconductors [71–74]. To overcome the problem of poor charge transport mechanism in CPs, inorganic nanocrystals (NCs) have been blended with CPs to integrate the desirable characteristics of organic and inorganic materials within a single device [75–80]. Photovoltaics devices based on CPs and inorganic NCs have the advantages attributes of organics, such as low cost, lightweight, large area, and flexibility [81–83], with the size-dependent electrical and optical properties, a broad absorption spectrum, excellent carrier mobility, and good environmental stability of inorganic [84, 85]. Hybrid device based on CPs/NCs have the same photon to electric current generation mechanism as in OPV: excitons created upon absorption of photon by organic semiconductors. These excitons are separated into free charge carriers at CPs/NCs junction where they move toward the respective electrodes [86]. A wide range of inorganic NCs such as CdS [87, 88], CdSe [89], Sb_2S_3 [90], PbS [91], Bi_2S_3 [92], $CuInS_2$ [93], and ZnO [94] have been used as an acceptor materials along with donor polymers [95–98].

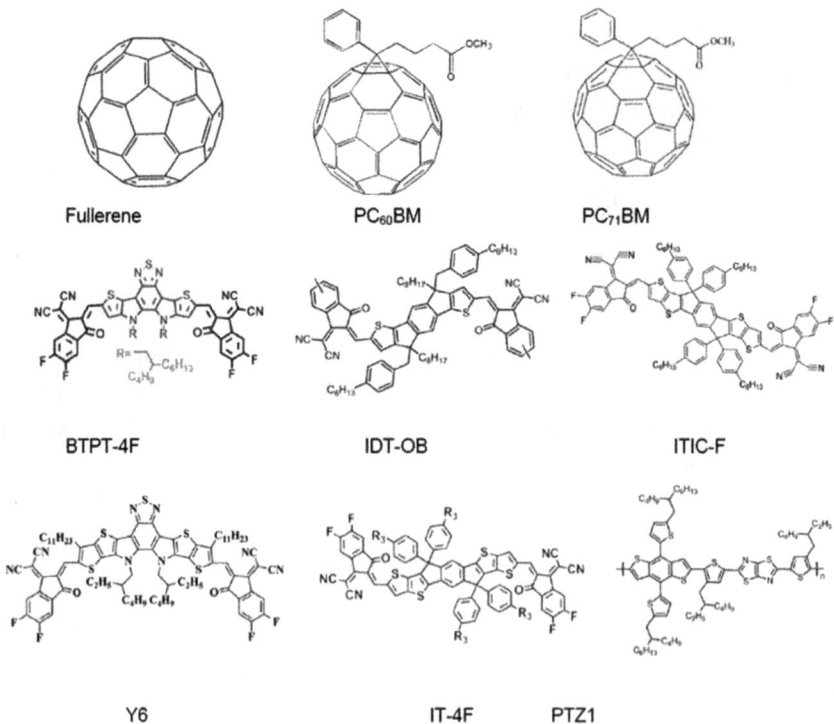

Fullerene   PC$_{60}$BM   PC$_{71}$BM

BTPT-4F   IDT-OB   ITIC-F

Y6   IT-4F   PTZ1

FIGURE 7.4   (Top) Fullerene (middle and bottom) NFAs used in BHJ PSCs.

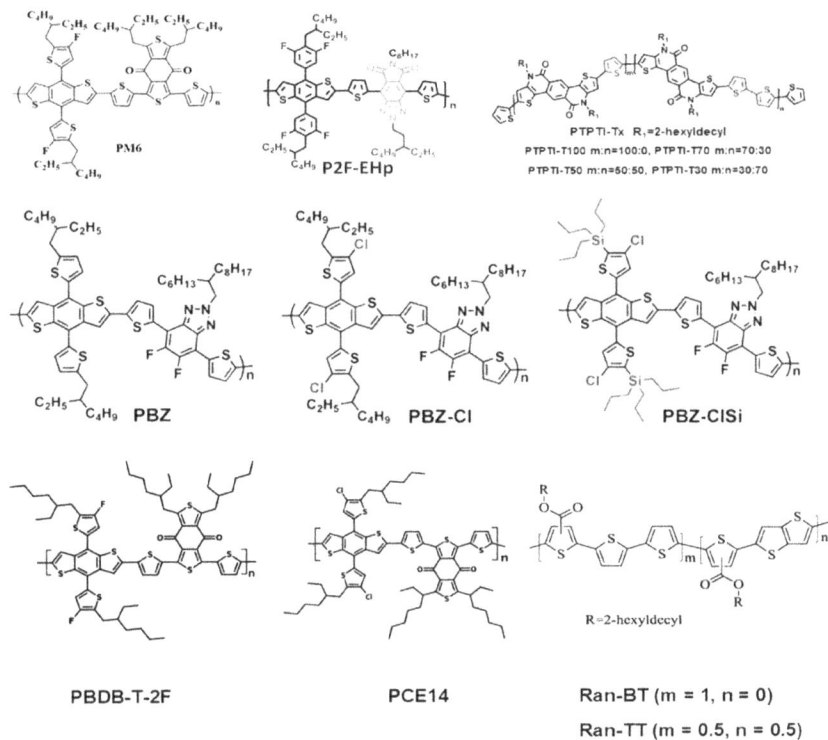

FIGURE 7.5 Nonfullerene-based donor molecules used in BHJ PSCs.

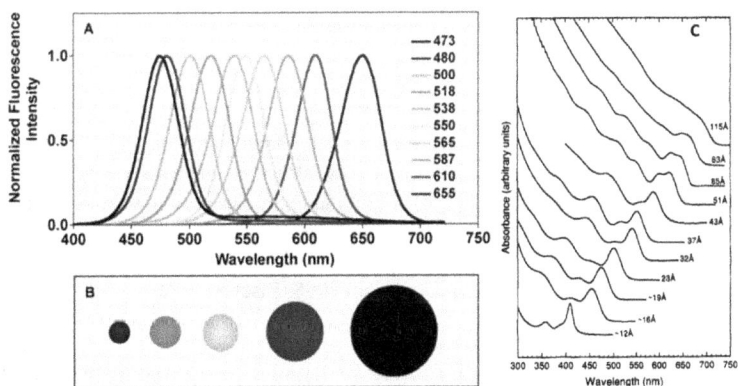

FIGURE 7.6 Size-tunable fluorescence spectra of CdSe quantum dots (A), and illustration of the relative particle sizes (B). From left to right, the particle diameters are 2.1 nm, 2.5 nm, 2.9 nm, 4.7 nm, and 7.5 nm. Reprinted with permission from Smith [130]. Copyright (2004) Royal Society of Chemistry (C) Absorption spectra of CdSe NCs dispersed in hexane and ranging in size from 1.2 to 11.5 nm. (A) Reprinted with permission from Murray et al. [131]. Copyright (1993) American Chemical Society.

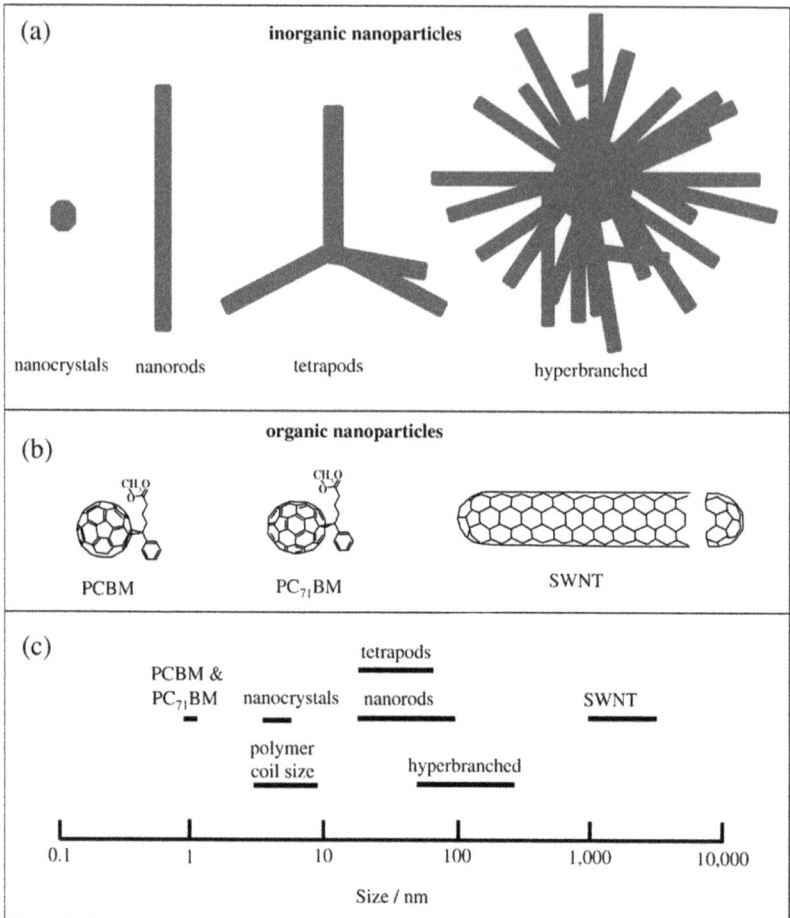

FIGURE 7.7 Structural depictions and approximate dimensions for selected nanoparticles are used in nanoparticle–polymer PV cells. The structures shown in (a) are for CdSe nanoparticles. The structures shown in (b) are for nanoparticles used in organic nanoparticle–polymer PV cells. The size ranges shown in (c) are estimates based on literature reports where the nanoparticles have been used in nanoparticle–polymer PV cells. Reprinted with permission from Saunders et al. [132]. Copyright 2008, Elsevier.

The first breakthrough work related to organic/inorganic hybrid solar cells was reported by Greenham et al. [99]. They have studied the optoelectrical properties of hybrid system based on poly(1-methoxy-4-(2-ethylhexyl oxy-2,5-phenylenevinylene)) and NCs of cadmium chalcogenides (CdS or CdSe). It has been found that with the increase of concentrations of inorganic NCs in polymer, the photoconductivity of the hybrid nanocomposite increases.

The first hybrid solar cell based on polymer/NCs was demonstrated by Alivisatos et al. [100]. They have achieved a PCE as high as 1.7% in the hybrid device based on regioregular P3HT as donor and CdSe nanosphere as acceptor. In 2003, Sun et al. [101] did a comparison of the performance of solar cells based on CdSe nanorods and tetrapods. They found that the device based on CdSe tetrapods shows better performance as compared to the CdSe nanorods due to more connecting pathways that are available for charge transport in tetrapods. Hindson et al. [102] did a comparative study of CdSe nanosphere and nanorods. They observed that the later have highly connected network of particles that are homogeneously distributed throughout the polymer film. Zhou et al. [108] reported a PCE of 2% with J_{sc} of 5.8 mA/cm^2 and V_{oc} of 0.67 V in a hybrid device fabricated using rr-P3HT and CdSe QDs. The hybrid device based on P3HT:CdSe/ZnS quantum dots shows a PCE of 5.1% and FF of 45% under the illumination of 8.19 mW/cm^2 [112]. The size of the nanoparticles also plays an important role in the performance of the device because it is expected that the nanorods or tetrapods might improve the charge transport of the inorganic network. It is also considered that the mixing of spherical nanoparticles with nanorods reduce the nanorods–nanorods horizontal aggregation, which improve the charge transport in the vertical direction.

The solvent vapor annealing process is an effective way to enhance the crystallinity of polymer thin films and improve their optoelectrical properties [133, 134]. Wu et al. [135] have used a chemical vapor annealing process to improve the PCE of a P3HT/CdSe nanorods based hybrid solar cell. After spin casting the P3HT/CdSe active layer on top of ITO/PEDOT:PSS, the substrates were transferred to a glass vessel containing benzene-1,3-dithiol. The whole vessel was kept at 120 °C for 20 min on a preheated hot plate and then Al contacts were applied via thermal evaporation. The device undergone through benzene-1,3-dithiol chemical vapor annealing process exhibits an improved device performance with V_{oc} of 0.553 V, J_{sc} of 9.7 mA/cm^2, and PCE of 2.65%, compared to without annealed device with V_{oc} of 0.577 V, J_{sc} of 5.823 mA/cm^2, and PCE of 1.56% (Figure 7.8).

To further improve the efficiency of hybrid devices, some other CPs have been employed in devices besides P3HT. In 2010, Jilian et al. [109] have studied the effect of incorporation of CdSe QDs in poly(9,9-*n*-dihexyl-2,7-fluorenilenevinylene-*alt*-2,5-thienylenevinylene) (PFT)/PCBM system. In this work, they found that the incorporation of CdSe QDs in the mixture of PFT/PCBM changes the film morphology, and this is responsible for the improvement in device photocurrent and efficiency. Dayal et al. [136] have used low band gap polymer PCPDTBT as a donor in combination with CdSe tetrapods in BHJs hybrid device. The hybrid device exhibits a record PCE of 3.13% with V_{OC} of 0.674 V, J_{SC} of 9.015 mA/cm², and FF of 51.47%. The improved device performance was due to the enhanced charge transfer pathway provided by CdSe tetrapods and improved solar photon absorption of PCPDTBT polymer. Celik et al. [137] have used CdSe nanorods instead of the tetrapods with PCPDTBT polymer and achieved even a higher efficiency of 3.42%. They also did a series of washing steps with methanol and *n*-hexane, to remove the capping ligand that results in better charge transfer between polymer and NCs and hence increase of device performance. Zhou et al. did hexanoic acid treatment to the trioctylphosphine/oleic acid (OA)-capped CdSe QDs. The P3HT:CdSe device exhibited a PCE of 2.1% with V_{OC} of 0.628V, J_{SC} of 6.0 mA/cm², and FF of 56%, whereas the device with PCPDTBT:CdSe shows higher PCE of 2.7% with V_{OC} of 0.591V, J_{SC} of 8.30 mA/cm² [138] (see Figure 7.9a). The higher short-circuit current in PCPDTBT:CdSe device is attributed to broader external quantum efficiency (EQE) spectrum from 300 to 850 nm in comparison with the P3HT:CdSe device, as shown in Figure 7.9b. Later, they used mixtures of nanorods and QDs of CdSe instead of nanorods or QDs only in PCPDTBT:CdSe hybrid device. The hybrid device exhibits a PCE above 3% [139]. Kuo et al. [140] have incorporated the CdSe tetrapods in PDTTTPD and investigated the effects of thermal annealing on the morphology and photovoltaic device performance, as prepared hybrid device exhibits a PCE of 1% with V_{OC} of 0.89 V, J_{SC} of 3.16 mA/cm², and FF of 36%. When the device was annealed at 130 °C for 20 min, the device performance improved to 2.9% with V_{OC} of 0.88 V, J_{SC} of 7.26 mA/cm², and FF of 46%. The improved performance of annealed device attributed to the removal of pyridine ligands from the surfaces of the CdSe tetrapods results in a decrease in the thickness of the active layer as revealed from synchrotron X-ray reflectivity.

The solvent plays a vital role to control the morphology of active layer and phase separation between donor/acceptor and hence the performance of the device. Huynh et al. [103] used binary solvent, a mixture of pyridine/chloroform, and found that by varying the concentration of the pyridine to chloroform, phase separation was tuned from the micrometer scale to the nanometer scale. In 2005, Sun et al. [104] did a similar study between 1,2,4-trichlorobenzene and chloroform. They observed that the device prepared from 1,2,4-trichlorobenzene solvent delivered efficiencies of 2.4%, as compared to the of 0.7% obtained with films deposited from chloroform solutions. Other studies also reported that by optimizing the solvent mixture used during film deposition, the donor/acceptor phase separation and hence efficiencies of the device could be improved [105–107].

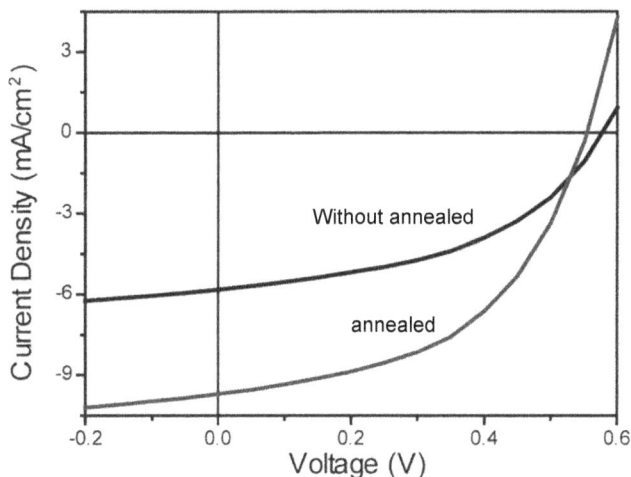

FIGURE 7.8   *J–V* characteristics of hybrid solar cell devices that have undergone benzene-1,3-dithiol chemical vapor annealing and without annealed. Reprinted from Wu et al. [135] with the permission of ACS publishing. Copyright 2010, ACS.

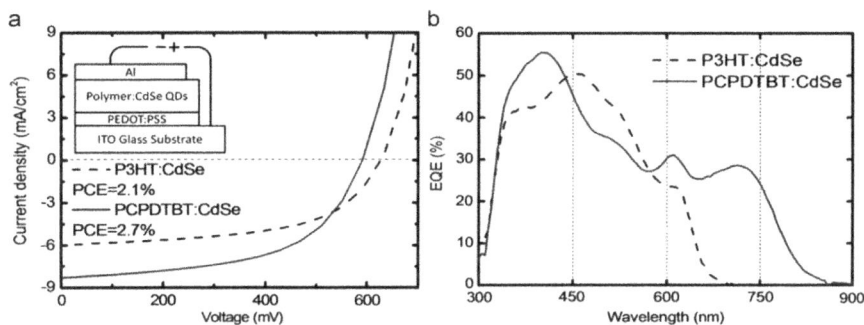

FIGURE 7.9   (a) *J–V* characteristics of P3HT:CdSe and PCPDTBT:CdSe hybrid photovoltaic devices. (b) EQE spectra of the P3HT:CdSe and the PCPDTBT:CdSe devices. Reprinted from Zhou et al. [138] with the permission of ACS publishing. Copyright 2011, Elsevier.

The higher electron mobility, good electron affinity, and wide band gap (2.4 eV) of CdS attracted many researchers to use it as an acceptor in organic/inorganic hybrid solar cells [117, 118]. Jiang et al. have fabricated hybrid device based on vertical aligned CdS nanorod P3HT via surface modification of CdS nanorod using aromatic acid as surface ligands [119]. The device exhibits a PCE of 0.25% with J_{sc} 1.56 mA/cm², V_{oc} of 0.34 V, and FF of 47%. Ren et al. [120] have controlled the P3HT/CdS interface and CdS QD interparticle distance via chemical grafting and ligand exchange methods, respectively. The device exhibits a record PCE of 4.1% with J_{SC} of 10.9 mA/cm^2, V_{OC} of 1.1 V, and FF of 35%. We have studied the effect of CdS QDs in P3HT:PCBM-based device [121]. The incorporation of CdS QDs in the P3HT:PCBM results in the enhancement in the device performance from 0.45% to 0.87%. The postproduction thermal annealing at 150 °C for 30 min further improved PCE to 0.95%. In a similar work based on P3HT: C$_{60}$:CdTe device, a PCE of 0.47%, with J_{SC} of 2.775 mA cm^{-2}, V_{OC} of 0.442 V, and FF of 0.38 was obtained [123].

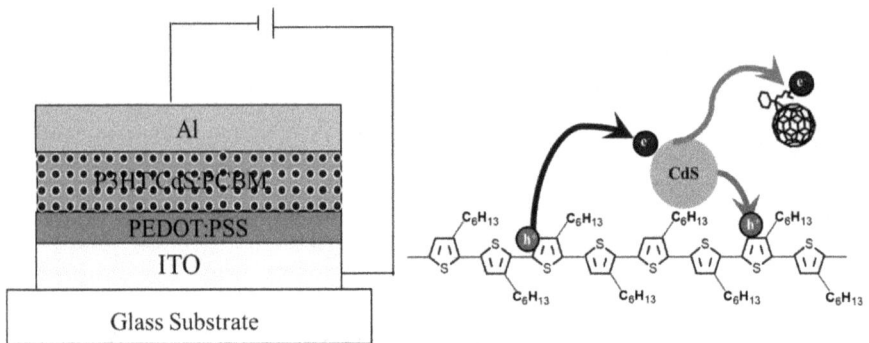

FIGURE 7.10 (Left) Device schematic of P3HT:CdS:PCBM hybrid solar cells and (right) charge transfer between P3HT, CdS, and PCBM in the same device.

To reduce the aggregation of NCs generally, they are capped with organic ligands. These organic ligands hinder the efficient charge transfer between photoexcited polymer and inorganic NCs [113]. To remove the organic ligands, polymer–NCs hybrid system was treated with pyridine. As pyridine is an immiscible solvent for the polymer and flocculation of the P3HT chains in an excess of pyridine lead to reduce the donor/acceptor phase separation, which resulting in poor photovoltaic performance [114]. Fu et al. [115] have used *n*-butanethiol to remove the OA on the surface of CdSe nanoparticles. The postproduction treatment of P3HT:CdSe thin films with *n*-butanethiol enhance the device performance as high as 3.09%. Later, they used acetic acid instead of *n*-butanethiol for the postdeposition treatment of P3HT:CdSe films, but lower efficiency about ~2% was obtained because the acetic acid is not strong enough to replace all of the original ligands [116]. To overcome the effects of the capping ligands on charge transport, the NCs have been grown *in situ* in the polymer matrix. The *in-situ* growth of the NCs in polymer templates controls the dispersion of the inorganic phase in the organic one, thus ensuring a large, distributed surface area for charge separation. Moreover, the *in-situ* growth also ensure the uniform distribution of NCs to the entire device thickness and thus contains a built-in percolation pathway for transport of charge carriers to the respective electrodes. In surfactant-assisted synthesis, nanoparticles growth is controlled by electrostatic interactions induced by the surfactant functional group and steric hindrance from the surfactant side alkyl chains. P3HT provides a combination of both effects, as it contains an electron-donating sulfur functionality, a potential anchorage for the nucleation, and growth of nanoparticles along with steric hindrance due to long hexyl side chains. Liao et al. have directly synthesize CdS single-crystal nanorods in P3HT matrix, where the P3HT is acting as a molecular template for geometrical manipulation of CdS NCs and, in the meantime, as an efficient charge conductor in composite form. The photovoltaic device of *in-situ* grown CdS nanorods in the matrix showed a PCE as high as 2.9% [122]. In a similar work, Dayal et al. have directly synthesized CdSe QDs in P3HT solution by using a high boiling point solvent, 1-octadecene [124]. Time-resolved microwave conductivity of fabricate thin films of *in-situ* synthesized hybrid system shows that photoinduced charge separation occurs at the polymer/CdSe interface. Stavrinadis et al. have synthesized the PbS nanorods/MEH–PPVcomposites at low temperature and surfactant-free method [125]. Khan et al. have also synthesized NCs of CdTe in P3HT matrix, as shown in Figure 7.11 [126]. The photovoltaic performances of P3HT:PCBM as well as P3HT-CdTe:PCBM have been investigated in the device configuration, viz., ITO/PEDOT:PSS/P3HT:PCBM/Al and ITO/PEDOT:PSS/P3HT-CdTe:PCBM/Al. The P3HT:PCBM device showed J_{SC} of 2.25 mA cm^{-2}, V_{OC} of 0.58 V, FF of 0.44, and PCE of 0.72%. The device with *in-situ* grown CdTe NCs in P3HT matrix (P3HT:-CdTe:PCBM) exhibits an increased J_{SC} to 3.88 mA cm^{-2}, V_{OC} of 0.80 V, and PCE to 0.79% while diminishing the FF to 0.32. Haque et al. [127] have thermally decomposed the solution-processable metal xanthate precursor complex Cd(S$_2$COEt)$_2$(C$_5$H$_4$N)$_2$ in P3HT film to fabricate P3HT–CdS hybrid film. The photovoltaic devices based upon P3HT–CdS hybrid film in inverted architecture: ITO/TiO$_x$/active layer/PEDOT:PSS/Au exhibits a PCE of 0.7% under full AM1.5 illumination and 1.2% under 10% incident power. Later, they observed that the nanomorphology of the synthesized CdS–P3HT hybrid film and yield of charge photogeneration are strongly depend upon the annealing temperature [128]. In the above device, they have inserted CdS interface layer between TiO$_x$ and active layer and then annealed at 160 °C for 10 min, the device performance enhanced to 2.17% with J_{SC} of 4.848 mA cm^{-2}, V_{OC} of 0.842 V, and FF of 53.23%.

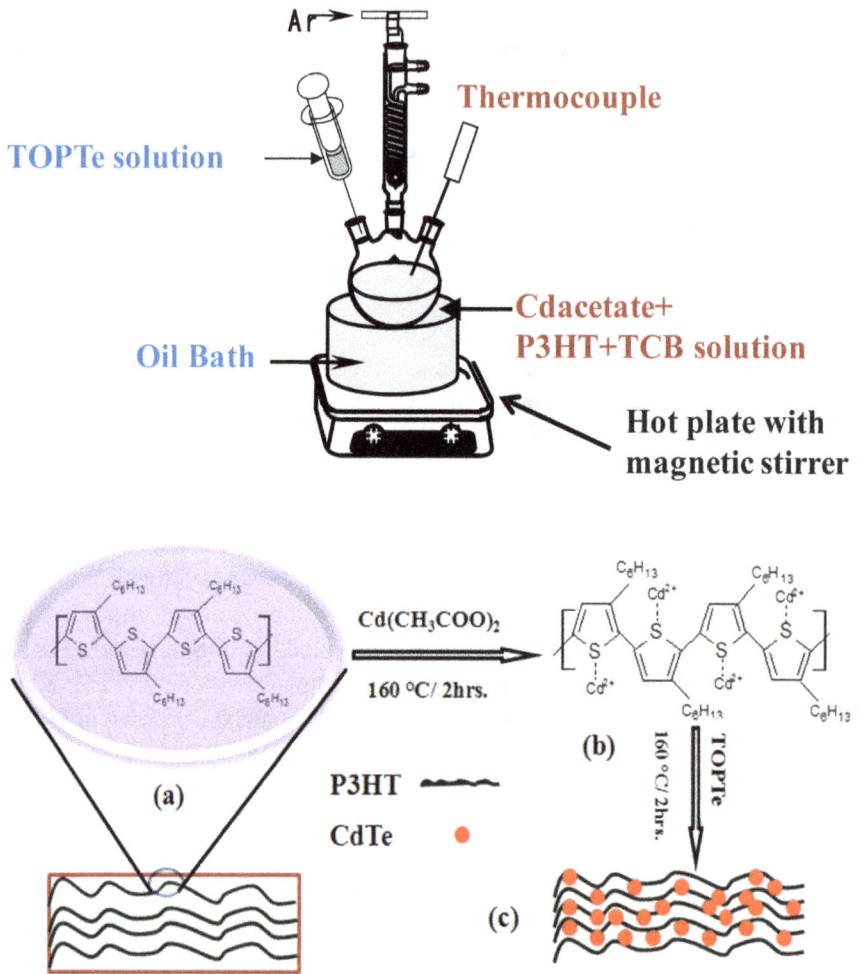

FIGURE 7.11   (Top) Experimental setup used for the *in-situ* synthesis of CdTe NCs in P3HT matrix, (bottom) *in-situ* growth of the CdTe NCs in the P3HT matrix. Reprinted from Khan et al. [126] with the permission of AIP Publishing.

Further, they have expanded this approach to fabricate the copper indium sulfide (CIS)/poly[(2,7-silafluorene)-*alt*-(4,7-di-2-thienyl-2,1,3-benzothiadiazole)] (PSiF-DBT) nanocomposite solar cells as CIS offers the advantage of a lower band gap (1.5 eV), compared to cadmium based NCs (CdSe ~1.74 eV, CdS ~2.4 eV). The photovoltaics device is fabricated in conventional configuration, viz., ITO/PEDOT:PSS/PSiF-DBT:CIS(1:7)/Al shows a efficiencies up to 1.7%. However, the device performance was improved to 2.8% when the dual layer of PSiF-DBT:CIS was used in the above configuration [129]. Zinc oxide NCs have been also used as acceptor materials in organic/inorganic hybrid solar cells. Beek et al. [110] have achieved a PCE of 0.9% with J_{SC} of 2.4 mA/cm^2 and a V_{OC} of 685 mV in a hybrid device based on P3HT:ZnO has been achieved. The incorporation of ZnO nanofibers in a blend of P3HT and PCBM enhance the efficiency up to 2.03% [111].

Ternary systems have recently attracted attention as candidates for traditional organic solar cells. Such systems can be comprised of polymer/inorganic NCs/fullerene [143–147], polymer/fullerene/fullerene [148, 149], polymer/polymer/fullerene combinations [150–1533], or small organic molecule/polymer/fullerene [154–156]. In 2008, Chen et al. [157] first demonstrated a high photoconductive gain in polymer/fullerene-based device on incorporation of CdTe NCs. An EQE of 8000% at 350 nm was achieved under the −4.5 V bias. Such a high photoconductive gain suggests that the CdTe NCs assist the hole injection into the polymer.

In 2010, De Freitas et al. [158] have studied a device based on PFT/PCBM/CdSe ternary systems. The PFT is a polymer that contains fluorine and thiophene units in the main chain [159]. It has been observed that both J_{SC} and V_{OC} increases after incorporation of CdSe NCs into the PFT/PCBM system, which results in an enhancement in PCE from 0.5% to 0.8%. In the same year, we have investigated devices based on P3HT:PCBM:CdSe in 1:0.8:1 ratio. The incorporation of CdS QDs results in the enhancement of efficiency from 0.45% to 0.87%, due to an increase in J_{SC} from 2.57 to 4.65 mA cm^{-2} [160]. We also observed that the postproduction thermal annealing improves device performance due to enhancement in the device parameters like FF, V_{OC}, and improvement in contact between active layer and electrodes. In 2011, Peterson et al. [7] have studied the performance of hybrid solar cells device based on P3HT/PCBM/CdSe ternary system. They used 2.5 nm CdSe NCs that were functionalized with an electron acceptor methyl viologen. An increase in the photocurrent was observed when a higher concentration of CdSe NCs incorporated into the P3HT/PCBM system. However, for a low concentration of CdSe, they observed that the photocurrent was decreased which decreases the efficiency of the hybrid devices. In 2012, Fu et al. [161] investigated ternary systems based on P3HT/PCBM/CdSe in an inverted solar cell configuration with TiO$_2$ as the electron collection layer. An increase in current density, from 7.51 mA cm^{-2} to 8.15 mA cm^{-2}, and V_{OC} from 0.57 V to 0.60 V as well as enhancement in PCE from 2.06% to 3.05% was observed in hybrid device as compared to the device without CdSe NCs.

In 2013, Huang et al. [162] investigated the effect of six different kinds of inorganic NCs (Au, CdS, and PbS) in P3HT/PCBM-based devices. The authors reported that the hybrid ternary-based P3HT/PCBM device showed no sign of efficiency improvement. Among all semiconducting nanoparticles investigated, the best device was obtained with CdS capped with an aromatic thiophenol molecule. The efficiency of 3.91% obtained was slightly lower than the 4.01% of the binary P3HT/PCBM based device. In the same year, photovoltaics performance of $P3HT/PC_{71}BM$ and $P3HT/PC_{71}BM/CdSe$ tetrapods have been investigated [163]. It has been observed that the device with CdSe tetrapods delivered a higher short-circuit current of (6.11 mA cm^{-2}) as compared to the device without CdSe NCs (4.78 mA cm^{-2}). Furthermore, the PCE has been also improved from 1.07% to 1.5% on the incorporation of CdSe tetrapods in $P3HT/PC_{71}BM$ hybrid system. This was explained by increased light absorption as well as the elongated shape of CdSe tetrapods. Tetrapods shape provide efficient charge transport inside the device also accept the electrons from P3HT and transferring them toward the PCBM.

7.3   DEVICE PHYSICS OF ORGANIC BULK-HETROJUNCTION SOLAR CELLS

7.3.1   BASICS OF MOLECULAR PHOTOPHYSICS

The knowledge of device physics is important to optimize the device structure or design of new materials to further push the performance of OPV devices. In this section, we are discussing various photophysical process that occurs from photoexcitation to electric current generation in OPV devices. The absorption of sunlight by CPs results in excitation of ground-state singlet electron (S_0) to upper excited singlet state referred as Franck–Condon transition [140]. There are various possibilities by which Franck–Condon state attain equilibrium state, which is shown in Figure 7.12a. As there are two different materials in OPV devices that differ in electron-donating and -accepting properties. Therefore, in addition to the transitions shown in Figure 7.12a, the photogenerated charges in CPs can participate in a number of inter- and intra-molecular processes, such as energy transfer or charge transfer to acceptor molecules. After photo-excitation of an electron from the HOMO to the LUMO, the electron can jump from the LUMO of the donor to the LUMO of the acceptor. However, this process, which is called photoinduced charge transfer, can lead to free charges only if the hole remains on the donor due to its higher HOMO level, as shown in Figure 7.7c. In contrast, if the HOMO of the acceptor is higher, the exciton transfers itself completely to the material of the lower band gap referred to as energy transfer shown by Figure 7.12b.

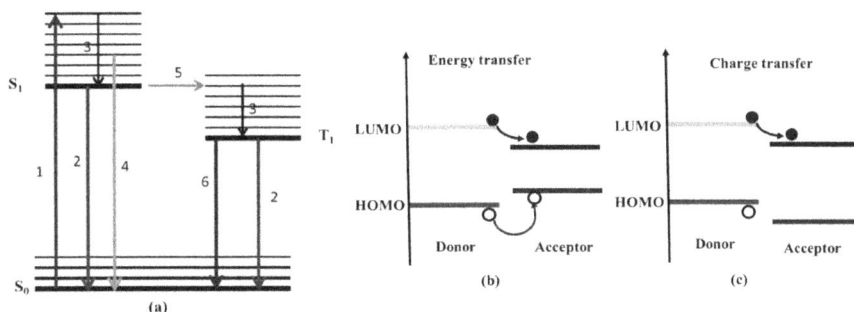

FIGURE 7.12 (a) Various process occurs after (1) absorption of sunlight by CPs, (2) internal conversion, (3) vibrational relaxation, (4) fluorescence, (5) intersystem crossing, and (6) phosphorescence. The interface between donor and acceptor can facilitate either (b) energy transfer or (c) charge transfer.

7.3.2   *FUNDAMENTAL PHYSICAL PROCESS IN BULK-HETROJUNCTIONS SOLAR CELLS*

The main physical processes that occur in the BHJ OPV devices are schematically illustrated in Figure 7.13. The absorption of sunlight leads to the creation of exciton in donor polymer depicted by process 1 in the figure. Because of the high exciton binding energy of CPs, the photogenerated exciton does not split into free charge carriers. The photogenerated excitons start to diffuse (2) within the donor phase and if they come across the interface with the acceptor then a fast dissociation takes place, leading to charge separation at donor–acceptor interface (3). Subsequently, the separated free charge carriers are transported toward the respective electrodes (4) with the aid of the internal electric field caused by the different work functions of electrodes. These dissociated charge carriers move toward the electrodes where they are collected and driven into the external circuit (5). However, the excitons can decay (6), yielding, for example, luminescence, if they are generated too far from the interface. Thus, the excitons should be formed within the diffusion length of the interface being an upper limit for the size of the CPs phase in the BHJ.

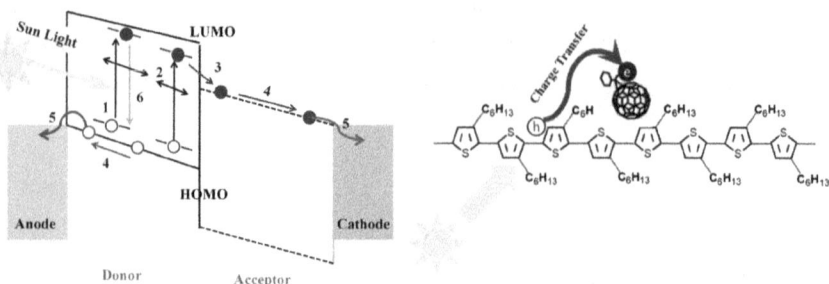

FIGURE 7.13 (Left) Schematic showing fundamental operation process in BHJs solar cells, the numbers refer to the operation processes explained in the text. (Right) Pictorial representation of charge transfer from P3HT to PCBM.

7.4 DEVICE FABRICATION

The combination of good electrical resistivity (<20 Ω/cm^2) and excellent optical transparency (>85%), the ITO-coated glass substrates are the most suitable substrate for OPV device fabrication. Moreover, the work function of ITO is also well matched with the energy levels of the donor polymers used in OPV device fabrication which facilitate the hole collection. The devices are generally fabricated on 2.0 × 2.0 cm^2 ITO-coated glasses substrates. To prevent short circuit between ITO and top metal electrode, part of the ITO is generally etched with aqua regia. The etching process takes place on clean substrates, patterned with tape. After taping, the substrates are dipped in HCl (1:1 of HCl:H$_2$O) and Zn dust was put in the solution. After waiting for 15–45 min, the ITO is removed from open region and remains under the tape, as shown in Figure 7.14. The laser patterning and photolithography are other important process for ITO etching, which remove the ITO completely without affecting the glass substrate. The etched substrates are cleaned twice with Hallmax solution, and then washed by deionized water. After washing with deionized water, the substrates are ultasonicated for 30 min in acetone at 50 °C, followed by boiling trichloroethylene and isopropanol for 20 min separately. Finally, these substrates were dried in vacuum oven at 120 °C for 2 h. Prior to use, the cleaned substrate was treated with oxygen plasma.

Poly(3,4-ethylenedioxythiophene):poly(styrene sulfonate)(PEDOT:PSS) is used as hole transport layer in OPV devices. PEDOT:PSS layers is spin-coated onto the precleaned ITO substrate and annealed to remove the moisture. The ITO-coated glass substrate with a layer of PEDOT:PSS serves as the transparent anode through which light is incident on the device. For the preparation of OPV devices, a solution of donor polymers and acceptor materials in the ratio of *x:y* with a concentration of *z* wt.% in solution are dissolved by ultrasonication. The active layer is spin casted from this solution on the top of PEDOT:PSS layer in the glove box, followed by annealing. Finally, aluminum or silver or Au electrodes are deposited via thermal evaporation through a shadow mask at 2×10^{-6} Torr.

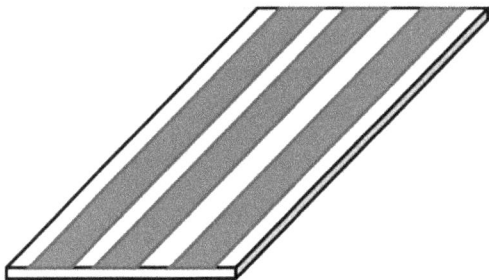

FIGURE 7.14 A patterned ITO coated glass substrate. White regions represent glass and gray represent ITO.

7.5 SUMMARY AND FUTURE OUTLOOK

This chapter provides insight into the solution-processed thin-film organic solar cells devices, and we hope that this will be worthwhile for the beginner of researcher in the OPV technologies. In the first part of the chapter, we have discussed the different device architecture and why they were developed. Polymer–fullerene bulk-hetrojunction is the most successful device configuration to date. At the earlier stage, BHJs devices were fabricated by using MDMO-PPV or P3HT as donor material and PCBM as acceptor in the device configuration viz. ITO/PEDOT:PSS/Active layer/Al and achieved a PCE of more than 6%. In the second stage, low band gap polymers such as PTB7, PCDTBT, PBTT, etc., have been used as donor polymer along with the PCBM in the same configuration and have shown a PCE of more than 11%. The devices with inverted structure viz. ITO/ZnO/Active layer/ MoO_3/Al have shown similar performance with improved environmental stability. Recently, NFA has emerged as very promising acceptor materials and device efficiency has been cross the 15% PCE target. To date, the most efficient OPV devices are based on (1) PM6-Y6 in the inverted configuration viz. ITO/ZnO/PM6:Y6/MoO_3/Ag with a PCE of 15.70%, (2) BTPTT-4F in conventional structure viz. ITO/PEDOT:PSS/BTPTT-4F/PNDI-Br/Ag with PCE of 16.02% and (3) D18:Y6 based solar cells with PCE of 18.22%. Besides organic acceptor molecules, inorganic NCs such as cadmium chalcogenides (CdS, CdSe, and CdTe) and lead chalcogenides (PbS, PbSe) have been also employed in OPV devices aims to integrate the desirable characteristics of organic and inorganic materials within a single device. At the end of the chapter, we discussed the device physics of OPV devices and fabrication methods. Finally, we conclude that OPV devices have a bright future due to its lightweight, flexibility, and transparent characteristics. There are many opportunities for OPVs to fall into a niche market category if the panels are able to be developed to be more efficient and environmental stable comparable to silicon-based photovoltaics.

CONFLICTS OF INTEREST

The authors declare no conflict of interest.

ACKNOWLEDGEMENTS

The authors would like to thanks to Deanship of Scientific Research at Islamic University of Madinah for funding this work through the 2nd Tamayyuz programme of the academic year 1441/1442 H (project No: 497). Author Mohd. Shkir would like to express his gratitude to King Khalid University, Saudi Arabia, for providing administrative and technical support.

KEYWORDS

- **organic solar cells**
- **donor/acceptor**
- **device physics**
- **device fabrication**

REFERENCES

1. Zhang S., Qin Y., Zhu J., Hou J., Over 14% efficiency in polymer solar cells enabled by a chlorinated polymer donor, Adv. Mater. **2018**, 30(20), 1800868.
2. Hou J., Olle I., Friend R. H., Gao F., Organic solar cells based on non-fullerene acceptors, Nat. Mater., **2018**, 17, 119–128.
3. Li S., Ye L., Zhao W., Yan H., Bei Y., Liu D., Li W., Ade H., Hou J., A wide band gap polymer with a deep highest occupied molecular orbital level enables 14.2% efficiency in polymer solar cells, J. Am. Chem. Soc., **2018**, 140(23), 7159–7167.

4. Liu, Q., Jiang Y., Jina K. Qin J., Xu J., Li W. Xiong J. Liu J., Xi Z., Sun K., Yang S., Xiaotao Z., Ding L., 18% Efficiency organic solar cells, Sci. Bull., **2020**, 65, 272–275.

5. Kallmann H., Pope M., J. Chem. Phys., **1959**, 30, 585.

6. Blom P. W. M., Mihailetchi V. D., Koster L. J. A., Markov D. E., Adv. Mater., **2007**, 19, 1551.

7. Khan M. T., Bhargav R., Kaur A., Dhawan S. K., Chand S., Thin Solid Films, Elsevier, **2010**, 519, 1007, ISSN: 0040-6090, citations: 44.

8. Khan M. T., Bajpai M., Kaur A., Dhawan S. K., Chand S., Electrical, optical and hole transport mechanism in thin films of poly(3-octylthiophene-*co*-3-hexylthiophene): synthesis and characterization, Synth. Met., **2010**, 160, 1530, ISSN: 0379-6779, citations: 3.

9. Khan M. T., Kaur A., Dhawan S. K., Chand S., Hole transport mechanism in organic/inorganic hybrid system based on in-situ grown cadmium telluride nanocrystals in poly(3-hexylthiophene), J. Appl. Phys., **2011**, 109, 114509, ISSN: 0021-8979 E-ISSN: 1089-7550.

10. Khan M. T., Agrawal V., Almohammedi A., Gupta V., Solid State Electron., **2018**, 145, 49–53, citations: 4.

11. Khan M. T., Almohammedi A., J. Appl. Phys., **2017**, 122, 075502. doi: 10.1063/1.4999316, citations: 4.

12. Choong V., Park Y., Gao Y., Wehrmeister T., Mullen K., Hsieh B. R., Tang C. W., Appl. Phys. Lett., **1996**, 69, 1492.

13. Halls J. J. M., Pichler K., Friend R. H., Moratti S. C., Holmes A. B., Appl. Phys. Lett., **1996**, 68, 3120.

14. Halls J. J. M., Friend R. H., Synth. Met., **1997**, 85, 1307.

15. Markov D. E., Amsterdam E., Blom P. W. M., Sieval A. B., Hummelen J. C., J. Phys. Chem. A, **2005**, 109, 5266.

16. Markov D. E., Tanase C., Blom P. W. M., Wildeman J., Phys. Rev. B, **2005**, 72, 045217.

17. Hertel D., Bassler H., Chem. Phys. Chem., **2008**, 9, 666.

18. Barth S., Bässler H., Phys. Rev. Lett., **1997**, 79, 4445.

19. Silinsh E. A., Capek V., Organic Molecular Crystals, AIP, New York, 1994.

20. Marks R. N., Halls J. J. M., Bradley D. D. C., Friend R. H., Holmes A. B., J. Phys. Condens. Matter., **1994**, 6, 1379.

21. Tang C. W., Appl. Phys. Lett., **1986**, 48, 183.

22. Yu G., Gao J., Hummelen J. C., Wudl F., Heeger A. J., Science, **1995**, 270, 1789.

23. Shaheen S. E., Brabec C. J., Sariciftci N. S., Padinger F., Fromherz T., Hummelen J. C., Appl. Phys. Lett., **2001**, 78, 841.

24. Brabec C. J., Shaheen S. E., Winder C., Sariciftci N. S., Denk P., Appl. Phys. Lett., **2002**, 80, 1288.

25. Reyes-Reyes M., Kim K., Carroll D. L., Appl. Phys. Lett., **2005**, 87, 3.

26. Kim J. Y., Kim S. H., Lee H. H., Lee K., Ma W. L., Gong X., Heeger A. J., Adv. Mater., **2006**, 18, 572.

27. Ma W. L., Yang C. Y., Gong X., Lee K., Heeger A. J., Adv. Funct. Mater., **2005**, 15, 1617.

28. Li G., Shrotriya V., Huang J. S., Yao Y., Moriarty T., Emery K., Yang Y., Nat. Mater., **2005**, 4, 864.

29. Kim K., Liu J., Namboothiry M. A. G., Carroll D. L., Appl. Phys. Lett., **2007**, 90, 163511.

30. Gadisa A., Mammo W., Andersson L. M., Admassie S., Zhang F., Andersson M. R., O. Inganas, Adv. Funct. Mater., **2007**, 17, 3836.

31. Zhang F., Jespersen K. G., Bjorstrom C., Svensson M., Andersson M. R., Sundstrom V., Magnusson K., Moons E., Yartsev A., Inganas O., Adv. Funct. Mater., **2006**, 16, 667.
32. Andersson L. M., Zhang F., Inganas O., Appl. Phys. Lett., **2007**, 91, 071108.
33. Coffin R. C., Peet J., Rogers J., Bazan G. C., Nat. Chem., **2009**, 1, 657.
34. Chen H.-Y., Hou J., Zhang S., Liang Y., Yang G., Yang Y., Yu L., Wu Y., Li, G., Nat. Photonics, **2009**, 3(11), 649–653.
35. Liang Y., Xu Z., Xia J., Tsai S.-T., Wu Y., Li G., Ray C., Yu L., Adv. Mater., **2010**, 22, 1.
36. Price S. C., Stuart A. C., Yang L., Zhou H., You W., J. Am. Chem. Soc., **2011**, 133, 4625.
37. Jain N., Chandrasekaran N., Sadhanala A., Friend R. H., McNeill C. R., Kabra, D., J. Mater. Chem. A, **2017**, 5(47), 24749–24757.
38. Zhou C., Zhang G., Zhong, C., Jia, X., Luo, P., Xu R., Cao, Y., Adv. Energy Mater., **2016**, 7(1), 1601081. doi:10.1002/aenm.201601081.
39. Zhao J., Li Y., Yang G., Jiang K., Lin H., Ade H., Ma W., Yanl H., Nature Energy, **2016**, 1, 15027.
40. Liao S. H., Jhuo H. J., Cheng Y. S., Chen S. A., Adv. Mater., **2013**, 25, 4766.
41. Chen J. D., Cui C., Li Y. Q., Zhou L., Ou Q. D., Li C., Li Y., Tang J. X., Adv. Mater., **2015**, 27, 1035.
42. Zhong H., Ye L., Chen J.-Y., Jo S. B., Chueh C.-C., Carpenter J. H., Ade H., Jen A. K. Y., J. Mater. Chem. A, **2017**, 5, 10517.
43. Shaheen S. E., Brabec C. J., Sariciftci N. S., Padinger F., Fromherz T., Hummelen J. C., Appl. Phys. Lett., **2001**, 78(6), 841–843.
44. Chirvase D., Chiguvare Z., Knipper M., Parisi J., Dyakonov V., Hummelen J. C., Synth. Met., **2003**, 138(1-2), 299–304.
45. Padinger F., Rittberger R. S., Sariciftci N. S., Adv. Funct. Mater., **2003**, 13(1), 85–88.
46. Reyes-Reyes M., Kim K., Dewald J., López-Sandoval R., Avadhanula A., Curran S., Carroll D. L., Org. Lett., **2005**, 7(26), 5749–5752.
47. Peet J., Kim J. Y., Coates N. E., Ma W. L., Moses D., Heeger A. J., Bazan G. C., Nat. Mater., **2010**, 6, 497–500.
48. Price S. C., Stuart A. C., Yang L., Zhou H., You W., J. Am. Chem. Soc., **2011**, 133(12), 4625–4631. doi:10.1021/ja1112595.
49. Zhou H., Zhang Y., Seifter J., Collins S. D., Luo C., Bazan G. C., Heeger A. J., Adv. Mater., **2013**, 25(11), 1646–1652.
50. Hu X., Yi C., Wang M., Hsu C.-H., Liu S., Zhang K., Cao Y., Adv. Energy Mater., **2014**, 4(15), 1400378. doi:10.1002/aenm.201400378.
51. Liu Y., Zhao J., Li Z., Mu C., Ma W., Hu H., Yan H., Nat. Commun., **2014**, 5(1), 5293. doi:10.1038/ncomms6293.
52. Jin Y., Chen Z., Dong S., Zheng N., Ying L., Jiang X.-F., Liu F., Huang F., Cao, Y., Adv. Mater., **2016**, 28(44), 9811–9818.
53. Liu X., Nian L., Gao K., Zhang L., Qing L., Wang Z., Ying L., Xie Z., Ma Y., Cao Y., Liu F., Chen J., J. Mater. Chem., **2017**, 5(33), 17619–17631.
54. Jin Y., Chen Z., Xiao M., Peng J., Fan B., Ying L., Cao Y., Adv. Energy Mater., **2017**, 7(22), 1700944.
55. Hu H., Chow P. C. Y., Zhang G., Ma T., Liu J., Yang G., Yan H., Acc. Chem. Res., **2017**, 50(10), 2519–2528.
56. Fan B., Du X., Liu F., Zhong W., Ying L., Xie R., Cao Y., Nat. Energy, **2018**, 3, 1051–1058, doi:10.1038/s41560-018-0263-4.
57. Kang Q., Ye L., Xu B., An C., Stuard S. J., Zhang S., Yao H., Ade H., Hou J., Joule, **2019**, 3, 227–239.

58. Fan B., Zhang D., Li M., Zhong W., Zeng Z., Ying L., Huang F., Cao Y., Sci China Chem, **2019**, 62, https://doi.org/10.1007/s11426-019-9457-5.

59. Li J. L., Dierschke F., Wu J. S., Grimsdale A. C., Müllen K., J. Mater. Chem., **2006**, 16, 96–100.

60. Lin Y., Wang J., Zhang Z.-G., Bai H., Li Y., Zhu D., Zhan X., Adv. Mater. **2015**, 27, 1170–1174.

61. Mo D., Wang H., Chen H., Qu S., Chao P., Yang Z., Tian L., Su Y.-A., Gao Y., Yang B., Chen W., He F., Chem. Mater. **2017**, 29, 2819−2830.

62. Feng S., Zhang C., Liu Y., Bi Z., Zhang Z., Xu X., Ma W., Bo Z., Adv. Mater. **2017**, 29, 1703527.

63. Guo B., Li W., Guo X., Meng X., Ma W., Zhang M., Li Y., Adv. Mater. **2017**, 29, 1702291.

64. Chen S., Cho H. J., Lee J., Yang Y., Zhang Z.-G., Li Y., Yang C., Adv. Energy Mater., **2017**, 7, 1701125.

65. Fan Q., Wang Y., Zhang M., Wu B., Guo X., Jiang Y., Li W., Guo B., Ye C., Su W., Fang J., Ou X., Liu F., Wei Z., Sum T. C., Russell T. P., Li Y., Adv. Mater., **2017**, 30, 1704546.

66. Fan Q., Su W., Wang Y., Guo B., Jiang Y., Guo X., Liu F., Thomas P. R., Zhang M. J., Li Y. F., Sci. China Chem., **2018**, 61, 531–537. doi:10.1007/s11426-017-9199-1.

67. Fan Q., Zhu Q., Xu Z., Su W., Chen J., Wu J., Guo X., Ma W., Zhang M., Li Y., Nano Energy, **2018**, 48, 413–420. https://doi.org/10.1016/j.nanoen.2018.04.002.

68. Zhao W., Li S., Yao H., Zhang S., Zhang Y., Yang B., Hou J., J. Am. Chem. Soc. **2017**, 139, 7148−7151.

69. Fan Q., Su W., Wang Y., Guo B., Jiang Y., Guo X., Liu F., Russell T. P., Zhang M., Li Y., Sci China Chem, **2018**, 61(5), 531–537.

70. Zhang S., Qin Y., Zhu J., Hou J., Adv. Mater. **2018**, 30, 1800868.

71. Greene L. E., Law M., Yuhas B. D., Yang P., J. Phys. Chem. C, **2007**, 111, 18451.

72. Ravirajan P., Peiro A. M., Nazeeruddin M. K., Graetzel M., Bradley D. D. C., Durrant J. R., Nelson J., J. Phys. Chem. B, **2006**, 110, 7635.

73. Zhu R., Jiang C. Y., Liu B., Ramakrishna S., Adv. Mater., **2009**, 21, 994.

74. Xin H., Reid O. G., Ren G., Kim F. S., Ginger D. S., Jenekhe S. A., ACS Nano, **2010**, 4, 1861.

75. Coakley K. M., McGehee M. D., Chem. Mater., **2004**, 16, 4533.

76. Kim B. G., Kim M. S., Kim J., ACS Nano, **2010**, 4, 2160.

77. Zhang S., Cyr P. W., McDonald S. A., Kostantatos G., Sargent E. H., Appl. Phys. Lett., **2005**, 87, 233101.

78. Groves C., Reid O. G., Ginger D. S., Acc. Chem. Res., **2010**, 5, 612–620.

79. Gunes S., Sariciftci N. S., Inorganica Chimica Acta, 2008, 361, 581.

80. Alivisatos A. P., Science, **1996**, 271, 933.

81. Steigerwald M. L., Eisrus L., Acc. Chem. Res. **1990**, 23, 283.

82. Empedocles S. A., Bawendi M. G., Acc. Chem. Res., **1999**, 32, 389.

83. Murphy C. J., Coffer J. L., Appl. Spectr., **2002**, 56, 16.

84. Weller H., Ang. Chem. Int. Eng. Ed., **1993**, 32, 41.

85. Arici E., Meissner D., Schäffler F., Sariciftci N. S., Int. J. Photoenergy, **2003**, 5, 199.

86. Ginger D. S., Greenham N. C., Phys. Rev. B, **1999**, 59, 10622.

87. Khan M. T., Bhargav R., Kaur A., Dhawan S. K., Chand S., Thin Solid Films, **2010**, 519, 1007.

88. Ren S., Chang L.-Y., Lim S.-K., Zhao J., Smith M., Zhao N., Bulovic V., Bawendi M., Gradecak S., Nano Lett., **2011**, 11, 3998.

89. Huynh W. U., Dittmer J. J., Alivisatos A. P., Science, **2002**, 295, 2425−2427.
90. Bansal N., O'Mahony F. T. F., Lutz T., Haque S. A., Adv. Energy Mater., **2013**, 3, 986.
91. Zhang Y., Li Z., Ouyang J., Tsang S.-W., Lu J., Yu K., Ding J., Tao Y., Org. Electron., **2012**, 13, 2773.
92. Martinez L., Higuchi S., MacLachlan A. J., Stavrinadis A., Miller N. C., Diedenhofen S. L., Bernechea M., Sweetnam S., Nelson J., Haque S. A., Tajima K., Konstantatos G., Nanoscale, **2014**, 6, 10018.
93. Rath T., Edler M., Haas W., Fischereder A., Moscher S., Schenk A., Trattnig R., Sezen M., Mauthner G., Pein A., Meischler D., Bartl K., Saf R., Bansal N., Haque S. A., Hofer F., List E. J. W., Trimmel G., Adv. Energy Mater., **2011**, 1, 1046.
94. Oosterhout S. D., Koster L. J. A., van Bavel S. S., Loos J., Stenzel O., Thiedmann R., Schmidt V., Campo B., Cleij T. J., Lutzen L., Vanderzande D., Wienk M. M., Janssen R. A. J., Adv. Energy Mater., **2011**, 1, 90.
95. Gao F., Ren S., Wang J., Energy Environ. Sci., **2013**, **6**, 2020.
96. Rath T., Trimmel G., Hybrid Mater., **2014**, 1, 15.
97. Freitas J. N., Goncalves A. S., Nogueira A. F., Nanoscale, **2014**, 6, 6371.
98. Moule A. J., Chang L., Thambidurai C., Vidu R., Stroeve P., J. Mater. Chem., **2012**, 22, 2351.
99. Greenham N. C., Peng X., Alivisatos A. P., Phys. Rev. B, Condens. Matter., **1996**, 54, 17628.
100. Huynh W. U., Dittmer J. J., Alivisatos A. P., Science, **2002**, 295, 2425−2427.
101. Sun B. Q., Marx E., Greenham N. C., Nano Lett., **2003**, 3, 961.
102. Hindson J. C., Saghi Z., Hernandez-Garrido J.-C., Midgley P. A., Greenham N. C., Nano Lett., **2011**, 11, 904.
103. Huynh W. U., Dittmer J. J., Libby W. C., Whiting G. L., Alivisatos A. P., Adv. Funct. Mater., **2003**, 13, 73.
104. Sun B. Q., Snaith H. J., Dhoot A. S., Westenhoff S., Greenham N. C., J. Appl. Phys., **2005**, 97, 014914.
105. Sun B. Q., Greenham N. C., Phys. Chem. Chem. Phys., **2006**, 8, 3557.
106. Han L. L., Qin D. H., Jiang X., Liu Y. S., Wang L., Chen J. W., Cao Y., Nanotechnology, **2006**, 17, 4736.
107. Wang L., Liu Y. S., Jiang X., Qin D. H., Cao Y., J. Phys. Chem. C, **2007**, 111, 9538.
108. Zhou Y., Riehle F. S., Yuan Y., Schleiermacher H.-F., Niggemann M., Urban G. A., Kruger M., Appl. Phys. Lett., **2010**, 96, 013304.
109. d. Freitas J. N., Grova I. R., Akcelrud L. C., Arici E., Sariciftci N. S., Nogueira A. F., J. Mater. Chem., **2010**, 20, 4845.
110. Beek W. J. E., Wienk M. M., Janssen R. A. J., Adv. Funct. Mater., **2004**, 16, 1009.
111. Olson D., Piris J., Tcollins R., Shaheen S. E., Ginley D. S., Thin Solid Films, **2006**, 496, 26.
112. Dixit S. K., Madan S., Madhwal D., Kumar J., Sihgh I., Bhatia C. S., Bhatnagar P. K., Mathur P. C., Org Electron., **2012**, 13, 710–714.
113. Huynh W. U., Dittmer J. J., Libby W. C., Whiting G. L., Alivisatos A. P., Adv. Funct. Mater., **2003**, 13, 73.
114. Cui D., Xu J., Zhu T., Paradee G., Ashok S., Gerhold M., Appl. Phys. Lett., **2006**, 88, 183111.
115. Fu W.-F., Shi Y., Qiu W. M., Wang L., Nan Y. X., Shi M.-M., Li H.-Y., Chen H.-Z., Phys. Chem. Chem. Phys., **2012**, 14, 12094–12098.

116. Fu W.-F., Shi Y., Wang L., Shi M.-M., Li H.-Y., Chen H.-Z., Sol. Energy Mater. Sol. Cells, **2013**, 117, 329–335.
117. Chen F., Qiu W., Chen X., Yang L., Jiang X., Wang M., Chen H., Sol Energy, **2011**, 85, 2122–2129
118. Mohamed N. B. H., Haouari M., Ebdelli R., Zaaboud Z., Habchi M. M., Hassen F., Maaref H., Ouada H. B., Phys. E, **2015**, 69, 145–152.
119. Jiang X., Chen F., Qiu W., Yan Q., Nan Y., Xu H., Yang L., Chen H., Sol. Energy Mater. Sol. Cells, **2010** 94, 2223–2229.
120. Ren S., Chang L.-Y., Lim S.-K., Zhao J., Smith M., Zhao N., Bulovic V., Bawendi M., Gradecak S., Nano Lett., **2011**, 11, 3998–4002.
121. Khan, M. T., Bhargav, R., Kaur, A., Dhawan, S. K., Chand S., Thin Solid Films, **2010**, 519(3), 1007–1011.
122. Liao, H.-C., Chen, S.-Y., Liu, D.-M., Macromolecules, **2009**, 42(17), 6558–6563. doi:10.1021/ma900924y.
123. Li Y., Mastria R., Li K., Fiore A., Wang Y., Cingolani R., Manna L., Gigli1 G., Appl. Phys. Lett., **2009**, 95, 043101.
124. Dayal S., Kopidakis N., Olson D. C., Ginley D. S., Rumbles G. J. Am. Chem. Soc., **2009**, 131(49), 17726–17727. doi:10.1021/ja9067673.
125. Stavrinadis A., Beal R., Smith J. M., Assender H. E., Watt A. A. R., Adv. Mater., **2008**, 20(16), 3105–3109. doi:10.1002/adma.200702115.
126. Khan M. T., Kaur A., Dhawan S. K., Chand S., J. Appl. Phys., **2011**, 110(4), 044509. doi:10.1063/1.3626464.
127. Leventis H. C., King S. P., Sudlow A., Hill M. S., Molloy K. C., Haque S. A., Nano Lett., **2010**, 10,1253–1258, ISSN: 1530-6984.
128. Dowland S., Lutz T., Ward A., King S. P., Sudlow A., Hill M. S., Haque S. A., Adv. Mater., **2011**, 23(24), 2739–2744. doi:10.1002/adma.201100625.
129. Rath T., Edler M., Haas W., Fischereder A., Moscher S., Schenk A., Haque S. A., Trimmel G., Adv. Energy Mater., **2011**, 1(6), 1046–1050. doi:10.1002/aenm.201100442.
130. Smith A. M., Nie S., Analyst, **2004**,129, 672–677
131. Murray C. B., Norris D. J., Bawendi M. G., J. Am. Chem. Soc., **1993**, 115(19), 8706–8715. doi:10.1021/ja00072a025.
132. Saunders B. R., Turner M. L., Adv. Colloid Interface Sci., **2008**, 138(1), 1–23. doi:10.1016/j.cis.2007.09.001.
133. Mascaro D. J., Thompson M. E., Smith H. I., Bulovic V., Org. Electron., **2005**, 6, 211–220.
134. Dickey K. C., Anthony J. E., Loo Y. L., Adv. Mater., **2006**, 18, 1721–1726.
135. Wu Y., Zhang G., Nano Lett. **2010**, 10, 1628–1631.
136. Dayal S., Kopidakis N., Olson D. C., Ginley D. S., Rumbles G., Nano Lett., **2010**, 10, 239.
137. Celik D., Krueger M., Veit C., Schleiermacher H. F., Zimmermann B., Allard S., Dumsch I., Scherf U., Rauscher F., Niyamakom P., Sol. Energy Mater. Sol. Cells, **2012**, 98, 433.
138. Zhou Y., Eck M., Veit C., Zimmermann B., Rauscher F., Niyamakom P., Scherf U., Sol. Energy Mater. Sol. Cells, **2011**, 95(4), 1232–1237. doi:10.1016/j.solmat.2010.12.04.
139. Zhou Y., Eck M., Men C., Rauscher F., Niyamakom P., Yilmaz S., Krüger M., Sol. Energy Mater. Sol. Cells, **2011**, 95(12), 3227–3232. doi:10.1016/j.solmat.2011.07.015.
140. Franck J., Elementary processes of photochemical reactions, Trans. Faraday Soc., **1926**, 21, 536.
141. Hal P. A. V., Wienk M. M., Kroon J. M., Verhees W. J. H., Sloff L. H., Gennip W. J. H. V., Jonkhejm P., Janssen R. A. J., Adv. Mater., **2003**, 15, 118.

142. Oosterhout S. D., Wienk M. M., van Bavel S. S., Thiedmann R., Koster L. J. A., Gilot J., Loos J., Schmidt V., Janssen R. A. J., Nat. Mater., **2009**, 8, 818.
143. Wang D. H., Kim D. Y., Choi K. W., Seo J. H., Im S. H., Park J. H., Par O. O., Heeger A. J., Angew. Chem., Int. Ed., **2011**, 50, 5519.
144. Wang D. H., Park K. H., Seo J. H., Seifter J., Jeon J. H., Kim J. K., Park J. H., Park O. O., Heeger A. J., Adv. Energy Mater., **2011**, 1, 766.
145. Kim C. H., Cha S. H., Kim S. C., Song M., Lee J., Shin W. S., Moon S. J., Bahng J. H., Kotov N. A., Jin S. H., ACS Nano, **2011**, 5, 3319.
146. Spyropoulos G. D., Stylianakis M. M., Stratakis E., Kymakis E., Appl. Phys. Lett., **2012**, 11, 213904.
147. Xu X. Y., Kyaw A. K. K., Peng B., Zhao D. W., Wong T. K. S., Xiong Q. H., Sun X. W., Heeger A. J., Org. Electron., **2013**, 14, 2360.
148. Street R. A., Davies D., Khlyabich P. P., Burkhart B., Thompson B. C., J. Am. Chem. Soc., **2013**, 135, 986.
149. Li H., Zhang Z. G., Li Y. F., Wang J. Z., Appl. Phys. Lett., **2012**, 101, 163302.
150. Seo J., Cho M. J., Lee D., Cartwright A. N., Prasad P. N., Adv. Mater. **2011**, 23, 3984–3988.
151. Zhang Y., Li Z., Ouyang J., Tsang S.-W., Lu J., Yu K., Ding J., Tao Y., Org. Electron., **2012**, 13, 2773–2780.
152. Ameri T., Heumuller T., Min J., Li N., Matt G., Scherf U., Brabec C. J., Energy Environ. Sci., **2013**, 6, 1796.
153. An Q. S., Zhang F. J., Zhang J., Tang W. H., Wang Z. X., Li L. L., Xu Z., Teng F., Wang Y. S., Sol. Energy Mater. Sol. Cells, **2013**, 118, 30.
154. Min J., Ameri T., Gresser R., Lorenz-Rothe M., Baran D., Troeger A., Sgobba V., Leo K., Riede M., Guldi D. M., Brabec C. J., ACS Appl. Mater. Interfaces, **2013**, 5, 5609.
155. Lyons D. M., Kesters J., Maes W., Bielawski C. W., Sessler J. L., Synth. Met., **2013**, 178, 56.
156. Derouiche H., Mohamed A. B., Sci. World J., **2013**, 2013, 914981.
157. Chen H.-Y., Lo M. K. F., Yang G., Monbouquette H. G., Yang Y., Nat. Nanotechnol., **2008**, 3, 543.
158. de Freitas J. N., Grova I. R., Akcelrud L. C., Arici E., Saricifci N. S., Nogueira A. F., J. Mater. Chem., **2010**, 20, 4845.
159. de Freitas J. N., Pivrikas A., Nowacki B. F., Akcelrud L. C., Saricifici N. S., Nogueira A. F., Synth. Met., **2010**, 160, 1654.
160. Peterson E. D., Smith G. M., Fu M., Adams R. D., Coffin R. C., Carroll D. L., Appl. Phys. Lett., **2011**, 99, 073304.
161. Fu H., Choi M., Luan W., Kim Y.-S., Tu S.-T., Solid-State Electron., **2012**, 69, 50.
162. Huang Y.-J., Lo W.-C., Liu S.-W., Cheng C.-H., Chen C.-T., Wang J.-K., Sol. Energy Mater. Sol. Cells, **2013**, 116, 153.
163. Ahmad R., Arora V., Srivastava R., Sapra S., Kamalasanan M. N., Phys. Status Solidi A, **2013**, 210, 785.

CHAPTER 8

Optoelectronics for Biomedical Applications

MOHD IMRAN[1*], MOHAMMAD SHARIQ[2], and MD. MOTTAHIR ALAM[3*]

[1]*Department of Chemical Engineering, Faculty of Engineering, Jazan University, Jazan 45142, Saudi Arabia*

[2]*Department of Physics, Faculty of Science, Jazan University, Jazan 45142, Saudi Arabia*

[3]*Department of Electrical & Computer Engineering, Faculty of Engineering. & Technology, King Abdulaziz University, Jeddah 21589, Saudi Arabia*

Corresponding author. E-mail: imranchaudhary0@gmail.com; mohammad.mottahir@gmail.com

ABSTRACT

This chapter covers the basics of optoelectronics and general optoelectronic properties applied in the development of equipment used in biomedical fields. This chapter starts with the introduction of photonics and electromagnetic wave. Semiconductor energy band diagram and its interaction with the light are mentioned afterward. Some of the important optoelectronic materials are also introduced and their properties and applications, especially in the fields of biomedical, are explained. Recently explored optoelectronic materials such as conducting polymer, phosphorene, carbon nanotubes, and other metal oxides nanoparticles are also mentioned in the optoelectronic materials section. Optoelectronic devices starting from simple diode to complex solar cells have been added and presented well. Special attention has been given to the application of optoelectronics to biomedical equipment. The well-known optoelectronic sensors have been described and classified according to their nature of invasiveness and noninvasiveness. The design and the components of these sensors as well as working principles are also part of the discussion carried out in this chapter. Other familiar techniques such as endoscopic imaging techniques and other important medical equipment are also included. Some of the important diagrams or schemes are included and explained according to their function. The future of optoelectronics and its application in the biomedical field are discussed. The future development in optogenetics and retinal prosthesis is included in the biomedical application section as well as in the future prospective section. A conclusion of all the components covered in this chapter is given as the closing of this chapter.

8.1 INTRODUCTION

Photonics is a branch of science concerning the conduct of photons (light). Electronics is a section of science and technology concerning the behavior of electrons. Optoelectronics is the study and use of electronic gadgets interacting with light. It depends on the quantum mechanical impacts of light on electronic materials, particularly semiconductors. Optoelectronic devices utilize electronic technology in which light is emitted, adjusted, or converted. Fundamental optoelectronic devices based on semiconductor junction electronics include photodiodes (PDs) in which optical radiation is converted to an electrical signal, light-emitting diodes (LEDs) in which electrical energy is changed into an optical signal, and laser diodes (LDs) that involve the transformation of electrical energy into optical energy in the form of laser. In this section, we will introduce some basics of electromagnetic wave (light), the interaction of light with matter, and the idea of analysis of interacted radiation (after interaction with sample) for medical measurement.

8.1.1 ELECTROMAGNETIC WAVES

Electromagnetic (EM) waves are a movement of coupled electric and magnetic oscillations with the speed of light exhibiting wave-like behavior. Electromagnetic waves are considered waves because under appropriate conditions they show wave properties of diffraction, interference, and polarization. There are some examples of other circumstances in which electromagnetic waves behave as though they comprise streams of particles. Wave-particle duality along with special relativity is central to a comprehension of modern physics. British physicist James Clerk Maxwell in the year 1864 suggested that accelerating electric charges produced coupled electric and magnetic oscillation that can move continuously in the space. Periodically oscillating charges produce disturbances in the form of waves whose magnetic and electric components are perpendicular to each other and propagation direction; Maxwell demonstrated that the velocity of electromagnetic waves (v) is given by

$$v = \frac{1}{\sqrt{\varepsilon_0 \mu_0}} = 2.99 \times 10^8 \text{ m/s} \tag{8.1}$$

where μ_0 is the magnetic permeability of free space and ε_0 is the electric permittivity. This value is equal to the speed of light. Based on this extraordinary correspondence, Maxwell reasoned that light comprises electromagnetic waves. Finally, in 1888, physicist Heinrich Hertz demonstrated the existence of EM waves and the EM waves behaved the same as Maxwell had anticipated. Not only light, but also every such wave has similar fundamental nature. Numerous features of EM waves' interaction with matter rely on their frequencies. Our eye responds to light waves (EM waves) that have a span of small frequency intervals, from about 4.3×10^{14} Hz (red light) to about 7.5×10^{14} Hz (violet light). EM wave spectrum (10^3–10^{22} Hz) covers radiation from the low frequencies including electrical power transmission radio frequencies, microwaves, infrared, visible, ultraviolet to the high frequencies found in X-rays and gamma rays. A trademark feature of all waves is that they follow the principle of superposition [1]. All phenomena of diffraction, interference, and polarization are completely explained by the wave nature of light. Following Hertz's experiments, the question of the fundamental nature of light seemed clear: light consisted of EM waves that obeyed Maxwell's theory. This fact lasted a few years as some strongly observed phenomena could not explain the wave nature of light. For example, the origin of black body radiation was explained only by the quantum theory of light. In 1900, German physicist Max Planck was surprised to find that energy was always emitted in the form of tiny discrete packets called the quantum of energy. He experimentally determined the equation for the spectral energy density of blackbody radiation that explained the whole spectral emission of the black body (Equation (8.2))

$$u(v)\,dv = \frac{8\pi h}{c^3} \frac{v^3\,dv}{e^{hv/kT-1}} \tag{8.2}$$

Here, h is a Plank's constant whose value is 6.626×10^{-34} J s and c is speed of light.

Not only blackbody radiation but the photoelectric effect, Compton effect, and pair production were failed to explain by considering light as a wave. In the photoelectric effect, electrons are emitted from the metal surface when the light of adequately high frequency is incident on it and the emitted electrons are called photoelectrons. In the year 1905, Einstein explained the photoelectric effect by assuming energy of light is not extended over wavefronts (as Huygens principle) but it is a form of small energy packets or photons. A photon of frequency v has energy hv, equivalent to Planck's quantum energy. Compton effect further confirms the photon model of light. As indicated by the quantum hypothesis of light, photons act like particles except for zero rest mass (i.e., absence of rest mass). Compton in the early 1920s, Compton observed scattering of the photon that established solid proof in support of the quantum hypothesis of radiation [1]. The scattering of a photon by an electron is called the Compton effect. It pursues conservation of energy, momentum, and energy of the scattered photon (longer wavelength) is less than that of the incident photon. Another evidence of photon model is pair production in which a photon materializes into an electron and a positron (a positively charged electron). Now the question comes that what is the nature of light? The idea that light moves as a progression of small energy packets is completely contradicted to the wave hypothesis of light and both perspectives have experimental evidence, as we have mentioned. In the consideration of wave theory of light, each point of a wavefront is assumed as a source of secondary wavelets that spread out continuously in the forward direction at speed of light. As indicated by the quantum hypothesis, light comprises discrete energy packets (photons), which can be absorbed by a single electron in an atom. However, in spite of the particle nature the light displays, the frequency of light is required to know the energy of a photon in quantum theory. We can consider light having a double character. The wave hypothesis and the quantum hypothesis complement each other. The "true nature" of light incorporates both wave and particle characters.

There are certain energy levels that electrons can have in an isolated atom. Atoms are brought closer together to form a solid and atoms are so close to each other that their valence electron wave functions overlap. Greater the number of interacting atoms, the more the number of levels produced by the mixing of their respective valence wave functions. In a solid, splitting into many levels (equal to the number of atoms present) makes levels very close to each other to form an energy band comprising nearly continuous spread of allowed energies. The energy band structure of a solid determines whether it is a conductor, an insulator, or a semiconductor. An electron in a conductor can only have energies that fall within its energy bands. In a conductor, valence and conduction bands (CBs) may overlap and valence electrons have a continuous distribution of allowed energies. In other solids (semiconductors and insulators), the bands may not overlap and the intervals between them are called energy band gaps. In insulators, the energy band is generally high so that valence electrons do not have enough thermal energy (KT0.025 eV) to jump into the CB where they can move about freely. The energy band gap of a semiconductor is small enough. The small number of valence electrons in a semiconductor at room temperature has enough thermal energy to jump the forbidden band and enter the CB. These few electrons are sufficient to enable a small amount of electric current to flow when an electric field is connected.

8.1.2 OPTICAL ABSORPTION

Optoelectronic gadgets depend on the interaction of light with matter, commonly semiconductor materials. Some basic examples of devices are solid-state semiconductor laser, LEDs, photodetectors, and solar cells. Mostly light or EM radiation is absorbed by the semiconductor and transformed into electrical signals (electron–hole pairs) or electric current into light energy. It is important to know the interaction of light with semiconductor material for understanding the working of these devices. Consider light of wavelength λ incident on a semiconductor with band-gap E_g. The energy of the radiation E is given by hc/λ. Based on the relation between E and E_g there can be two conditions.

1. $E < E_g$—for this condition semiconductor is transparent for light; however, it can be scattered at the interfaces.
2. $E > E_g$—light radiations are absorbed by the semiconductor. Absorption of light creates electrons and holes in the conduction and valence bands (VBs), respectively. Light gets absorbed for photon energies larger than band gap in direct band-gap semiconductors. Absorption of light is normally little for indirect semiconductors, except if the photon energy is higher than the direct gap value. Produced electrons and holes are not stable rather thermalize material by losing energy to the lattice and occupy the valence and CB edges [2]. This is clear by looking at the occupation probability ($f(E):g(E)$) and this probability is maximum near to the band edges, as shown in Figure 8.1. Excess energy ($E-E_g$) of the light photon is lost in the form of heat by increasing lattice vibrations of semiconductor material.

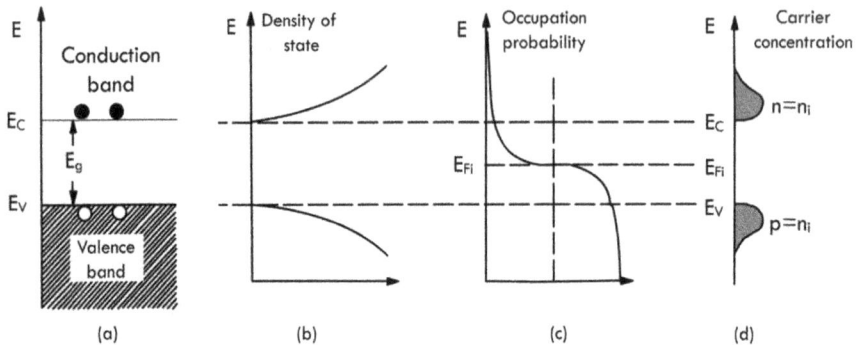

FIGURE 8.1 (a) Diagram of an energy band for a semiconductor displaying edges of conduction and VB edges. (b) Monotonical increment of the density of states from the band edges. (c) Occupation probability of electrons and holes in the CB and VB, calculated with respect to the intrinsic Fermi function. (d) Product of density of states and occupation probability result in the concentration of electrons in CB and holes in the VB.

Source: Reprinted with permission from Ref. [2]. © 2006 John Wiley.

The photoelectric effect, Compton scattering, and pair production are three fundamental ways in which photons of light, X-rays, and gamma rays have interaction with matter. In all different ways, the energy of a photon is transmitted to electrons that thus lose energy to atoms in the absorbing material. Energy may be lost in a media in which light is propagating. The fractional energy loss−dI/I radiation while going through a thickness dx of absorbing material is observed to be proportional to dx. If $I(x)$ is the intensity of the radiation at depth (x) within the absorber, then fractional change in intensity for gradual change in distance (dx) is given by

$$-dI/I = \alpha dx$$

$$I = I_0 e^{-\alpha x} \tag{8.3}$$

This is called the Beer–Lambert law (Equation (8.3)) where α is the absorption coefficient (or attenuation coefficient), with unit m^{-1}. I_0 is the value of Intensity at the position $(x = 0)$. The absorption coefficient is positive if energy is lost in the medium, zero for a lossless media, and has a negative value for a media with gain. As per the equation, the power lessens exponentially as the depth inside the semiconductor increases. Intensity becomes I_0/e at penetration depth equal to $1/\alpha$ [1]. The absorption coefficient is a function of the wavelength (energy) of the photon. Absorption is actually small if the energy band gap is greater than photon energy, while for photon energies above the band gap, photon absorption creates electron–hole pairs. This relies upon the distribution of density of states in the valence and CBs. The optical attenuation coefficient also relies upon the types of semiconductors (i.e., direct or indirect band gap).

8.1.3 POST-INTERACTION RADIATION ANALYSIS

Optoelectronic sensors (OS) are utilized in different needs of human life and various scientific activities like environmental monitoring, industry, chemical examination, science, and medicine. Some present medical usage incorporates heart-beat oximetry, pulse checking, estimating the oxygen level in the blood, observing blood glucose, urine analysis, breathed out biomarkers checking, and dental color matching, These sensors should follow requirements of safety, simple maintenance, biocompatible, reliable, stable, suitable for sterilization, and immune for biologic refusal [3]. The working of optoelectronic devices depends on examinations of light-matter interactions. The main processes that occur in the matter illuminated by radiation at a given wavelength are reflection, absorption, transmission, and emission (in the form of fluorescence and phosphorescence). The type of interaction of light with matter relies upon peculiar features of the sample. The outcome of the interaction can be utilized as an issue signature (marker). For instance, scattered light is utilized for fluid examination (observation of bloodstream or glucose identification). The absorption spectrum utilized for substance identification has a high importance for therapeutic estimations. Optoelectronic devices based on absorption generally evaluate the change of intensity and absorption spectrum of light that is passed through the sample. Reflectance sensors are additionally intended for such explicit investigation, however, for low transparency substances. The excited molecule of the sample while coming back to the ground state (singlet state S_0) can emit photons of lower energy (longer wavelength) than the absorbed one. The absorbed radiation can likewise create fluorescence at a wavelength not quite the same as the incident wavelength. Because of inner transformation, the molecule can likewise come back to the ground (S_0) state through triplet state (T_1, with longer life) if there should be an occurrence of phosphorescence. Phosphorescence with lower energy radiation endures longer than fluorescence [4]. For both phenomena, the intensity of emitted radiation depends proportionally on the concentration of excited molecules. Parameters estimated ordinarily in medical diagnosis can be separated into physical (e.g., pressure temperature, humidity) and chemical (lipoproteins, lipids, pH, oxygen partial pressure) one [5]. Since stimuli are not electrical, medical sensors use different steps of energy conversion before giving an electrical signal, which is measured and interpreted for symbolizing medical parameters of interest.

Photo-detector is used in optoelectronic sensors for the detection of photon flux that provides conversion of an electric signal. For example, periodic stress working on a fiber-optic pressure sensor first produces strain in the fiber, and this strain creates a change in refractive index and optical transmission, which at last outcomes modulation of the photon flux recorded by a photodiode. In an optical fiber-based optoelectronic sensor, the light might be directly modulated either by the investigating parameter or by exceptional reagent associated with the fiber [6]. The optical response of a reagent is altered by changing the stimulation agent of a measured medium. Such a probe is generally known as an optrode. The operation of this probe involves physical phenomena of fluorescence, absorption, Raman scattering, evanescent wave, and plasmon. Most of the optoelectronic sensors utilize monitoring of light intensity. However, its measuring parameters response is very sensitive to changes in light source parameters, fiber couplings, fiber lessening, signal noises, and so on. These factors reduce the accuracy of the measurement. In this way, such interferometric arrangements as Sagnac, Michelson, Mach–Zehnder, and Fabry–Perot are utilized for minimization of error [7].

8.2   OPTOELECTRONICS MATERIALS

There are a lot of optoelectronic materials that are used in medical and biomedical applications with different requirements and challenges. Here, we will discuss some materials with recent trends in biomedical applications.

8.2.1   CONDUCTIVE POLYMER

Polymers are materials that can be molded into additional complex structures compared to metals and ceramics [8]. However, a large portion of these applications was nonelectronic before the revelation of conducting polymers. From that point forward, researchers from various disciplines have given attention to them and helped in the improvement of conducting polymers. Research attraction of conducting polymers is that they have a wide scope of electrical conductivity like metals while they keep up their inherent polymeric properties in the meantime [9]. The conductivity of a conducting polymer can be changed by controlling the doping level and utilizing distinctive dopant ions. Additionally, the conductivity of leading polymers is not just controlled by the sort of monomers and dopants yet in addition dictated by the conditions under which the polymers are manufactured [10].

Straight backbone polymers such as polyacetylene, polypyrrole, polyindole, and polyaniline and their copolymers are the prime category of conductive polymers. Some sufficiently investigated category of polymers are poly(pyrrole)s, polyanilines, poly(thiophene)s, poly(3,4-ethylenedioxythiophene), poly(*p*-phenylene sulfide), poly(acetylene)s, and poly(*p*-phenylene vinylene). Chemical synthesis and electro (co) polymerization are two leading processes to synthesize conductive polymers. Some review reports propose their additional encouraging in natural solar cells, printing electronic circuits, organic light-emitting diodes, actuators, electrochromism, supercapacitors, chemical sensors and biosensors, adaptable transparent displays, and electro-magnetic protecting. Organic LEDs and organic polymer solar cells are the latest accentuations [11]. Polymer Electronics Research Center at the University of Auckland is building a scope of novel DNA sensor advances based on conducting polymers, photoluminescent polymers, and inorganic for easy, fast, and delicate gene identification. There are several biomedical applications of conducting polymers fields because of their remarkable properties: (1) their organic nature, (2) receptivity to electrical stimuli, (3) compatibility with biomolecules, and (4) appearance of ionic as well as electronic conductivity. Conducting polymers can be utilized in biosensors, neural prosthetic gadgets, drug delivery, and actuators. However, the application of conducting polymers in biomedical fields is restricted because of their typical elastic moduli in the order of 1–8 GPa [12]. This issue has attracted attention to the improvement and structure of new crossover or composite conducting polymer-based materials with soft, progressively strong mechanics. Biosensors are expository gadgets used to recognize explicit analytes. Generally, a biosensor is made out of an organic detecting component and a transducer [13]. The organic detecting component can be enzymes, antibodies, cell receptors, and DNA tests [13], which is to be interacted with a particular analyte. The transducer made of piezoelectric or optical material [13] makes an interpretation of biological signals into electrical or optical signals. Different kinds of organic material can be detected by using various biological sensing elements within the conducting polymers. Abidian and colleagues revealed another sort of glucose biosensor by utilizing distinctive conducting polymers [14]. Conducting polymers have been utilized for neural chronicle and incitement of neural prostheses, for example, cochlear implants and deep brain stimulators. Conducting polymer can likewise be utilized as biomedical actuators because of the volume change amid decrease or oxidation reactions. Additionally, Conducting polymer-based actuators have been utilized as artificial muscles and biomedical gadgets [15].

8.2.2   *PHOSPHORENE*

Black phosphorus (BP) has started gigantic research enthusiasm since its disclosure in 2014 because of its distinctive structures and valuable properties. Single-layer or few-layer BP, known as phosphorene is a sort of two-dimensional (2D) nanomaterial with peculiar physical and substance properties because of the dimensional impact. Having a band gap, in contrast to graphene, phosphorene is anticipated as a substitution of graphene. Phosphorene can be exfoliated by mechanical exfoliation [16], liquid-phase exfoliation [17], or some different technique from bulk phosphorous because of the way that the BP has a layered structure, where weak interaction favors stacking of layers and covalent bonding puts atoms in one plane [18]. Phosphorene belongs to direct band gap that can be tuned from 0.3 to 2.0 eV. It has large carrier mobility, emphatically anisotropic character, and valuable saturation current in field-effect type gadgets [19]. These amazing features create phosphorene an incredible possibility for electronic, photonic, medical, and thermoelectric devices. Red phosphorus and some phosphorus-containing mixes have been utilized as fire retardants in the ahead of schedule past, in light of the fact that they can be shaped into phosphoric corrosive by high-temperature thermal decay that elevates the polymers to frame into a heat-resistant carbonaceous defensive layer, therefore obstructs with the movement of O_2 to the burning zone [20].

Feature of 2D phosphorene nanomaterial is the same as nano flame retardants such as graphene and carbon nanotubes prompting high fire-safe effectiveness with a low added substance. Distribution of these materials into matrix materials is consistent and qualifies a decent similarity with polymers because of the little size. Along these lines, phosphorene is required to be effective fire-resistant. Being a segment of nucleic acids, phosphorus is essential in keeping up human health, prompting a biocompatible material with broad application potential in the biomedical field [21]. phosphorene has been utilized as a biosensing material for the identification target analytes (e.g., immunoglobulin G (IgG) [22] and myoglobin (Mb) [23] because of its inalienable electrochemical properties. phosphorene is potentially used for medication conveyance and against tumor treatment inferable from its high drug loading proficiency, great biocompatibility, and fantastic photothermal and photodynamic properties [24]. Despite the fact that phosphorene-based biomedical application is still in its earliest stages with various specialized difficulties staying to be tackled, it might bring novel open doors for future medical determination and treatment. So, a broad examination of the capability of phosphorene in biomedical applications like biosensor and regenerative medication would be raising interest for BP as an option to graphene-based materials. A comprehensive review with time is strongly needed on a wide scope of biomedical utilization of BP, including colorimetric detecting, fluorescent sensing, electrochemical sensing, field-effect transistor detecting, malignant growth imaging, disease treatment, and medication conveyance.

8.2.3 CARBON NANOTUBES

Carbon nanotubes (CNTs) are carbon-based nanomaterials with three-dimensional hexagonal sp^2 hybridized arrangement and consist of graphite or fullerene sheets having C–C separation of ~1.4 Å [25]. CNTs are the third allotropes of carbon in which thin graphite sheets are rolled up into hollow cylinders having nano-needle shape and size. Based on dimension and number of graphite sheets, CNTs are arranged chiefly into four kinds

1. Single-walled carbon nanotubes (SWCNTs)
2. Double-walled carbon nanotubes (DWCNTs)
3. Triple-walled carbon nanotubes (TWCNTs)
4. Multiple-walled carbon nanotubes (MWCNTs).

CNTs have developed as captivating nanomaterials for various fields along with biomedical employments. With excellent properties like high viewpoint proportion, high electrical and thermal conductivity, nonimmunogenicity, and so on, CNTs open another vista in the field of nanobiotechnology. CNTs offer numerous advantages in different applications such as targeted drug delivery, imaging and analysis, and photothermal treatment.

8.2.3.1 BIOMEDICAL APPLICATIONS OF CNTS

1. Nanotubes are utilized as a transporter to convey quantum spots (QDs) and proteins into malignant growth cells in light of the fact that QDs have photoluminescent property that is advantageous in imaging [26].
2. The achievement of bone uniting depends on the capacity of scaffold that helps the normal healing process. In any case, the scaffold might be related with few weaknesses like low strength and body refusal [27]. Recuperating procedure can be enhanced by giving a CNTs platform to new bone material. An examination uncovered that CNTs

could emulate the role of collagen as a platform for the development of hydroxyapatite into bones [28].

3. CNTs have been utilized for Gene treatment and undifferentiated stem cell therapy [29].

4. Adequate contractility of CNTs makes them possible replacement of harmed muscle tissue [26].

5. Working of CNTs with polyethylene glycol (PEG) causes them stealth that halts white blood cells from perceiving the nanotubes as outside materials, in this manner enabling them to circulate in the circulatory systems for longer length of time [26].

6. CNTs are utilized widely for active medication conveyance and imaging/analysis for sickness treatment and health checking [26].

7. Surface engineered CNTs have developed as nanocarriers for the therapeutics conveyance. CNTs have appeared potential in explicit cells focusing without influencing the normal cells and at a dosage lower than that of traditional medication [26].

8. CNTs can be utilized for tissue designing through tracking and naming of cells and by improving cell execution. CNTs may improve the tracking of cells, biosensing, and conveyance of transfecting operators.

9. Photothermal treatment utilizing CNTs is a nonintrusive therapeutic methodology for malignancy treatment in which photon energy demolishes cancer cells by heat conversion.

10. CNTs have been additionally utilized as a contrast specialist in imaging and identification of malignant cells inferable from their momentous optical properties. In that capacity, the electrical conductance of CNTs is delicate to the environment and furthermore, changes upon the surface adsorption attributes of various molecules. The CNTs are recommended as a sensing array component for reorganization of cancer markers in liquids.

8.2.4   SILICA-BASED NANOCAPSULES

Silica cases are perfect competitors in biomedical applications since they couple benefits of silica and capsular nanostructure. Silica is a nonlethal and biocompatible material and intact for normal utilization in the food added substances and nutrient enhancements. It has benefits of cheap production, good mechanical and chemical stability, and optical transparency. Different methods have been used to synthesize various kinds of silica capsules. The significant change in the scope of silica capsules in sizes, shapes, and surface science offered attention to new applications and basic research on silica capsules. In particular, silica-based nanocapsules (SNCs) are increasing attention, particularly in the scope of biomedical research.

8.2.4.1 FLUORESCENT PROBES

SNCs have been utilized for bioimaging applications by the inclusion of different kinds of fluorescent probes and magnetic nanocrystals due to their high colloidal and chemical stability in the physiological media, protection for photosensitive compounds with the desired image contrast, controllable nanoscale size with adequately long blood flow, and transparency so that they cannot absorb sufficient light in the near-infrared (NIR), visible, and ultraviolet regions or meddle with magnetic fields. A substantial number of fluorescent SNCs have been created by either exemplifying fluorescent dyes into the pit of SNCs or combining dye molecules to the outside of SNCs by means of synthetic holding. Liu et al. detailed the utilization of single micelle-templated SNCs for the conveyance of Nile red to DU-145 human prostate cancer cells [30].

8.2.4.2 MAGNETIC RESONANCE IMAGING

Relaxation of water protons in the presence of an external magnetic field and a radio-frequency pulse provides magnetic resonance imaging (MRI) contrast. There are two ways of relaxation: longitudinal (T1) and transverse (T2) relaxations. Recently, utilization SNCs to encapsulate MRI T1 contrast agents (e.g., MnO NPs) has been observed enhancement to these "positive" contrast agents to apply a more hyperintense region and consequently better relaxivity. Silica is a suitable candidate to coat T1 contrast agents since it is bio-inert and nonpoisonous. In such a manner, SNCs provide magnetic nanocrystals not only great biocompatibility but also improve chemical and mechanical stability.

8.2.4.3 MULTIMODAL IMAGING

SNCs have likewise been widely investigated for various imaging modalities, for example, a mix of fluorescence imaging and MRI.

8.2.4.4 SILICA-BASED NANOCAPSULES FOR DRUG DELIVERY

Improvement of SNCs has attracted great consideration for drug delivery. They have been generally utilized to make soluble and convey drugs that show moderately poor dissolvability, low soundness, and unwanted pharmaceutical properties. Exceptional colloidal and chemical stability and their little sizes enable them to flow in the body for a significant time. Henceforth, they can enter the flawed vasculature of tumors and aggregate in the influenced zone all the more effectively.

8.2.5 METAL OXIDES FOR BIOMEDICAL

Metal oxides have an important role in different chemical and physical areas. Many researchers are working on metal oxides due to their various applications in catalysis, gas sensing, field-effect transistors, microelectronics, ecological cleaning, ceramics synthesis, biomedical, and biosensors. Some broadly utilized nanoparticles metal oxides are TiO_2, ZrO_2, CeO_2, Fe_2O_3, and Fe_3O_4. These oxides behave as catalyst, antioxidant, and show great steadiness, bacterial activities, and biocompatibility. There are different useful biomedical applications of these materials for example, medication conveyance, therapeutic and diagnostic agents, and parts in medicinal implants [31]. For instance, TiO_2 being biocompatible surface for cells, and their multiplication is frequently utilized in therapeutic implants [32]; MRI, cell labeling, and focused drug delivery are reported to utilize magnetic iron oxides [33]; CeO_2 is well-established catalytic and cancer prevention agent [34]. MnO nanoparticles synthesized by the thermal processes are utilized as fluorescence probes for marking and optical imaging of organic tissues [35].

8.2.5.1 MRI AND CANCER TREATMENT

The working principle of MRI is the same as nuclear magnetic resonance analysis that is standout among the most valuable diagnostic imaging systems due to its great spatial resolution and noninvasive nature. Improvement in imaging quality includes sensitivity, resolution, tissue, or organ specificity and use of contrast agents. Cure of malignant growth through hyperthermia by tissue warming at 42–45°C, related to radiation and chemotherapy, has brought about synergistic impacts to expand the adequacy of ordinary medicines and has permitted to decrease poisonous symptoms [36]. The alternating magnetic field is utilized for tissue heating with the help of magnetic nanoparticle in magnetically interceded hyperthermia. Surface-modified magnetic nanoparticles with PEG are helpful for biological systems with good colloidal stability. In addition, flexible surface functioning with antibodies aptamers peptides [37] permits targeting distinctive receptors overexpressed on cells. Superparamagnetic iron oxides are the most utilized in MRI among the numerous kinds of magnetic nanoparticles.

8.2.5.2   ANTIBACTERIAL PROPERTIES

Various nanomaterials are well known for powerful restraint wide scope of bacteria, for instance, Fe_3O_4, TiO_2, CuO, and ZnO. These oxides have the advantage of toughness, lower poisonous quality, higher strength, and selectivity, over established natural antibacterial agents. The efficiency of nanoparticles diameter around 20 nm or less is commonly more effective than the bigger one. In the real sense, small nanoparticles are more effective to infiltrate into bacterial cells and may discharge lethal metal ions upon disintegration [38].

8.2.5.3   BIOMEDICAL IMPLANT

Thermal and chemical inertness of bioceramics makes it important for biomedical application and surgical implants. Friction coefficients of bioceramics are small and they are able to stimulate bone growth. They do not create strong biologically relevant interfaces with bones, but they elevate strong adhesions to bones [39]. Bioceramics are not only resorbable; they can likewise be bioactive, biodegradable, and dissolvable and are additionally accessible as composites with a polymer segment and microspheres [40]. Al_2O_3 and ZrO_2 metal oxides are used very commonly for the biomedical implant. Al_2O_3 is exceptionally inert and hard with excellent wear resistance, even under physiological conditions.

8.3   PRINCIPLES OF OPTOELECTRONIC DEVICES

8.3.1   INTRODUCTION

Optoelectronic devices are extensively used in areas such as consumer electronics, telecommunication, biomedical devices, solar energy, imaging, optical fiber systems, military applications, and so forth, and therefore, have become crucial in our society's infrastructure equipment and systems. Most optoelectronic devices are based on semiconductor PN-junction diodes, and their function relies upon the properties of the semiconductor material as well as the PN-junction. The most common optoelectronic pn-junction devices are laser diodes, LEDs, and photodiodes. In order to know the functioning in a better way, it is essential to learn the basics of the processes associated.

This section describes the important concepts of optoelectronic p–n junction devices in terms of the materials' properties, operating principles, and applications in the realization of electro-optical hybrid logic circuits. They include light sources like LEDs, three-terminal light detectors such as phototransistors, and integrated light sources and detectors such as optocouplers.

8.3.2   RADIATIVE TRANSITIONS

The interaction between a photon and an electron in a solid can be broadly categorized into three different processes [41], namely:

1. absorption (stimulated generation);
2. spontaneous emission (spontaneous recombination); and
3. stimulated emission (coherent photon emission).

To illustrate these processes, we consider a simplistic system of an atom with electrons having two energy levels E_1 and E_2, where E_1 and E_2 are the ground state and an excited state, respectively. At the normal room temperature, all the atoms in the solid are supposed to be in the ground state. Any transition between these energy states will involve the emission or absorption of a photon (Equation (8.4)) with frequency v_{12} is given by

$$hv_{12} = E_2 - E_1 \qquad\qquad (8.4)$$

8.3.3   ABSORPTION (STIMULATED GENERATION)

If a photon of energy hv_{12} falls on a semiconducting material, the electron in the ground state E_1 absorbs the photon and goes to an excited state E_2. This is known as the absorption process [42]. It results in the addition of an electronic charge carrier in the CB and a hole in the VB. This phenomenon plays a vital role in devices such as solar cells and photodetector devices. The absorption process is illustrated in Figure 8.2.

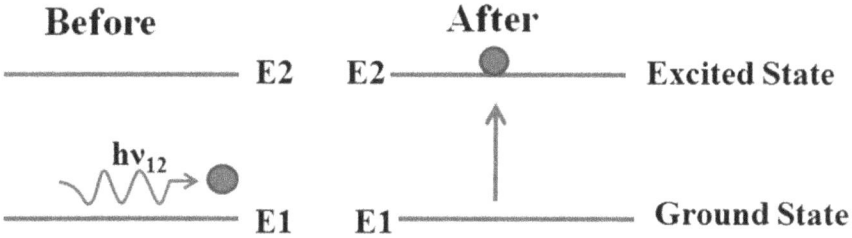

FIGURE 8.2   Absorption process [42].

8.3.4   *SPONTANEOUS EMISSION (SPONTANEOUS RECOMBINATION)*

When an electron goes to the excited state, it becomes unstable, and therefore, it spontaneously returns to the ground state emitting a photon of energy hv_{12}. This process is known as spontaneous emission. When this process occurs for a large number of atoms, the emission time and direction will be different and the photons will be unable to add to a coherent radiation field. Hence, it results in the substantially incoherent emission of photons. This is the basic mechanism on which an LED works, in which photon feedback is not provided. It is illustrated in Figure 8.3.

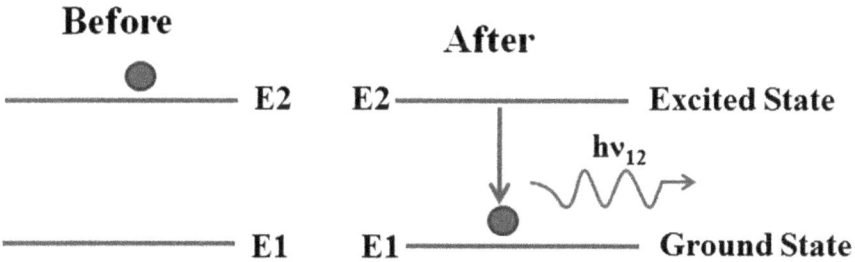

FIGURE 8.3   Spontaneous emission [42].

8.3.5   *STIMULATED EMISSION (COHERENT PHOTON EMISSION)*

If a photon of energy hv_{12} strikes an electron while it is in the excited state, the atom is additionally excited to return to the ground state and in the process, emits a new photon of energy hv_{12} that is in phase with the incident radiation. This process is known as stimulated emission. It is similar to the absorption process where the sign of the excitation is reversed. In this process, the incident photon excites the system resulting in the electron–hole recombination and simultaneously producing a new photon. The operation of the LASER is based on this mechanism. The phenomenon is illustrated in Figure 8.4.

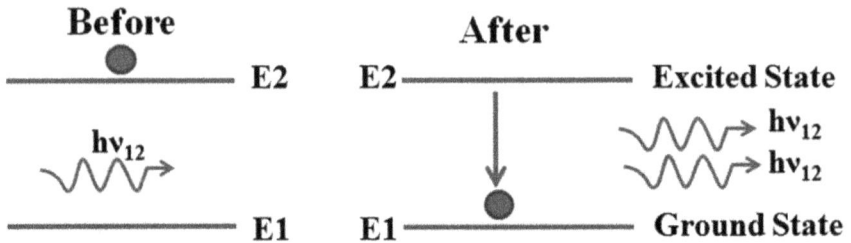

FIGURE 8.4   Stimulated emission [42].

8.3.6   INTRODUCTION TO LIGHT SOURCES

The two widely elements used to make optoelectronic devices used are LED and LASER diodes. The main operating mechanism for the LED is spontaneous emission while it is stimulated emission for the LASER diode.

    The charge carriers, in a p–n junction, acquire sufficient energy to overcome the junction potential barrier. Under a forward applied voltage, minority charge carriers cross the p–n junction, and recombination with majority charge carriers takes place. These electronic recombinations are of two types: radiative and nonradiative.

    In radiative recombination, a charge carrier makes a transition from an initial to a final state and in the process emits a photon. This is the basis of all optoelectronic devices which operate based on radiative transitions and semi-conductors. Essentially, the main goal of generating light-emitting devices is to maximize the probability of radiative transitions while minimizing nonradiative transitions [43]. The materials fit for radiative transition are the direct band-gap semiconductors. In the direct band-gap semiconductors, the recombining electron and hole have the same value of momentums. Some of the popular direct band-gap semiconductors used in optoelectronic devices are compounds formed by group III and group V elements of the periodic table, namely, GaAs, GaSb, InAs, and InSb. Bandgap energies (E_g) of GaAs, GaSb, InAs, and InSb are 1.43, 0.73, 0.35, and 0.18 eV, respectively.

    In nonradiative transitions, the electron makes the transition without a photon and the extra energy goes out as heat. The materials that undergo nonradiative transitions are indirect band-gap semiconductors. For indirect band-gap semiconductors, the minima of the CB energy and the maxima of the VB energy occur at different values of momentum (Figure 8.5). In such cases, recombinations involve a particle called phonons (i.e., crystal lattice vibrations) to conserve momentum, as the momentum of the photon will be quite small. Deep levels and Auger recombination are the main types of non-radiative recombinations. In Auger recombination, when an electron and a hole recombine, any extra energy goes into knocking a secondary electron or hole deeper into the band. It is clearly undesirable and may become a major source of loss for lasers. Examples of indirect band-gap semiconductors are Si, Ge, GaP, AlAs, and AlSb.

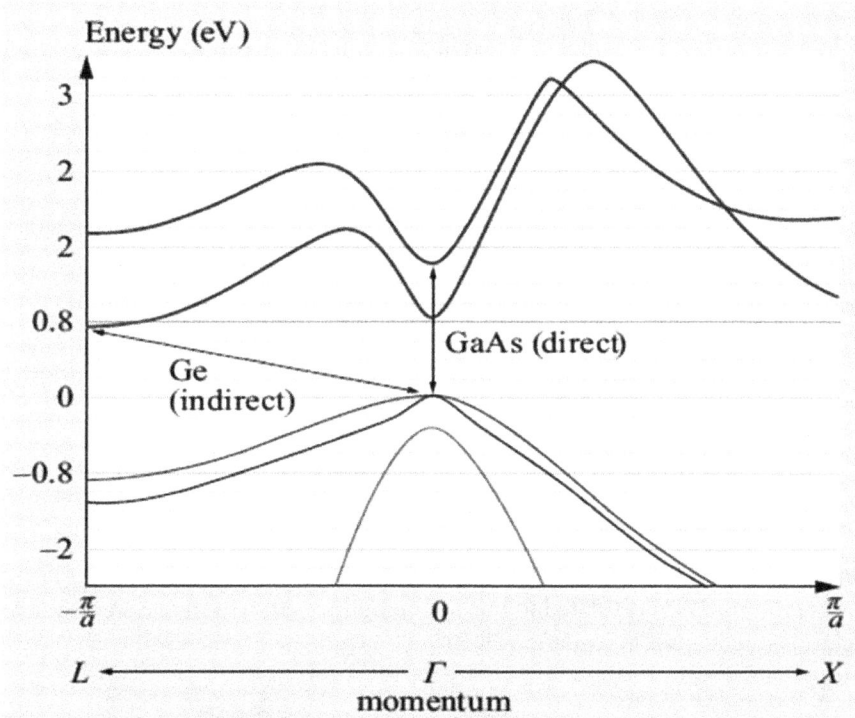

FIGURE 8.5   Direct-gap semiconductor and indirect semiconductor band structure.
Source: Reprinted with permission from Ref. [43]. © 2017 Springer Nature.

The rate of radiative recombination R_r is defined in terms of the number of photons emitted per volume per second, and is proportionate to the charge carrier concentration as follows:

$$R_r = B_{rec}pn \qquad (8.5)$$

where,

p = concentration of p-type carrier,
n = concentration of n-type carrier,
B_{rec} = recombination coefficient.

From the above Equation (8.5), it is obvious that the radiative recombination rate increases with the increase in the concentration of charge carriers that can be achieved by adding impurity.

TABLE 8.1   A Few Properties of Common Semiconductor Materials [42, 44]

Semiconductor Type	Bandgap (eV)	Bandgap Type	Electron Mobility (cm²/V s)	Hole Mobility (cm²/V s)	Lattice Constant (Å)	Recombination Coefficient
InSb	0.18	Direct	77,000	1000	6.4794	4.58×10^{-11}
InAs	0.35	Direct	33,000	460	6.0584	8.5×10^{-11}
Ge	0.67	Indirect	3900	1900	5.64613	5.25×10^{-15}
GaSb	0.73	Direct	5000	1000	6.0959	2.39×10^{-10}
Si	1.12	Indirect	1500	470	5.43095	1.79×10^{-15}
InP	1.35	Direct	4600	150	5.8693	1.26×10^{-9}
GaAs	1.43	Direct	8500	400	5.6533	7.21×10^{-10}
GaP	2.26	Indirect	110	75	5.4512	5.37×10^{-14}

An electron lying in the CB in an atom can fall into a hole in the VB, thereby the atom returning to its ground state. This process is called recombination of electron–hole pair that liberates energy in the form of a photon and is the basis by which a source emits light. The energy E emitted during this recombination is related to the wavelength of emitted light λ through the relationship shown in Equation (8.6)

$$E = 1.240/\lambda \qquad (8.6)$$

where λ is the wavelength of light in micrometers (μm) and E is the band-gap energy in electron volts (eV).

As every material has different band-gap energy, therefore, electron–hole recombination in different semiconductor materials leads to the emission of distinct wavelengths, and hence distinct colors. The wavelength at which the maximum light is produced is called the peak wavelength, denoted by λP. It is calculated based on the energy band-gap (E_g) of the semiconductor material.

It is to be noted that the band gaps of the semiconductor compounds can be manipulated through alloying. Bandgap engineering facilitates the generation of heterojunctions that is crucial for the design of high-end optoelectronic devices [45].

8.4 OPTOELECTRONIC DEVICES

Optoelectronics is a very broad term that refers to the study and use of electronic devices that emit, detect, and control light. There are many examples of optoelectronic devices in everyday life, such as the LEDs in my computer monitor and the photodetector in my computer mouse. Let us take a moment and examine the electronics that we own. We can hardly find any device that does not have any optoelectronic components (i.e., do not emit or detect light). Most of the devices that we own have at least an indicator light to tell us whether the device is on or off. This activity gives us a solid idea of why we should use optoelectronics. The main reason why optoelectronic devices are so prevalent is that most people depend greatly on vision to observe the world around them. Because of this, being able to manipulate light is extremely useful—the capability has allowed us to see at night, record images, and videos of things we observe, and easily share information with others. There are many important optoelectronic applications such as solar cells, optical fiber communication, lasers, and more. Optoelectronic devices find wide applications in industry, telecommunications, military, medical equipment, aerospace, control systems, etc.

Optoelectronic devices produce electrical energy when the light is incident on it. These devices use light in visible or infrared ranges of the electromagnetic spectrum. Solid-state devices such as light emitters, IR sensors, lasers, and LEDs are extensively used in optoelectronics [46]. Optoelectronic devices can be broadly divided into two categories, namely: (1) photoconductive and (2) photovoltaic devices.

8.4.1 PHOTOCONDUCTIVE DEVICES

These are devices that sense the changes in the light intensity in order to trigger or inhibit a circuit. They operate in the reverse-biased mode to produce current when illuminated. The important components belonging to this category are LDR, photodiodes, phototransistors, etc. These devices generate a voltage and current through acquiring energy from light. The intensity of light falling on the p–n junction of these devices controls the amount of voltage and current that gets generated. Examples of photovoltaic devices are photovoltaic cells and solar cells. The electrical conductivity of the semiconductor materials employed in photoconductive devices alters as per the intensity of light falling on it. As it operates on the principle of photo-resistivity, it is also called as photoresistive cell or photoresistor. The resistance of these devices gets increased in dark surroundings and gets extremely small resistance under bright light. The typical photoconductive devices are:

1. LDR
2. LEDs
3. Photodiode
4. Phototransistors
5. Infrared diodes.

8.4.1.1   LIGHT-DEPENDENT RESISTOR

LDR is a small device used to detect variation in light intensity. It is a kind of variable resistor and the resistance depends on the intensity of light falling on it. In the complete darkness condition, its resistance is as high as 10 mΩ that reduces to a few Ohms in bright light. LDR is called CdS cells because its semiconductor is cadmium sulfide. Its light-sensitive area is doped with impurities such as silver (Ag), antimony (Sb), or indium (In). When the light falls on an LDR, the flow of electron–hole pairs occur and electrical conductivity takes place. In structure, the LDR consists of a wafer-thin, zig-zag pattern semiconductor line covered in a transparent case (Figure 8.6A).

LDR can handle very high current and even AC passes through it without harming the device. Two leads are arising from the two ends of the semiconductor. Figure 8.6B represents the working of LDR that shows that as the intensity of light sensed varies, there is a change in the brightness of the light. LDR is available in sizes 3 mm, 5 mm, 10 mm, 20 mm, etc.

8.4.1.2   LIGHT-EMITTING DIODES

It is another light source that can be used in optical communication. It has a wider spectral width but operates at optical frequency.

An LED is operated in forward bias mode, the same as a common PN junction diode for it to conduct. In forward bias mode, the free electrons from the CB jump the junction and drop into the holes of the VB. Since these electrons drop from a higher to lower energy level, they emit energy as light. The breakdown voltage of LEDs is very small, usually in the 3–5 V range. Figure 8.7 shows the symbol for LED.

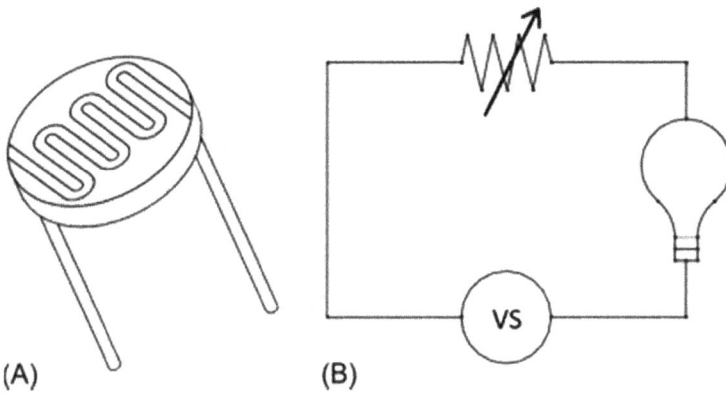

(A)   **(B)**

FIGURE 8.6   (A) Example of LDR, (B) circuit diagram with a light-dependent resistor. *Source*: Reprinted with permission from Ref. [47]. © 2019 Elsevier.

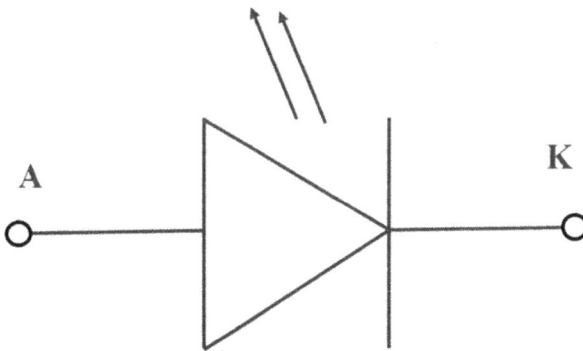

FIGURE 8.7   Symbol for LED [48].

The color of the light emitted by the LEDs is dependent on the energy band gap of the material used. LEDs emitting different light colors can be produced with the compounds of elements like gallium (Ga), arsenic (As), and phosphorous (P). For example, an LED made of gallium arsenide phosphide (GaAsP) can emit both red and yellow light, LED with gallium phosphide (GaP) can emit red or green light, and gallium nitrate ($Ga(NO_3)_3$)-based LED can emit blue light.

The degree of illumination of an LED depends on the amount of current. When the applied voltage is much greater than the diode voltage VD, the brightness of an LED gets almost constant. It has various applications such as a seven-segment display, remote control transmitter, photodetectors, and display devices such as LED TVs.

8.4.1.3 PHOTODIODE

A photodiode, as the name implies, is a PN junction that operates due to the light. Photodiode can be said to be a light operated switch. It is an LED type, a high-speed device employed in reverse bias mode to generate photocurrent (in µA) when a light is incident on it [49]. A photodiode has a relatively large p–n junction, so the photons of light striking the depletion region results in the breaking of covalent bonds of the semiconductor. Consequently, electron–hole pairs are created and the flow of electron happens. This conduction of electrons depends upon the intensity of the directed light. If the intensity of incident light increases, extra electron–hole pairs are generated and the flow of electron (current) increases.

The operating mechanisms of photodiodes depend on the electrical as well as the optical properties of the semiconductor material of which the pn-junction is made of. The functions of photodiodes are, however, quite different from those of LEDs and laser diodes, as the optical absorption processes are used here. The operating wavelength of the light in case of photodiodes depends on the band-gap energy of the semiconductor material. Moreover, the light intensity also affects the magnitude of conduction in the photodiode. The photodiode current is directly proportional to the number of photons falling at the junction.

Silicon (Si), germanium (Ge), and compounds of groups III--V and II--VI semiconductors such as GaAs, InP, and CdTe are mostly used in the fabrication of photodiodes [4]. Photodiodes made of Si are mostly used in consumer electronics, while those manufactured from Ge and InGaAs(P)/InP are mainly used in optical fiber communication systems. For the higher wavelength range, InAs and InSb can also be used as photodiode materials. Figure 8.8 displays a practical photodiode (A) and the symbol for a photodiode (B).

Photodiodes have wide-ranging applications such as object detection, high stability, and speed circuits, switching circuits, demodulation, and optical communication equipment. It is also used in remote control sensor, light operated switch, designing of optocouplers, to read the audio track recorded on motion picture film [50].

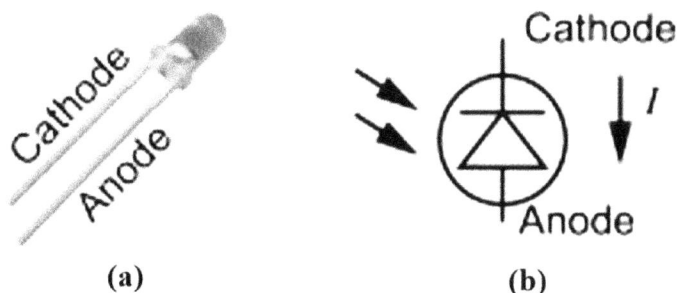

FIGURE 8.8   (A) Practical photodiode. (B) Symbol for photodiode.
Source: Reprinted with permission from Ref. [48]. © 2016 Elsevier.

8.4.1.4   PHOTOTRANSISTORS

Phototransistors are similar to photodiodes but they also amplify the generated current. It is like a normal bipolar junction transistor except that it is light-sensitive, and has no connection to its base terminal (common base mode). It operates on the basis of the photodiode that exists at the collector–base (CB) junction. The input to a phototransistor is supplied by the incident light, instead of the base current. Only the emitter and collector terminals are connected to the circuit. Each phototransistor consists of a see-through window that has a focusing lens to incident light at its junction. The symbol of a phototransistor is given in Figure 8.9.

A phototransistor has two junctions (n–p and p–n) that are separated by a large base region. The n–p junction is somewhat forward biased whereas the p–n junction is reverse biased. When photons of light strike the n–p junction, electron–hole pairs are created. Then, these electrons of the p-region diffuse out while the holes reside in the p-region that builds a positive charge. The accumulation of positive charge results in the increase of the forward bias voltage of n–p junction that in turns raises current flow.

The applications of phototransistors are similar to the photodiodes, the difference lies only in the response time and current capacity. Phototransistors have better current capacity and response time, that is, sensitivity than photodiodes. However, photodiodes are faster than phototransistors having switching speed of less than a nanosecond.

Phototransistors can also work as optoelectronic mixers, where it performs photo-detection as well as frequency up-conversion at the same time in the same device [52]. Darlington phototransistors use an extra bipolar NPN transistor to perform additional amplification. This extra transistor is also useful when higher sensitivity of a photodetector is needed due to low light levels. However, the response time of a Darlington phototransistor is higher than that of normal NPN phototransistors.

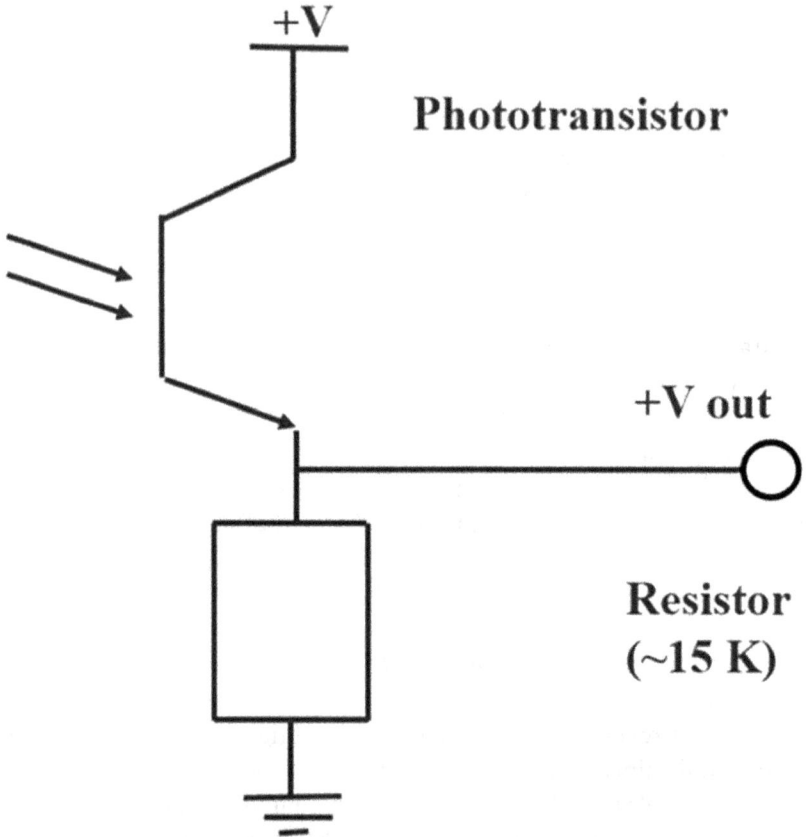

FIGURE 8.9 Symbol for phototransistor [51].

8.4.1.5 INFRARED DIODES

An IR diode is a photodiode that transmits light in the infrared region of the electromagnetic spectrum that is undetectable to the human eye. The p–n junctions in the infrared diodes are formed of gallium arsenide. When a voltage is applied across an IR diode, electrons in the n-region moves to combine with the holes in the p-region. This recombination of electrons and holes occurs in the recombination region leading to photon emission in the infrared region. The wavelength of these rays is generally in the range of 700 nm to 1 mm. Figure 8.10 displays an infradiode image.

FIGURE 8.10   Infrared diode [53].

IR LEDs find very common application in several types of remote controls. If IR LED is used with infrared cameras, it may be used as a spotlight that is invisible to the human eye. Since IR diodes can work in combination with varieties of sensors, they are getting increasingly popular in Internet of Things uses and machine-to-machine environments.

8.5   PHOTOVOLTAIC DEVICES

Photovoltaic devices are made of semiconductor materials of p-type and n-type materials. When sunlight energy in the form of photons strikes n-type material, free electrons are produced. These free electrons are accelerated toward p-type semiconductor causing a voltage difference that leads to the generation of direct current (DC).

8.5.1   SOLAR CELL

Photovoltaic devices or cells are fabricated from single-crystal silicon p–n junctions, similar to photodiodes, having a very wide light-sensitive area.

They do not operate in the reverse biasing mode. Their characteristics are similar to very large photodiodes placed in the dark. A silicon (Si) solar cell is demonstrated in Figure 8.11.

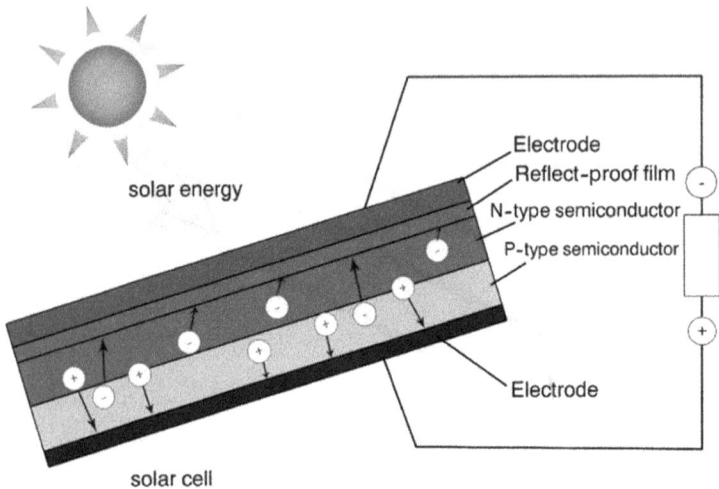

FIGURE 8.11 A silicon solar cell.

Source: Reprinted with permission from Ref. [54]. © 2018 Elsevier.

Each photocell consists of many p–n junctions joined in series. One of these junctions is very narrow to allow the passage of light energy. When sunlight strikes on the p–n junction, electron–hole pairs are created that cause current flow proportionate to the incident sunlight energy. Besides, the incident sunlight on the junction of solar cell strikes the valance electrons, causing the formation of charge carriers that traverse the p–n junction in an opposite direction and build a voltage across the p–n junction.

Photovoltaic cells and solar cells are extensively used in numerous applications to generate electricity. When a photovoltaic cell is lightened the light photons cause electrons to flow across the PN junction. A single solar cell can produce a voltage of ~0.6 V. It consists of a "positive" and a "negative" side the same as in a battery.

8.5.2 SOLAR PANEL

A solar panel consists of several units of photovoltaic cells. These photovoltaic cells absorb sunlight and generate direct current (DC), which is then converted into useful alternating current (AC) by an inverter.

8.6   BIOMEDICAL APPLICATIONS

Biomedical devices have changed the way of diagnosis, treatment, and patient care because of the passionate development in the optoelectronics. Recently, the immense development has taken place in the field of diagnostic tools, illuminating imagery, endoscopic medicine, ophthalmology, surgery, and biotechnology. The most common optoelectronic devices in the field of biomedical are the sensor or optoelectronic sensors. In recent years, optoelectronic sensors have emerged as a part of the design of medical devices and are very familiar to the clinicians or physicians with more accuracy and faster results without the fear of any error such as human error. The more sophisticated, small size and almost noninvasive sensors are in high demand because of their easy to use and handling in the diagnosis as well as in the treatment process. Even patients are able to handle and may operate at home to avoid the rush and emergency in the hospitals. We are dealing here briefly about few of the optoelectronics devices that have emerged in recent years. Optoelectronic sensors for biomedical applications have changed the method of diagnosis and treatment by invasive or noninvasive methods recently. An optoelectronic sensor is based on the principles of the amount of light incident on an active area and produces electrical signal. Optoelectronic sensors are more reliable, real-time performer, small in size, easy packaging and handling, light in weight, and accurately measured devices that are required by medical practitioners, doctors, and even patients. A few examples of optoelectronic sensors for biomedical applications are mentioned below. Some of the application of optoelectronics for biomedical applications is mentioned below.

8.7   NONINVASIVE OPTOELECTRONIC SENSORS (NIOS) FOR BIOMEDICAL APPLICATION

8.7.1   PULSE OXIMETRY

Pulse oximetry is classified as a noninvasive optoelectronic sensor that determines the percentage of oxygen saturation associated with hemoglobin (Hb). Hemoglobin is a protein associated with red blood cells (RBCs) of the blood that is responsible to carry the oxygen from the lungs to the rest of the body and return carbon dioxide from the body to the lungs that is exhaled through respiration. When Hb attached with oxygen is called oxygenated hemoglobin, and absorbs light in the infrared (IR) range while deoxygenated hemoglobin (Hb not attached with oxygen) absorbs light in the visible range.

In pulse oximetry, two organic LEDs (OLEDs) are connecting with an organic polymer photodiode presented by Lochner et al. [55]. When the light sensor is placed on a patient's finger, one LED emits green light (532 nm) and the other LED emits red light (626 nm). This is attributed to the emission of light while pulsating the arterial blood, nonarterial blood, venous blood, and tissue. There is a change in the volume of blood when systolic and diastolic pulsation (heart contraction; systolic, heart relaxation; diastolic) takes place that results in a change in the photodiode (Figure 8.12). This change in the light is because of the absorption of light by oxygenated and deoxygenated Hb. Thus oxygen saturation percent in blood can be determined.

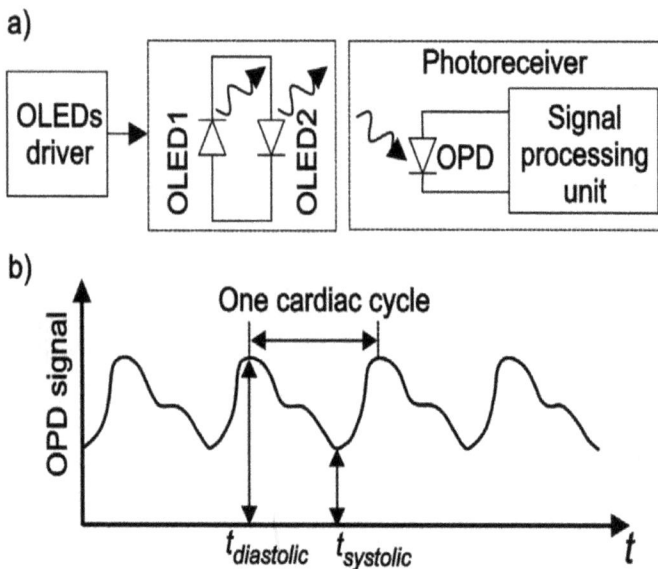

FIGURE 8.12   Pulse oximetry sensor (a) and an example of the output signal (b).
Source: Reprinted with permission from Ref. [3]. © 2018 Elsevier.

8.7.2   BLOOD GLUCOSE MONITORING

Blood glucose is simple glucose that is present in humans' and other animals' blood. Blood glucose is a metabolite and is regulated by the body as a part of metabolic functions. Higher or lower blood glucose level than normal blood glucose level is a medical condition and need to be checked regularly. Higher blood glucose leads to diabetes mellitus, the most common disease nowadays. Therefore, a routine checkup of blood glucose is the need of the patient. There are several traditional methods for testing blood glucose by taking the blood samples and testing them in the laboratory with the help of an instrument called glucometer. Even though glucometers are small-sized devices, they are so common and can be kept by a patient even at home for regular checkup of the blood glucose. Recently, a noninvasive method of testing of blood glucose has been developed. In the research of Ergin et al. [56], blood glucose concentration is determined by Raman spectroscopy of the eye lymph. In Raman spectrum, a laser light incident on a sample interacts with the matter and a shift in the wavelength is detected [57]. In a typical setup an optical fiber probe is for both excitation and collection, connected to a PC and a spectrometer for data measurement. Raman spectrum is characterized by a unique spectrum for each eye and metabolite structure. Different eye structure and metabolite present in aqueous solutions can be identified by the Raman spectrum. These features of identification make this characterization useful for the detection of metabolites. The physicochemical changes in biological specimens can be understood by this technique easily because of its good sensitivity [58–61].

8.7.3 BLOOD FLOW METER

Blood flow is important within a human body because of the function of the body, various metabolites, nutrients, and oxygen supply to the various organs get affected by an improper blood flow rate. Therefore, it is essential to monitor the blood flow rate for an early finding of a disease. Blood flow rate monitoring has been developed based on the laser Doppler flowmetry (LDF) technique and to be considered a noninvasive optoelectronic technique. It is a simple technique as it comprises two components, one is laser light source guided by optical fiber probe and the second is photodetector that measures the changes in the electrical signal. Briefly, the light is incident on the vascular network and or the tissue by the optical fiber probe where it is diffusely scattered and partially absorbed by the sample. The scattering of light is due to the movement of RBCs in the blood vesicles results in a shifting of the frequency of light while in the tissue the shifting of light frequency remains unaltered. The shifting in frequency depends on the cell velocity and the direction of the incident light beam. Therefore, the interference patterns arising from shifted and unshifted light frequency are measured by a photodetector [62]. Stern [63] performed blood flow measurement by illuminating a small area of skin using a low power collimated beam. The advantage of LDF technique is that the continuous measurement of microvascular blood perfusion can be done (Figure 8.13).

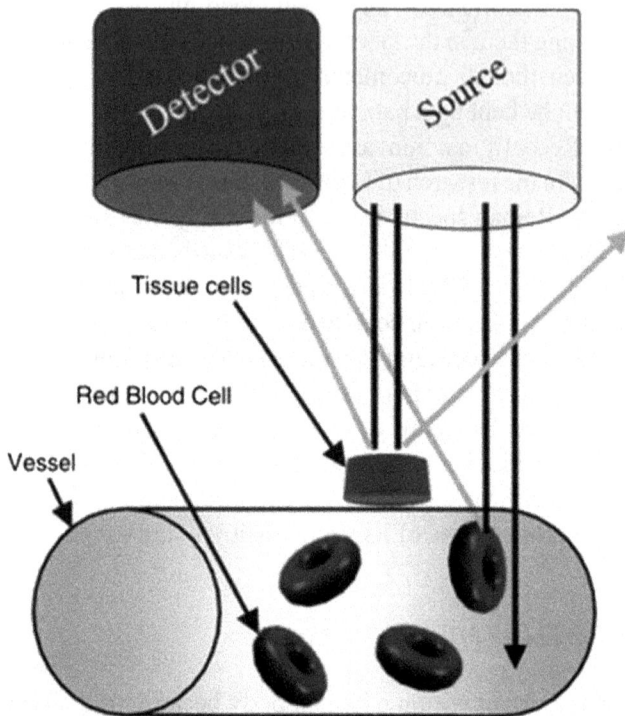

FIGURE 8.13   Optical fiber-based laser Doppler flowmeter.

Source: Reprinted with permission from Ref. [64]. © 2019 John Wiley and Sons.

8.7.4   DENTAL MATCH COLOR SENSOR

Optoelectronic sensors based on light to frequency and light to voltage converter both have played a major role in detecting the color. So a color-based sensor has been developed by the optoelectronic method where light is objected on a surface and the frequency of the light reflected is changed according to the color of the object. The reflected light frequency is measured by a photodiode and hence is used in matching the color in dentistry to eliminate the human error. This is a totally noninvasive method of detecting color. The color sensor provides accurate and fast results and therefore makes optoelectronic sensor the choice for dental color matching and urine analysis. This has minimized the human error that was before based on the visual judgment of the human eye. Kim et al. [65] recently has developed a device for dental color matching with the help of support vector machine algorithm.

8.7.5 *MICROCANTILEVER SENSOR FOR PROSTATE CANCER*

Prostate cancer is the most common cancer in males of older age. Four of 10 males may have this critical problem in older age. Therefore, it is required to diagnose this dangerous disease early. A noninvasive method of detection of this disease has been developed recently. A microcantilever sensor has been demonstrated by Wu et al. [66] for the diagnosis of prostate cancer. According to this research, on the surface of a microcantilever beam, there is binding of prostate-specific antigen (PSA). PSA is found in the serum of a prostate cancer patient; hence, PSA is an extremely useful marker in the early diagnosis of prostate cancer. This can be helpful in monitoring disease progression and in monitoring patients for disease progression and the effects of the treatment. In Figure 8.14, the schematic diagram has shown the basic idea of the working of the microcantilever sensor. The surface of the cantilever binds with a coating that is specific to analyte adsorption and leads to the bending of the cantilever. With the time of adsorption of analytes, more bending of the microcantilever is observed that can be detected optically. The magnitude of angular deflection (ΔH) of the laser beam can be measured.

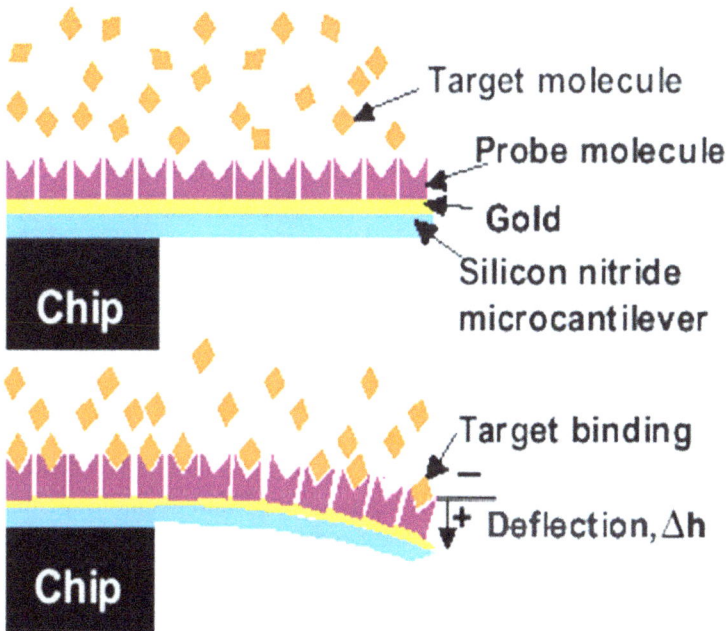

FIGURE 8.14 Working of microcantilever based sensor for prostate cancer.

Source: Reprinted with permission from Ref. [66]. © 2001 Springer Nature.

8.7.6 HUMAN BREATHE ANALYZER

In human exhaled breathe air, several organic and inorganic compounds have been already detected. The composition of exhaled air is carbon dioxide, nitrogen, water, oxygen, and other volatile organic compounds (VOCs) [67]. The VOCs are necessary to detect as these compounds may provide some important information about human health. Buszewski et al. [68] and Stacewicz et al. [69] have operated multipass spectroscopy (MUPASS) (Figure 8.15) and cavity-enhanced absorption spectroscopy (CEAS) (Figure 8.16) in the detection of VOCs and organic matters. For detailed information of their working ideas and method of operation, refer to Buszewski et al. [68] and Stacewicz et al. [69].

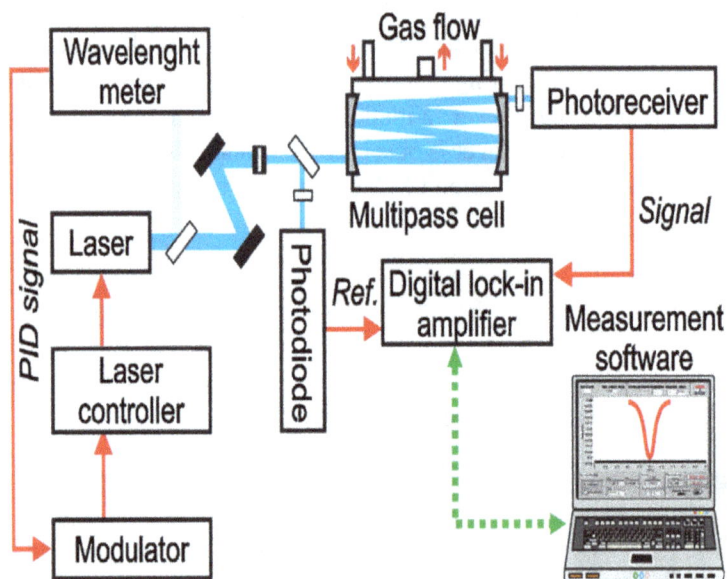

FIGURE 8.15   Scheme of MUPASS sensor.

Source: Reprinted with permission from Ref. [3]. © 2018 Elsevier.

They have also developed a more equipped and accurate technique of breathe air analysis by extending their work of MUPASS and CEAS. Laser absorption spectroscopy (LAS) plays a significant role in the detection of exhaled air biomarkers. This technique is highly sensitive and selective toward such biomarkers. This technique enables the detection of the small concentration of selective gas in situ or in tedlar bag, where the gas is collected. Bielecki et al. [3] have developed LAS-based sensors for the detection of the concentration of ammonia, nitrous oxide, carbon monoxide, and carbonyl sulfide.

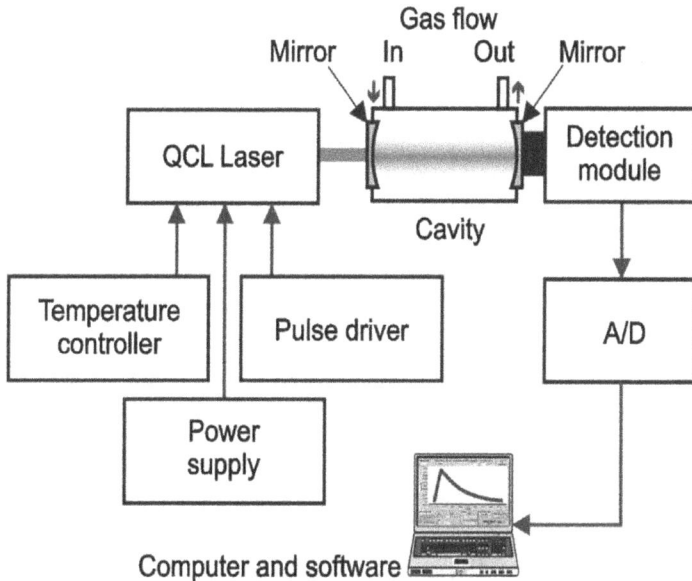

FIGURE 8.16 The idea of CEAS setup.
Source: Reprinted with permission from Ref. [3]. © 2018 Elsevier.

8.8 INVASIVE OPTOELECTRONIC SENSORS

8.8.1 *ENDOSCOPIC IMAGING*

The breakthrough has revived the development of optical fiber-based endoscopic imaging in the biomedical field. Endoscopy is used to check the gastrointestinal tract, confirm the diagnosis, usually with the biopsy, as well as the state of control for bleeding, inflammation, polyping, or removing foreign bodies. The endoscopic imaging system [70] guides a physician by illuminating a foreign object or tissue within the cavity and help in the removal of that object easily. Nowadays, there has been significant progress in the development of modern endoscopy since it was developed initially. A modern endoscopy comprises a light source, optical fiber, CCD unit, image processing unit, and instrument channel (Figure 8.17).

The endoscope inserted through the body and illuminates the cavity by endoscope fiber carried form a light source. An electrical signal originated form from a CCD unit that was the result of reflection of light from the body part, sent to an image processing unit. With the help of image processing unit, various kind of enlargement of the image can be done.

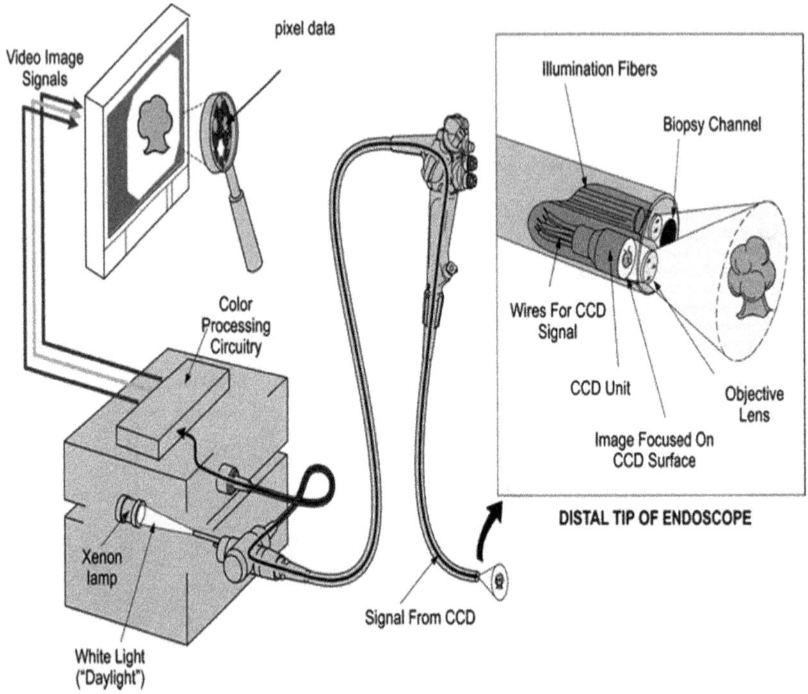

FIGURE 8.17   System design of modern video endoscopy.

Source: Reprinted with permission from Ref. [70]. © 2013 Springer Nature.

8.8.2   BILE SENSOR

Bile or bile juice is the secretion from the liver used to digest lipids in our body. The bile is stored in gall bladder. Bile can be detected by the optical fiber sensor designed for the detection of bile during the enterogastric and nonacid gastroesophageal refluxes [71]. These refluxes may leads to the numerous types of infection such as gastric ulcer, "chemical" gastritis, upper dyspeptic syndromes, and severe esophagitis. Gastric cancer may be the result of the enterogastric refluxes. The optical property of bile make enables the sensing of bile in gastrointestinal and other parts within it. Optoelectronic based bile sensors have been developed by Cecchi srl [72] and commercialized by Medtronic [73]. The sensor consist of two light emitting sources having wavelength (λ) at 465 and 570 nm [3], a signal-processing unit, and an optical probe. The light is supplied to the probe through optical fiber from the light sources to the stomach or the esophagus and returns the signals back to the photodiode [74–76]. The reflected light gives the difference in the absorbance at these wavelengths after absorption by bilirubin concentration. The bile reflux is related to the bilirubin concentration and can be determined by the difference in the absorbance of light. A photo of an optical probe for bile detection is shown in Figure 8.18.

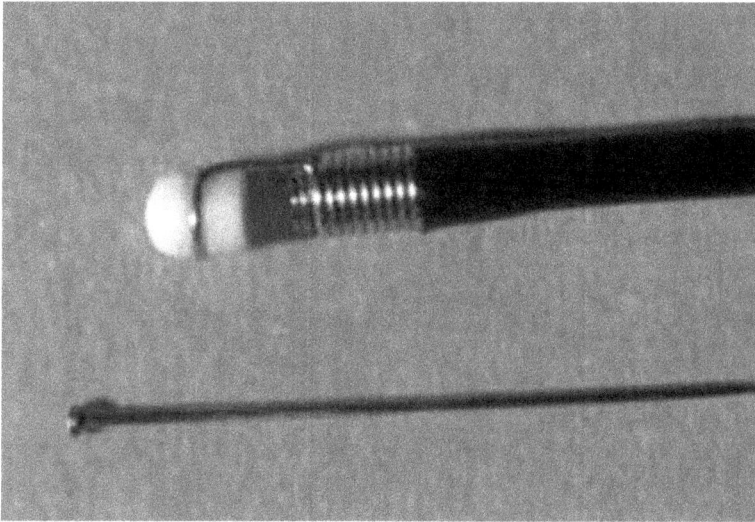

FIGURE 8.18   Photo of an optical probe for bile recognition.
Source: Gastric pH optical fiber probe (reprinted with permission from Ref.

8.8.3   pH OPTICAL SENSOR

Power/potential of hydrogen ion (pH) regulation is very important within the body as it help in the analysis of several human body organs functions. Many body fluids like blood, saliva, gastric juice, bile juice, and urine have different pH values for need to monitor. Therefore, an optical pH sensor is used for the measurement of pH within the body. Different pH indicator dyes are used to measure the hydrogen ion concentration in the sample. These dyes are typically weak organic acids or bases with their protonated (acidic) and deprotonated (basic) forms [71]. The working principle of a pH sensor is based on the interaction of hydronium ion (H_3O^+) and indicator dyes. These indicators intermingle with hydronium ion, H_3O^+, and investigate the variation in the absorption and reflection of light.

A pH optical sensor can monitor real time pH of blood, gastric juice in stomach, and other fluids. A miniaturized probe sensor is inserted into the stomach or esophagus through the nostril and measures the changes in the fluid (Figure 8.19). The range of pH from 1 to 3.5 is investigated by one channel while 3.5–8 pH range can be investigated by other channel because of the different composition of each channel. The individual channel equipped two LEDs in which one LED is for the signal and the other LED is for reference. Two types of dyes such as bromophenol blue and thymol blue are used for all pH ranges. [3,70,71].

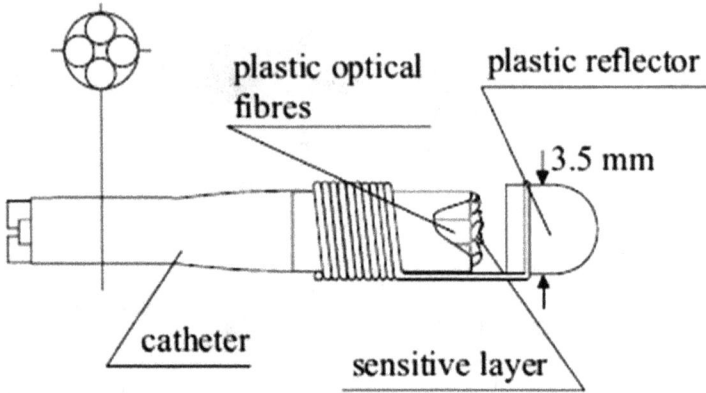

FIGURE 8.19   Gastric pH optical fiber probe.

Source: Reprinted with permission from Ref. [71]. © 2016 Springer Nature.

8.8.4   *OXYGEN AND CARBON DIOXIDE OPTICAL SENSOR*

Oxygen is the most essential element in the body. Its combination with blood makes oxygen such an important entity of investigation that is used in the diagnosis of cardiovascular and cardiopulmonary systems. This can be done with the help of spectroscopy by measuring the optical properties of hemoglobin that carry the oxygen to the tissues in the body or by using fluorophore. A fluorophore is a fluorescent chemical compound that re-emits light after light excitation. Therefore, real-time monitoring of oxygen and CO_2 can be done by using optical fiber sensor probe. The working principles of oxygen content are based on the oxygen saturation measuring technique by oximeter discussed in earlier Section 8.8.1. Although the hemoglobin as indirect indicator measurement of oxygen has some disadvantages. Therefore, fluorophore and perylene-dibutyrate based sensor was introduced first by Peterson [77]. The monitoring of CO_2 helps in the evaluation of the disturbance in tissue perfusion. Partial pressure of CO_2 (PCO_2)

has been investigated by an optical fiber sensor developed recently [78]. This sensor working principle is based on the measurement of pH change brought in optode following the diffusion of carbon dioxide [79–81].

8.9 OPTICAL COHERENCE TOMOGRAPHY

Optical coherence tomography (OCT) is an imaging technique that is non-destructive in nature and provides high axial resolution, say up to 5–7 μm range. Huang et al. first reported optical coherence tomography in 1991. OCT has emerged with very fast development and today it is a part of clinical practice. OCT provides in vivo "optical biopsy" of retina. Optical fibers based probes are used in the technique and are able to probe within gastrointestinal tracts. [82]. Brain cells and mammalian embryos have gone far imaging in animal studies [83,84]. The working principle of OCT is likely to ultrasound tomography imaging. The working principle of OCT is that light is produced from a broadband light source. This light is now divided into two different beams, a reference beam and a sample beam. When the light falls off to the retina, it is backscattered and interferes with the reference beam. This interference outline is measured to light resonances versus the depth profile of tissue in retina in vivo [85, 86].

8.10 LASER SURGERY

Nowadays, laser surgery is growing very enormously. In surgery, many lasers are used that make the surgery a hassle-free clinical practice. Ophthalmology is involving the laser light in the surgery of eye or retina. Eye surgery is possible due to the laser sight. Also, glaucoma is treated with the help of laser surgery. When eye medications do not have sufficient effect, then laser surgery is used in case of glaucoma. Laser surgery or the use of light in the surgery of retinal prosthesis is also possible with the help of optoelectronic technology. A brief discussion of retina prosthesis is mentioned in Section 8.13.

8.11 OPHTHALMOLOGY/OPTOMETRY

Ophthalmology or optometry is related to the functions, structures, disorders, and treatment of eyes. Both physicians and surgeons deal with ophthalmology. As we have said, in ophthalmology, laser light is used in treating the disorders or in the surgery of the eyes. Ophthalmology is related to eyesight restoration, degenerative diseases such as retinitis pigmentosa and macular degenerative, and so forth. Preoperative care and postoperative care are also a concern of ophthalmologists. Details of the retinal prosthesis are mentioned in the next section. Laser surgery is a well-known method in ophthalmology in which optical fibers are used. The laser that is used by ophthalmologists includes argon laser (blue, green), krypton laser (yellow, red), dye laser (blue, green, yellow, orange, red), and CO_2 laser [87].

8.12   RETINAL PROSTHESIS

Retinal prosthesis is the technique to implant the device for the partial restoration of sight and improved vision in the patients. The retinal prosthesis is the technology to improve the vision and understanding of electrical stimulation of the brain and retina. The retinal prosthetics promise to treat retinal degenerative diseases such as retinitis pigmentosa [88] and age-related macular degeneration [89]. In such diseases, the retinal photoreceptors deterioration averts the effective translation of light into electrical light that is construed in the visual cortex as visual function [90]. The example of the implantable devices such as the Argus I and II can be given under the retinal prosthesis. The Argus is an epiretinal prosthesis that comprises internally fixated electronic circuits and externally a camera mounted on a pair of sunglasses. Figure 8.20 shows the components of Argus II devices wear by a patient that consists both implanted and external components [91, 92].

The Argus II comprises external components as well as implanted devices. A small camera is mounted on the frame of a pair of sunglasses in the external portion of the system. The camera is connected through a cable to a video processing unit worn on a shoulder belt. In the video processing unit, a real-time image is collected by the camera upon turning on the system [91,92].

8.13   OPTOGENETICS

Optogenetics is the research technology in which neuroscientist is trying to mapping the brain cells and to control the neurons using light. The very first concept of Optogenetics was given by Andrew Hodgkin [93] when he decided to write about the history of the tools designed for the development of the field of neuroscience. Optogenetics is a set of tools that have encoded genetic molecules. When these genetic-encoded molecules are targeted to specific neurons in the brain, they enable their activity to be driven or silenced by light (blue color). This because to understand the function of each neuron in the brain to get the specific knowledge of feelings, emotions, judgments, and actions. This is one of the major issues in neuroscience to find the abnormality of neuron that leads to the brain disease. Biophysical proteins, known as "opsins" that is illuminated when targeted by light and translocate the specific ions through a membrane in which they are fascinating. Optogenetic tools have explored the use of opsins to facilitate either the activity or silencing of specified neuron classes with the effect of light. This study of neurons activity or silencing was done in the organism such as *Caenorhabditis elegans*, *Drosophila*, zebrafish, mouse, rat, and nonhuman primates to see their behavior [94]. Figure 8.21 shows the response of channelrhodopsin-2-YFP in a neuron when targeted via light and exhibiting light-activated spikes.

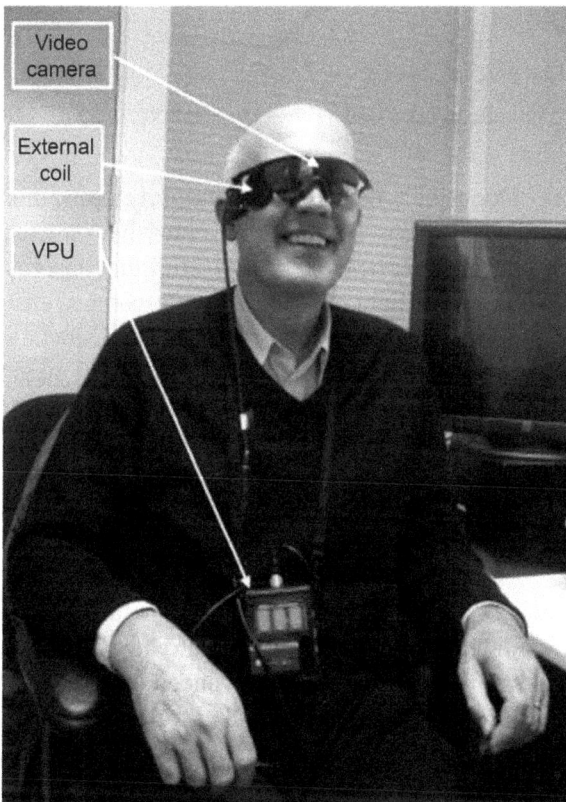

FIGURE 8.20 Photograph of a patient fitted with the Argus® II Retinal Prosthesis System.

Source: Reprinted with permission from Ref. [92]. © 2016 Elsevier.

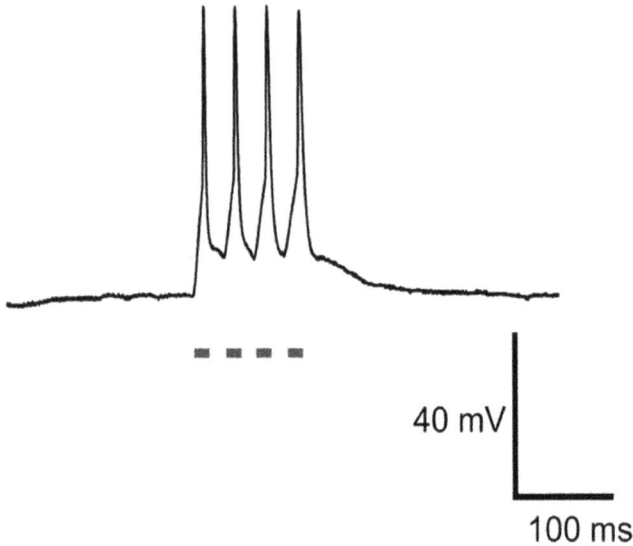

FIGURE 8.21   Response of channelrhodopsin-2 in light spiking in neurons.

Source: Reprinted from Ref. [95]. © 2011 Faculty of 1000 Ltd.

Moreover, in optogenetics, neuroscientists are developing more compatibile and reliabile specific neurons by using less invasive or even noninvasive optical fibers. Use of nanoparticles in optogenetics that absorbs near infrared laser light and emits visible light photons of blue/green wavelength that make ontogenetically modified neurons active or silence. Optogenetics are promising if succeed in treating chronic pain using switching on/off switching neurons. Optogenetics has also promised to know the role of neurons in the learning and memory, vision, and anxiety, and their effect in treating diseases [96].

8.14   FUTURE PROSPECTIVE

Optoelectronics is continuously developing the innovative ideas and skills to make new biomedical devices. Some of the already developed biomedical equipment are improved and in a more advanced and smart form, although there are many hurdles in improving and developing more accurate and real-time monitoring equipment for the body functions in humans. Restless work of research teams and scientists from various fields have boosted this technology and helped a lot in routine checkups of some of the metabolites or substances, which are responsible for various diseases in a human body.

The most common medical applications are in high demand for using NIOS, such as pulse oximetry, blood glucose monitoring, urine analysis, and dental diagnosis. However, invasive optoelectronic sensors such as endoscopic imaging, pH monitoring, bile sensor, oxygen, and carbon dioxide sensor are part of current clinical practice. Moreover, laser surgery, retinal prosthesis, ophthalmology, and other optoelectronic technologies are in regular development in the field of biomedical applications. The medical devices are complex but in high demand for the diagnosis of various invasive and noninvasive diseases. Such devices are sophisticated, small in size, reliable, easy to use, good in packaging, and available at home care. The director of sales and marketing at WZW optic in Balgach, Switzerland, David Varrie, said "This is the age of nanotechnology and we are starting to hear that an optoelectronic device inside a vitamin B12 capsules swallowed by human and devices are able to take pictures and even stream a video." "The whole digestive system or the esophageal pathways can be clearly seen using the optoelectronic devices," he added. The accumulation of optoelectronic devices with medical equipment has thundered the revolution in the field of biomedical applications. Optoelectronic devices have reduced human error, providing fast and more accurate results in real-time monitoring. Another area of innovation into the optoelectronic devices is the diagnostic systems with the use of personalized medicines. Also, the life-threatening diseases such as cancer, tumor, cardiovascular, respiratory disease, and others can be diagnosed in the early as well as the late stage. The fluorescent probes enables clinicians to see the growth of tissues, embryos, and so forth. The study and applications of optoelectronics may help in understanding the applications of tissue engineering technology in today's time. The most incredible use of optoelectronics in the field of optogenetics has grabbed attention of various researchers and neuroscientists by developing the response of neurons with the help of light that is responded to proteins such as rhodopsin. With the development of this technique, it will be easy to treat chronic pain, migraine, spinal cord injury, seizures, and brain-related diseases. The study, understanding, and improvement of learning memory, vision, emotions, grief, happiness, and other sensations can be measurable in the near future by the use of optogenetics. Retinal prosthetics have emerged recently for the treatment of retinitis pigmentosa and age macular degenerative. This technique has developed and is used in the treatment related to eyesight but is looking forward to a breakthrough in the restoration of eyesight and even blind persons can be treated. Recently organic optoelectronic devices have developed in the field of medicine using wearable optical sensors.

The organic optical devices have potential in the biomedical field such as biological activity, point of care diagnosis, lab-on-a-chip for biosensor, and flexible e-skin [97–100]. Improvements are also emerging in the sensors of veterinary medicine and food industry.

8.15   CONCLUSION

This chapter concludes the development of medical equipment using opto-electronic technology. We have tried to cover all the potential applications of optoelectronics for biomedical but the field has so vast that some of the applications or devices not part of this chapter. The recent research materials in the field of optoelectronic materials such as conducting polymer, organic polymer, metal oxide nanoparticles, carbon materials, and other nitrogen family member elements such as phosphorus have shown a great interest in optoelectronic application. We have given an introduction of design and application of the equipment and technology to the biomedical field. We have also discussed the working principles in some part but only a formal descrip-tion is available. The main thrust of this chapter is to show the application of optoelectronic-based medical equipment and their use in the biomedical field. A description of noninvasive and invasive optical sensors described and organized well on the basis of their sensing elements and methods. Ongoing research and development have been discussed in the area of optogenetics, retinal prosthesis, and endoscopic imaging. The challenges in optical sensors are to provide fast and accurate reading, real-time monitoring, and easy to handle the equipment. The medical technology needs to be advanced with high potential equipment and smart technology, small size, robust design, and reduced cost.

ACKNOWLEDGMENTS

The authors are thankful to the Department of Chemical Engineering, Faculty of Engineering and Department of Physics, Faculty of Science, Jazan University, Jazan, Kingdom of Saudi Arabia, for providing the necessary facilities. The authors are also thankful to the Department of Electrical & Computer Engineering, Faculty of Engineering. & Technology, King Abdulaziz University, Jeddah KSA, for all the resources and the facilities.

CONFLICT OF INTEREST

The authors declare that there are no conflicts of interest.

KEYWORDS

- **optoelectronics**
- **optoelectronic materials**
- **optoelectronic properties**
- **invasive and noninvasive optoelectronic sensors**
- **laser surgery**
- **optogenetics**
- **retinal prosthesis and ophthalmology**

REFERENCES

1. Beiser, A. Concepts of Modern Physics, 6th ed. McGraw-Hill Higher Education: New York, NY, 1994.
2. Sze, S. M.; Ng, K. K. Physics of Semiconductor Devices, 3rd ed. John Wiley & Sons, Inc: Hoboken, NJ, 2006.
3. Bieleckia, Z.; Stacewicz, T.; Wojtas, J.; Mikołajczyk, J.; Szabra, D.; Prokopiuk, A. Selected Optoelectronic Sensors in Medical Applications, Opto-Electron. Rev. 2018, 26, 122–133.
4. Perez, M.A; Gonzalez, O.; Arias, J.R. Optical Fiber Sensors for Chemical and Biological Measurements, InTech: London, UK, 2013, http://dx.doi.org/10.5772/52741(Chapter 10).
5. Mohamad, M.; Manap, H. An overview of optical fiber sensors for medical applications, Int. J. Eng. Technol. Sci. 2014, 1, 9–11.
6. Lechuga, L.M.; Calle, A.; Prieto, F. Optical sensor based on evanescent field sensing. Part I: surface plasmon resonance sensors, Anal. Chem. 2000, 19 (Suppl), 54–60.
7. Silvestri, S.; Schena, E. Optical-Fiber Measurement Systems for Medical Application, InTech: London, UK, 2013, http://dx.doi.org/10.5772/18845 (Chapter 11).
8. Callister W.D.; Rethwisch D.G. Materials Science and Engineering: An Introduction. New York: Wiley, 2018.
9. H.S. Nalwa, ed. Handbook of Nanostructured Materials and Nanotechnology. 5th ed. Academic Press: New York, USA, 2000. pp. 501–575.
10. Gardner, J.; Bartlett, P. Application of conducting polymer technology in microsystems Sensors Actuat. A Phys. A 1995, 51, 57e66.
11. Overview on Organic Electronics. Mrs.org. Retrieved on 2017-02-16.

12. Wang, X.S.; Tang, H.P.; Li, X.D.; Hua, X. Investigations on the mechanical properties of conducting polymer coating-substrate structures and their influencing factors, Int. J. Mol. Sci. 2009, 10(12), 5257e5284.
13. Gerard, M.; Chaubey, A.; Malhotra, B.D. Application of conducting polymers to biosensors, Biosens. Bioelectron. 2002, 17(5), 345e359.
14. Yang, G.; Kampstra, K.L.; Abidian, M.R. High performance conducting polymer nanofiber biosensors for detection of biomolecules, Adv. Mater. 2014, 26(29), 4954e4960.
15. Mazzoldi, A.; De Rossi, D. Conductive polymer based structures for a steerable catheter, Smart Struct. Mater. 2000, 3987, 273e280.
16. Koenig, S.P.; Doganov, R.A.; Schmidt, H.; Neto, A.H.C.; Ozyilmaz, B. Electric field effect in ultrathin black phosphorus. Appl. Phys. Lett. 2014, 104, 10451.
17. Coleman, J.N.; Lotya, M.; O'Neill, A.; Bergin, S.D.; King, P.J.; Khan, U.; Young, K.; Gaucher, A.; De, S.; Smith, R.J.; et al. Two-dimensional nanosheets produced by liquid exfoliation of layered materials. Science 2011, 331, 568–571.
18. Mao, D.; Li, M.; Cui, X.; Zhang, W.; Lu, H.; Song, K.; Zhao, J. Stable high-power saturable absorber based on polymer-black-phosphorus films. Opt. Commun. 2018, 406, 254–259.
19. Chen, Y.T.; Ren, R.; Pu, H.H.; Chang, J.B.; Mao, S.; Chen, J.H. Field-effect transistor biosensors with two-dimensional black phosphorus nanosheets. Biosens. Bioelectr. 2017, 89, 505–510.
20. Kuo, P.L.; Chang, J.M.; Wang, T.L. Flame-retarding materials—I. Syntheses and flame retarding property of alkylphosphate-type polyols and corresponding polyurethanes. J. Appl. Polym. Sci. 1998, 69, 1635–1643.
21. Lee, H.U.; Park, S.Y.; Lee, S.C.; Choi, S.; Seo, S.; Kim H.; et al. Black phosphorus (BP) nanodots for potential biomedical applications. Small. 2016, 12, 214–219.
22. Chen, Y.: Ren, R.; Pu, H.; Chang, J.; Mao, S.; Chen, J. Field-effect transistor biosensors with two-dimensional black phosphorus nanosheets. Biosens Bioelectron. 2017, 89, 505–510.
23. Kumar, V.; Brent, J.R.; Shorie, M.; Kaur, H.; Chadha, G.; Thomas, A.G.;. et al. A nanostructured aptamer-functionalised black phosphorus sensing platform for label-free detection of myoglobin, a cardiovascular disease biomarker. ACS Appl. Mater. Interfaces. 2016, 8, 22860–22868.
24. Tao, W.; Zhu, X.; Yu, X.; Zeng, X.; Xiao, Q.; Zhang, X. Black phosphorus nanosheets as a robust delivery platform for cancer theranostics. Adv. Mater. 2017, 29. doi: 10.1002/adma.201603276.
25. Mehra, N.K.; Verma, A.K.; Mishra, P.; Jain, N.K. The cancer targeting potential of D-α tocopheryl polyethylene glycol 1000 succinate tethered multi walled carbon nanotubes. Biomaterials 2014, 35(15), 4573–4588.
26. Singh, B.; Baburao, C.; Pispati, V.; Pathipati, H.; Muthy, N.; Prassana, S.; Rathode, B.G. Carbon nanotubes. A novel drug delivery system. Int. J. Res. Pharm. Chem. 2012, 2(2), 523–532.
27. Fujitani, T.; Ohyama, K-I.; Hirose, A.; Nakae, D.; Ogata, A. Teratogenicity of multi-wall carbon nanotube (MWCNT) in ICR mice. J. Toxicol. Sci. 2012, 37(1), 81–89.
28. Dey, P.; Das, N. Carbon nanotubes: it's role in modern health care. Int. J. Pharm. Pharm. Sci. 2013, 5(4), 9–13.
29. Chavan, R.; Desai, U.; Mhatre, P.; Chinchole, R. A review: carbon nanotubes. Int. J. Pharm. Sci. Rev. Res. 2012, 3(1), 125–134.
30. Huo, Q.; Liu, J.; Wang, L.-Q.; Jiang, Y.; Lambert, T.N.; Fang, E. J. Am. Chem Soc. 2006, 128, 6447–6453.

31. Campbell, C.T.; Peden, C.H. Chemistry. Oxygen vacancies and catalysis on ceria surfaces. Science 2005, 309(5735), 713–714.
32. Kim, S.E.; Lim, J.H.; Lee, S.C.; Nam, S.C.; Kang, H.G.; Choi, J.; Anodically nano-structured titanium oxides for implant applications. Electrochim. Acta 2008, 53(14), 4846–4851.
33. K. Hola, Z. Markova, G. Zoppellaro, J. Tucek, R. Zboril, Tailored functionalization of iron oxide nanoparticles for MRI, drug delivery, magnetic separation and immobilization of biosubstances, Biotechnol. Adv. 33 (6 Pt 2) (2015) 1162–1176.
34. Heckert, E.G.; Karakoti, A.S.; Seal, S.; Self, W.T. The role of cerium redox state in the SOD mimetic activity of nanoceria. Biomaterials 2008, 29(18), 2705–2709.
35. Moger, J.; Johnston, B.D.; Tyler, C.R. Imaging metal oxide nanoparticles in biological structures with CARS microscopy, Opt. Express 2008, 16(5), 3408–3419.
36. Genet, S.C.; Fujii, Y.; Maeda, J.; Kaneko, M.; Genet, M.D.; Miyagawa, K.; Kato, T.A. Hyperthermia inhibits homologous recombination repair and sensitizes cells to ionizing radiation in a time- and temperature-dependent manner. J. Cell. Physiol. 2013, 228(7), 1473–1481.
37. Laurent, S.; Stanicki, D.; Boutry, S.; Roy, J.C.; Vander Elst, L.; Muller, R.N. Development of a new molecular probe for the detection of inflammatory process. J. Mol. Biol. Mol. Imaging 2015, 2(1), 4 (id1013).
38. Sirelkhatim, A.; Mahmud, S.; Seeni, A.; Kaus, N.H.M.; Ann, L.C.; Bakhori, S.K.M.; Hasan, H.; Mohamad, D. Review on zinc oxide nanoparticles: antibacterial activity and toxicity mechanism. Nano-Micro Lett. 2015, 7(3), 219–242.
39. Chevalier, J.; Gremillard, L. Ceramics for medical applications: a picture for the next 20 years, J. Eur. Ceram. Soc. 2008, 29, 1245–1255.
40. Jayaswal, G.; Dange, S.; Khalikar, A. Bioceramic in dental implants: a review, J. Indian Postho. Soc. 2010, 10, 8–12.
41. Stateikina, I. Optoelectronic Semiconductor Devices—Principals and Characteristics. Concordia University: Quebec, Canada 2002.
42. http://shodhganga.inflibnet.ac.in/bitstream/10603/4556/13/13_chapter%202.pdf
43. Sweeney, S.J.; Mukherjee, J. Optoelectronic devices and materials. In: Springer Handbook of Electronic and Photonic Materials. Springer: Cham 2017.
44. Schroder, D.K. Semiconductor Material and Device Characterization. John Wiley & Sons: Hoboken, NJ 2015.
45. Mokkapati, S. and Jagadish, C., 2009. III–V compound SC for optoelectronic devices. Materials Today, 12(4), pp. 22–32.
46. Chen, C.-L. 'Elements of Optoelectronics and Fiber Optics. School of Electrical and Computer Engineering, Purdue University, Irwin, Chicago, Bogota, Boston 1996
47. Christenson, J. Sensors and Transducers, Handbook of Biomechatronics, Academic Press: New York, 2019, 61–93.
48. Lynch, K.M.; Marchuk, N.; Elwin, M.L. Sensors, Embedded Computing and Mechatronics with the PIC32, Newnes: Oxford, 2016, pp. 317–340.
49. Holonyak Jr, N. Is the light emitting diode (LED) an ultimate lamp? Am. J. Phys. 2000, 68(9), 864–866.
50. Lai, K.S.; Huang, J.C.; Hsu, K.Y.J. Design and properties of phototransistor photodetector in standard 0.35-μ m SiGe BiCMOS technology. IEEE Trans. Electron Dev. 2008, 55(3), 774–781.
51. https://www.elprocus.com/phototransistor-basics-circuit-diagram-advantages-applications/

52. Khan, H. 2010. Spectral Response Modelling and Analysis of Heterojunction Bipolar Phototransistors (Doctoral dissertation, The University of Manchester (United Kingdom)).
53. https://learn.sparkfun.com/tutorials/ir-communication/all
54. Husain, A.A.F.; Hasan, W.F.W.; Shafie, S.; Hamidon, M.N.; Pandey, S.S. A review of transparent solar photovoltaic technologies. Renew. Sustain. Energy Rev. 2018, 94, 779–791.
55. Lochner, C.M.; Khan, Y.; Pierre, A.; Arias, A.C. All-organic optoelectronic sensor for pulse oximetry. Nat. Commun. 2014, 5, 5745.
56. Ergin, A.; Thomas, G.A. Non-invasive detection of glucose in porcine eyes, Bioengineering Conference, Proc. IEEE 31st Annual Northeast 2005.
57. http://www.microspectra.com/support/the-science/raman-science.
58. Camerlingo, C.; Delfino, I.; Perna, G.; Capozzi, V.; Lepore, M. Micro-Raman spectroscopy and univariate analysis for monitoring disease follow-up. Sensors 2011, 11, 8309–8322.
59. Gnyba, M.; Wróbel, M.S.; Karpienko, K.; Milewska, D.; Edrzejewska-Szczerska, M.J. Combined analysis of whole human bloodparameters by Raman spectroscopy and spectral-domain low-coherenceinterferometry, in: Proc. SPIE 9537, Clinical and Biomedical Spectroscopy and Imaging IV, 15 July, 2015,
60. Smulko, J.M.; Dingari, N.C.; Soares, J.S.; Barman, I. Anatomy of noise in quantitative biological Raman spectroscopy. Bioanalysis 2014, 6, 411–421.
61. Wróbel, M.; Gnyba, M.; Jedrzejewska-Szczerska, M.; Myllyla, T.; Smulko, J.; Barman, I. Sensing of anesthetic drugs in blood with Raman spectroscopy, in: Advanced Photonics 2015, Optical Society of America, 2015, OSA Technical Digest (online), paper SeS1B.4.
62. Fredriksson, I.; Fors, C.; Johansson, J. Laser Doppler Flowmetry – A Theoretical Framework. Department of Biomedical Engineering, Linköping University, 2007
63. Stern, M.D. In vivo evaluation of microcirculation by coherent light scattering. Nature 1975, 254, 56–58.
64. Jafarzadeh, H. Laser Doppler flowmetry in endodontics: a review. Int. Endodontic J. 2009, 42(6), 476–490.
65. Kim, M.; Kim, B.; Park, B.; Lee, M.; Won, Y.; Kim, C.-Y.; Lee, S. A digital shade-matching device for dental color determination using the support vector machine algorithm. Sensors 2018, 18, 3051.
66. Wu, G.; Datar, R.H.; Hansen, K.M.; Thundat, T.; Cote, R.J.; Majumdar, A. Bioassay of prostate-specific antigen (PSA) using microcantilevers. Nat. Biotechnol. 2001, 19(9), 856.
67. Ulanowska; Ligor, T.; Michel, M.; Buszewski, B. Hyphenated andunconventional methods for searching volatile cancer biomarkers. Ecol. Chem. Eng. 2010, 17(1), 9–23.
68. Buszewski, B.; Grzywinski, D.; Ligor, T.; Stacewicz, T.; Bielecki, Z.; Wojtas, J. Detection of volatile organic compounds as biomarkers in breath analysis bydifferent analytical techniques. Bioanalysis 2013, 5(18), 2287–2306.
69. Stacewicz, T.; Wojtas, J.; Bielecki, Z.; Nowakowski, M.; Mikolajczyk, J.; Medrzycki, R.; Rutecka, B. Cavity ring down spectroscopy: detection of traceamounts of matter. Opto-Electron. Rev. 2012, 20, 34–41.
70. Dremel, H.W. General principles of endoscopic imaging, in: A. Ernst, F.J.F.Herth (Eds.), Principles and Practice of Interventional Pulmonology, vol. 15, Springer Science+Business Media: New York, 2013.
71. Baldini, F. Invasive sensors in medicine, in: Nato Sci Ser II Math, Springer: Berlin, 2006, pp. 417–435.
72. Cecchi srl, Viadotto Indiano 50145, Firenze, Italy.

73. Medtronic Functional Diagnostics, Tonsbakken 16–18, Skovlunde, Denmark, http://www.mfd.medtronic.com

74. http://www.cecchi.com/apparecchi-medicali/.

75. Falciai, R.; Scheggi, A.M.; Baldini, F.; Bechi, P. Method of detecting enterogastricreflux and apparatus for the implementation of this method. European Patent Number 0323816 B1 6–11-91.

76. Bechi, P.; Falciai, R.; Baldini, F.; Cosi, F.; Pucciani, F.; Boscherini, S. New fiber opticsensor for ambulatory entero-gastric reflux detection, Proc. P Socphoto-Opt. Ins. 1992, 1648, 130–135. http://dx.doi.org/10.1117/12.58293.

77. Peterson, J.I.; Fitzgerald, R.V.; Buckhold, D.K. Fiber-optic probe for in vivo measurement of oxygen partial pressure. Anal. Chem. 1984, 56, 62.

78. Baldini, F.; Falai, A.; De Gaudio, A.R.; Landi, D.; Lueger, A.; Mencaglia, A.; Scherr, D.; Trettnak, W. Continuous monitoring of gastric carbon dioxide with optical fibres, Sens. Actuators B 2003, 90, 132–138.

79. Mills, A.; Chang, Q.; McMurray, N. Equilibrium studies on novel colorimetric plastic film sensors for carbon dioxide. Anal. Chem. 1992, 64, 1383–1389.

80. Mills A.; Chang Q. Fluorescent plastic thin-film sensor for carbon dioxide. Analyst 1993, 118, 839–843.

81. Weigl, B.H.; Wolfbeis O.S.. Sensitivity studies on optical carbon dioxide sensors based on ion–pairing, Sens. Actuators B 1995, 28, 151–156.

82. Rollins, A.M.; Ung-arunyawee, R.; Chak, A.; Wong, R.C.K.; Kobayashi, K.; Sivak, M.V.; Izatt, J.A. Real-time in vivo imaging of human gastrointestinalultrastructure by use of endoscopic optical coherence tomography with a novel efficient interferometer design. Opt. Lett. 1999, 24, 1358–1360.

83. Tamborski, S.; Lyu, H.C.; Dolezyczek, H.; Malinowska, M.;Wilczynski, G.; Szlag, D.; Lasser, T.; Wojtkowski, M.; Szkulmowski, Extended-focus optical coherencemicroscopy for high-resolution imaging of the murine brain, Biomed. Opt. Express 2016, 7 (11), 4400–4414.

84. Karnowski, K.; Ajduk, A.; Wieloch, B.; Tamborski, S.; Krawiec, K.; Wojtkowski, M.; Szkulmowski, M. Optical coherence microscopy as a novel, non-invasivemethod for the 4D live imaging of early mammalian embryos, Sci. Rep. 2017, 7(1), 4165, http://dx.doi.org/10.1038/s41598–017-04220–8.

85. Huang, D.; Swanson, E.A.; Lin, C.P.; et al. Optical coherence tomography. Science 1991, 254, 1178–1181.

86. Sull, A.C.; Vuong, L.N.; Price, L.L.; et al. Comparison of spectral/Fourier domain optical coherence tomography instruments for assessment of normal macular thickness. Retina 2010, 30, 235–245.

87. Waidelich, W.; Waidelich, R. eds. Laser optoelectronics in medicine: proceedings of the 7th congress International society for laser surgery and medicine in connection with laser 87 optoelectronics. Springer Science & Business Media: Berlin, 2012.

88. Hartong, D.T.; Berson, E.L.; Dryja, T.P. Retinitis pigmentosa. Lancet 2006, 368(9549), 1795–1809.

89. Gehrs, K.M.; Jackson, J.R.; Brown, E.N.; Allikmets, R.; Hageman, G.S. Complement, age-related macular degeneration and a vision of the future. Archiv. Ophthalmol. 2010, 128(3), 349–358.

90. Mills, J.O. et al. Electronic retinal implants and artificial vision: journey and present. Eye 2017, 31(10), 1383–1398.

91. Do, B.K.; Humayun, M.S.; Ameri, H. Eyesight to the blind: the promise of retinal prosthetics: the team that developed the Argus II implant explains the state of the art. Rev. Optometry 2017, 154(2), 46–51.

92. Luo, Y.H.-L.; Da Cruz, L. The Argus® II retinal prosthesis system. Progr. Retinal Eye Res. 2016, 50, 89–107.

93. Hodgkin, A.L. Chance and design in electrophysiology: an informal account of certain experiments on nerve carried out between 1934 and 1952. J. Physiol. 1976, 263, 1–21.

94. Boyden, E.S.; Zhang, F.; Bamberg, E.; Nagel, G.; Deisseroth, K. Millisecond-timescale, genetically targeted optical control of neural activity. Nat. Neurosci. 2005, 8, 1263–1268. doi: 10.1038/nn1525

95. Boyden, E.S. A history of optogenetics: the development of tools for controlling brain circuits with light. F1000 Biol. Rep. 2011, 3, 11.

96. https://www.scientifica.uk.com/neurowire/recent-advances-in-optogenetics-research.

97. Shinar, J.; Shinar, R. J. Phys. D.: Appl. Phys. 2008, 41, 133001.

98. Hofmann, O.; Wang, X.H.; deMello, J.C.; Bradley, D.D.C.; deMello, A.J., Lab Chip 2005, 5, 863.

99. Ramuz, M.; Leuenberger, D.; Buergi, L.; J. Polym. Sci., Part B: Polym. Phys. 2011, 49, 80.

100. Bansal, A.K.; Hou, S.; Kulyk, O.; Bowman, E.M.; Samuel. I.D.W. Wearable organic optoelectronic sensors for medicine. Adv. Mater. 2015, 27(46), 7638–7644.

CHAPTER 9

Challenges and Future Prospects

MOHD. SHKIR* and S. ALFAIFY

Advanced Functional Materials and Optoelectronics Laboratory (AFMOL), Department of Physics, Faculty of Science, King Khalid University, 9004, Abha, Saudi Arabia

Corresponding author. E-mail: shkirphysics@gmail.com

ABSTRACT

The current book is based on nanomaterials for optoelectronics and describes the robust information about nanomaterials in different forms and applications, including nanostructured thin films and organic photovoltaics. Several methodologies have been discussed to prepare nanomaterials and their development for future device applications.

Nanomaterials or nanostructured films and organic solar cells are playing a fascinating role in commercial progress. Undeniably, the expectation is there to yield several wonderful innovations and novel predictions at economic scale for world economy through development in nanotechnological devices. As nanomaterials possess wide scale of usages for future, they might be broadly employed in several areas of clinical analysis and tumour therapies. These applications are based on nanoscale, biocompatibility, surface morphology, stability, tough tissue replacement, bone tissues, dental, catalysis, sensor, water purification, fertilizers manufacturing, production of agriculture, inhibitor for growth of cancer cells, drug carrier, nontoxicity, or changeable toxicity in biological arrangements [1–6].

The toxicity is a major issue in nanomaterials to be employed in living things, hence it is very important to investigate at deep level before use. Hence, the transformation of nanomaterials/nanotechnology in biological as well as optoelectrical practice needs a multidisciplinary tactic directed by clinical, moral, and societal observations. In process of extensive research and development to be devoted in the field, we may consider that the humans will be momentously furthered from nanomaterials/nanotechnology in future in several filed at economic level.

Besides these, the nanomaterials-based nanotechnology has the potential to transform the way of normal devices used in several electro-optic–biological–medical fields [7–14].

Concerning this the authors' belief is to combine the nanomaterials-based nanotechnology along with nanobiology, biotechnology, nanomedicine, nanopharma, etc., to make the revolt in diverse areas with clear practical usages, such as nanooptics, nanoscale solar cells, nanoelectronics, nanoscale supercapacitors, and nanobioelectronics [15–20]. However, in the fields of typical applications named as healthcare in the form of diagnosis, therapeutics, also in paints, filters, insulating, and lubrications essences. In these the participation of human subjects makes the whole situation very complicated that causes the substantial task in the process to deliver these technologies to the world.

In upcoming time, the issues related to the development of novel class of nontoxic-biocompatible nanomaterials in pure or by using some dopants with high stability should be outlined/developed with fascinating features which can replace the existing one.

The developed devices based on nanomaterials for different applications as mentioned above should be tested in the well-developed laboratories or should be developed to see the compatibility and also an outreach program should be carried out at ground level to provide the basic information about the product and its uses. As most of the people do not have information about several techniques through which they can come to know about dangerous diseases and the early cure can be done.

ACKNOWLEDGMENT

The authors would like to express their gratitude to King Khalid University, Saudi Arabia, for providing administrative and technical support.

CONFLICT OF INTEREST

The authors declare that there is no conflict of interest in current work.

KEYWORDS

- **nanomaterials**
- **optolectronics applications**
- **biomedical applications**

REFERENCES

1. J. Gao, B. Xu, Applications of nanomaterials inside cells, Nano Today, 4 (2009) 37–51.
2. L.R. Khot, S. Sankaran, J.M. Maja, R. Ehsani, E.W. Schuster, Applications of nanomaterials in agricultural production and crop protection: a review, Crop Protection, 35 (2012) 64–70.
3. G.E. Fryxell, G. Cao, Environmental Applications of Nanomaterials: Synthesis, Sorbents and Sensors, World Scientific, 2012.
4. H. Li, S. Liu, Z. Dai, J. Bao, X. Yang, Applications of nanomaterials in electrochemical enzyme biosensors, Sensors, 9 (2009) 8547–8561.
5. A. Fernandez-Fernandez, R. Manchanda, A.J. McGoron, Theranostic applications of nanomaterials in cancer: drug delivery, image-guided therapy, and multifunctional platforms, Applied Biochemistry and Biotechnology, 165 (2011) 1628–1651.
6. G. Ghasemzadeh, M. Momenpour, F. Omidi, M.R. Hosseini, M. Ahani, A. Barzegari, Applications of nanomaterials in water treatment and environmental remediation, Frontiers of Environmental Science & Engineering, 8 (2014) 471–482.
7. V.-V. Truong, J. Singh, S. Tanemura, M. Hu, Nanomaterials for light management in electro-optical devices, Journal of Nanomaterials, 2012 (2012) 981703.
8. S. Choi, H. Lee, R. Ghaffari, T. Hyeon, D.H. Kim, Recent advances in flexible and stretchable bio-electronic devices integrated with nanomaterials, Advanced Materials, 28 (2016) 4203–4218.
9. B. Scrosati, Nanomaterials: Paper powers battery breakthrough, Nature Nanotechnology, 2 (2007) 598.
10. H. Zhang, Ultrathin two-dimensional nanomaterials, ACS Nano, 9 (2015) 9451–9469.
11. U. Tisch, H. Haick, Nanomaterials for cross-reactive sensor arrays, Bulletin MRS, 35 (2010) 797–803.
12. A. Chen, S. Chatterjee, Nanomaterials based electrochemical sensors for biomedical applications, Chemical Society Reviews, 42 (2013) 5425–5438.
13. C. Cha, S.R. Shin, N. Annabi, M.R. Dokmeci, A. Khademhosseini, carbon-based nanomaterials: multifunctional materials for biomedical engineering, ACS Nano, 7 (2013) 2891–2897.
14. A.P. Blum, J.K. Kammeyer, A.M. Rush, C.E. Callmann, M.E. Hahn, N.C. Gianneschi, Stimuli-responsive nanomaterials for biomedical applications, Journal of the American Chemical Society, 137 (2015) 2140–2154.
15. Y. Xia, Nanomaterials at work in biomedical research, Nature Materials, 7 (2008) 758.

16. R.R. Letfullin, T.F. George, Nanomaterials in nanomedicine, Computational Studies of New Materials II: From Ultrafast Processes and Nanostructures to Optoelectronics, Energy Storage and Nanomedicine, World Scientific 2011, pp. 103–129.
17. L. Yan, F. Zhao, J. Wang, Y. Zu, Z. Gu, Y. Zhao, A safe-by-design strategy towards safer nanomaterials in nanomedicines, Advanced Materials, 31 (2019) 1805391.
18. R. DeLong, Y.-H. Cheng, P. Pearson, Z. Lin, R. Wouda, E.N. Mathew, A. Hoffman, C. Coffee, M.L. Higginbotham, Translating nanomedicine to comparative oncology: the case for combining zinc oxide nanomaterials with nucleic acid therapeutic and protein delivery for treating metastatic cancer, Journal of Pharmacology and Experimental Therapeutics, (2019) jpet. 118.256230.
19. M. Galetti, S. Rossi, C. Caffarra, A.G. Gerboles, M. Miragoli, Innovation in nanomedicine and engineered nanomaterials for therapeutic purposes, Exposure to Engineered Nanomaterials in the Environment, Elsevier 2019, pp. 235-262.
20. R. Zingg, M. Fischer, The consolidation of nanomedicine, Wiley Interdisciplinary Reviews: Nanomedicine and Nanobiotechnology, 11 (2019) e1569.

Index

For Product Safety Concerns and Information please contact our EU
representative GPSR@taylorandfrancis.com
Taylor & Francis Verlag GmbH, Kaufingerstraße 24, 80331 München, Germany

www.ingramcontent.com/pod-product-compliance
Lightning Source LLC
Chambersburg PA
CBHW060334220326
41598CB00023B/2702

9 781774 638224